PROGRESS IN ADVANCED MATERIALS RESEARCH

PROGRESS IN ADVANCED MATERIALS RESEARCH

NICOLAS H. VOLER
EDITOR

Nova Science Publishers, Inc.
New York

For permission to use material from this book please contact us:
Telephone 631-231-7269; Fax 631-231-8175
Web Site: http://www.novapublishers.com

LIBRARY OF CONGRESS CATALOGING-IN-PUBLICATION DATA

Progress in advanced materials research / Nicolas H. Voler, Editor.
p. cm.
Includes index.
ISBN-13: 978-1-60021-893-4 (hardcover)
ISBN-10: 1-60021-893-8 (hardcover)
1. Materials--Research. I. Voler, Nicolas H.
TA404.2.P68 2008
620.1'1--dc22 2007032461

Published by Nova Science Publishers, Inc. ✦ New York

CONTENTS

PREFACE

Advanced materials can be defined in numerous ways; the broadest definition is to refer to all materials that represent advances over the traditional materials that have been used for hundreds or even thousands of years. From this perspective advanced materials refer to all new materials and modifications to existing materials to obtain superior performance in one or more characteristics that are critical for the application under consideration. A more insightful and focused approach to advanced materials is to consider materials that are early in their product and/or technology lifecycle. In other words, there is significant room for growth in terms of the improvement of the performance characteristics (technology lifecycle) and their product lifecycle. This book presents the latest research in the field from around the world.

Chapter 1 - In this work, the study of modifying the local optical and electrical properties of ZnO down to a sub-micron domain is reported. It is found that an electron beam can be used to change the light emission and charge transport properties of ZnO simultaneously provided an AlO_x cap layer is present. The cap layer, works like a magic layer, gradually enhances the emission while at the same time reduces the conductivity of ZnO when under low electron irradiation. On the contrary, this effect is not observable in bare ZnO. The authors find the dynamics of electron beam induced effects is strongly dependent on the electron as well as energy doses. By carefully controlling the doses, sub-micron patterns with tailored optical and electrical properties have been fabricated on 2-D and 1-D ZnO nanostructures. Simple functional devices with controllable quantum size effect have been fabricated and are demonstrated. Finally, possible mechanisms that lead to the observed phenomena will be discussed.

Chapter 2 - The superconductivity of MgB_2 has been hidden for nearly 50 years although it has the highest T_c among the intermetallics superconductors. Beside the high T_c, simple crystal structure, large coherence length, high critical field, transparency of grain boundaries to current flow and low normal state resistivity MgB_2 is a fascinating topic to study for both large scale application and electronic devices. Moreover, the presence of two-gap superconductivity (π and σ band) has been theoretically and experimentally established.

However, critical current density of pristine MgB_2 drops rapidly in the high magnetic field due to the weak pinning centres and low upper critical field. During the past three years, novel techniques and developments for fabrication of useable MgB_2 have been reported, including chemical alloying, irradiation, thermo-mechanical processing techniques and magnetic shielding to improve the critical density, upper critical field and the irreversibility field. Among the studies, atomic substitution, especially using nano-particles, may help in clarification of the superconductivity mechanism thus making it appropriate for practical application. On the other hand, the momentum of enhancing flux pinning using chemical doping is moving to a positive side. Nano-SiC powder is well known for effective improvement of the critical current density.

In this chapter, a brief introduction of superconductivity in MgB_2 will be introduced. Further, the authors review the basic electronic and magnetic properties of the MgB_2 as well as its crystal structure. The authors will also discuss the preparation method that has been well developed for experimentation on MgB_2. Along with that, we review the experimental results of the chemical alloying on MgB_2, particularly on the critical current density. In addition, the future prospect of MgB_2 and the development for the application in the current superconductivity market will be given.

Chapter 3 - Polyelectrolyte capsules, composed of oppositely charged polyelectrolytes using the alternating layer-by-layer (LBL) self-assembly onto colloidal templates followed by core removal, have received widespread interests owing to their ease of fabrication, versatility with regards to selection of the materials, well-defined shell structure and tunable functionality. Although these capsules are structurally controlled at the nanometer scale, their robustness and high tolerance to imperfections offer significant promises for many applications in optics, electronics, sensors and biotechnology. In this chapter, the authors have given an overview of the recent progress and highlighted the potential developments in this field. Firstly, the core removal mechanisms and the structure-property relationships of capsules are outlined and elucidated. A variety of sacrificed templates had been utilized for capsule preparation, including polymeric colloids (i.e. melamine formaldehyde and polystyrene colloids), inorganic particles (i.e. silica and metal carbonate particles), and living materials (i.e. cells) etc. It is well documented that the wall thickness and the core decomposition conditions are two critical factors to govern the stability, integrity, and mesh size of polyelectrolyte capsules templated on pH-degradable melamine formaldehyde microparticles. After optimization of the core removal conditions, the diffusion and permeability of polyelectrolyte capsules are shown to be strongly dependent on the thickness, which is controlled by the variation of solvent quality (i.e. salt concentration) or the number of bilayers. The mesh size of intact capsules is also shown to decrease from meso- (2 ~ 50 nm) to microporous (less than 2 nm) as thickness increases. On the contrary, polyelectrolyte capsules made by manganese carbonate particles exhibit different permeability behaviors. Secondly, since many smart materials in bioengineering, nanotechnology and medicine require a depot of encapsulated drugs that are released on demand at a specific location by an external stimulus, polyelectrolyte capsules are very promising systems in the development of microencapsulation technologies with permeation control by changing environments. For example, electrochemical stimuli are very promising, since e.g. the redox potential near a desired organ where a pharmaceutical drug should be released may differ from other locations in the body or also, a corrosion pit may self-anneal due to the release of an inhibitor by local potential change. However, conventional, organic polyelectrolyte based microcapsules have their limitations due to their slow response to trace amounts of trigger or restricted freedom of choice of stimuli. The authors present a novel redox-sensitive capsules which show the dramatically changes of their permeability and diffusion behaviors upon triggered by redox agents. Consequently, many efforts had been devoted to fabricate various novel intelligent micro- and nanocapsules with unique controlled, responsive and switchable permeability.

Although these responsive capsules, such as pH-, light-, or temperature-sensitive etc., had been developed, many strict barriers for their real applications still exist and have to be overcome in the future. Concerning to the practical issues, the comments and suggestions will be summarized and discussed in the end.

Chapter 4 - Here the authors present the preparation of nine 2D and 3D lead halide perovskite structures whose exciton energy is between 2.05 and 3.64 eV. The authors investigate in details the optical properties of these excitons, by means of transmission, photoluminescence, reflectivity, and ellipsometric spectroscopies, and by atomic force microscopy. The authors characterize planar microcavities in which strong exciton-photon coupling occurs between the lead halide based perovskites and the standing wave. The small linewidth and large oscillator strength characteristic of perovskites make them very interesting structures to allow the observation of strong coupling for hybrid semiconductors. In this chapter the authors investigate the potential for application of these materials in hybrid organic-inorganic devices.

Chapter 5 - Photochromic compounds are very interesting and potentially useful for optical data storage, optical switching, optical filter, and so on. In recent years, unique photoinduced surface relief formation caused by lateral material migration has been reported and has attracted great attention. Large amplitude surface relief structures can be inscribed on thin films containing the azobenzene moiety upon irradiation with patterned light. This article focuses to selected research topics on possible applications proposed in the author's group using reversible dynamic lateral motion in the formation and erasure process of surface relief structure. They involve i) new methodology for optical patterning and alignment of various light-inert materials, ii) organic distributed feedback laser with oscillation wavelength programmability. In addition to these topics on surface relief structures formed on azobenzene-containing materials, new class of surface relief formation using a photochromic compound other than the azobenzene derivatives is presented. Promotions of these research areas are anticipated to provide wide and new opportunities for material processing and creation of new class of photonic devices.

Chapter 6 - This chapter summarizes recent progress in growth and characterization of aligned-crystalline silicon (ACSi) films on polycrystalline metal-alloy substrates. The ACSi deposition process uses, as a key technique, ion-beam-assisted deposition (IBAD) texturing to achieve *near-single-crystalline* silicon films with good electrical properties on low-cost, large-area, non-single-crystalline substrates. First, we describe the general aspects of IBAD texturing technique, and the motivation for applying it to semiconductors. This is followed by a detailed description of multilayer buffer structures implemented in this work. We have performed materials characterization using scanning electron microscopy, atomic force microscopy, electron backscattering diffraction, transmission electron microscopy, and scanning tunneling microscopy, and combined them with electrical measurements, such as carrier mobility, to understand the materials-property relationships in these structures. Optimized ACSi films on metal-alloy substrates with insulating IBAD buffer layer stacks demonstrate out-of-plane and in-plane mosaic spreads of 0.8° and 1.3°, respectively, and a room-temperature Hall mobility of 90 $cm^2/V.s$ (~50% of what is achievable with epitaxial Si films, and ~1000 times that of amorphous Si films) for a p-type doping concentration of $\sim 4x10^{16}$ cm^{-3}. By using various microscopic techniques, we have confirmed the underlying near-single-crystalline order and the superior characteristics of low-angle (<5°) grain boundaries in ACSi films. Also, forming gas anneal experiments indicated that Si films with low-angle grain boundaries do not need to be passivated to demonstrate improved majority carrier transport properties. Moreover, we have demonstrated a new conducting IBAD material, TiN, in addition to the standard insulating IBAD MgO, that can be integrated with Si. This has led to demonstration of a conducting nitride buffer stack that could be especially suitable for photovoltaic applications of ACSi films. As a further demonstration of the photovoltaic potential, the authors have fabricated thick (~1-5-micron-thick) ACSi films with well-defined doping profiles on these buffer layers. In the electrical characterization area, in addition to having extensive studies on the majority carrier transport using Hall measurements, the authors have performed surface potential measurements using AFM and transport measurements based on STM that showed that carrier transport is not significantly impeded by low-angle grain boundaries in optimized ACSi films. Also, with an eye for thin film transistor (TFT) applications, we have prepared metal-insulator-semiconductor structures and measured C-V curves. These measurements yielded near-electronic-grade surface properties and low surface defect densities in the ACSi films. Currently, no other monolithic technology combines the high performance of epitaxial silicon films with the low cost of non-single-crystalline substrates. As such, the ACSi film technology holds promise to make a major impact in the semiconductor industry by providing a new means to improve the performance (or, reduce material and fabrication costs) of silicon-based devices, such as solar cells and flat-panel displays.

Chapter 7 - One-dimensional (1D) nanomaterials (such as nanowires, nanotubes, nanobelts and nanorods) have attracted considerable attention due to their potential applications as building blocks in nanoscale circuits, optoelectronic, electromechanical and sensing devices. 1D nanostuctures with controlled diameters can act as a model system for systematic studies on the size-dependent electrical, optical, and thermal transport phenomena. Precision engineering or tailoring of materials is always the goal of much of the effort driving synthetic technology of 1D nanomaterials. The synthesis and fabrication of 1D nanostuctures nanostructures can be approached from one of two directions. One is the “top down” approach in which nanostructures are generated using a number of nanolithographic techniques. The other is the “bottom up” approach in which the nanostructures are built up from individual atoms or molecules. This is the basis of most “cluster science” as well as synthesis of crystal materials, usually via chemical means. In contrast with the nanolithographic techniques, the chemical method provides an attractive route to produce 1D nanostuctures of various materials at large scale and reasonably low cost.

This chapter will focus on our current research activities on the fabrication and characterization of one-dimensional nanowire heterostructures. The authors’ endeavors over the recent years with respect to the chemical synthesis of novel inorganic 1D nanostructures via simple thermal evaporation, their detailed structural analysis using state-of-the-art electron microscopy techniques, and electrical property evaluation are presented. These nanostructures including carbon nanotube or silicon-based heterostructures, namely, gallium-filled carbon nanotubes, Ga_2O_3-C core-shell 1D nanostructures, Si-ZnS side-by-side composite nanowires, Si-Cd_4SiS_6 composite nanowires and Si-In end-to-end nanowire heterostructures. These materials are thoroughly analyzed by using Transmission electron microscopy (TEM), high-resolution TEM, scanning transmission electron microscopy (STEM) and related EDX mapping, field-emission measurements and cathodoluminescent spectroscopy. Potential applications of these novel nanostructures, using as nanothermometers, nanoscale optoelectronics, electronics, were also demonstrated.

Chapter 8 - With the explosive growth in data volume in this digital age, higher and higher requirements on information storage techniques have been put forward. In recent years, novel storage techniques have welled up continuously. Among them, SPM-based information storage techniques, including scanning tunneling microscopy (STM), atomic force microscopy (AFM), scanning near-field optical microscopy (SNOM), etc., have become one of the most active research fields. The development of data storage technique depends to a large degree on the development of storage medium. Storage medium with high storage density, fast response speed and long lifetime is the critical factors for ultrahigh-density data storage. This review gives an overview of SPM-based nanoscale data storage on different medium, including inorganic materials, polymers, and organic small molecules films. Particular emphasis is placed on the novel organic materials. The related storage mechanism and principals are introduced and the development foreground is predicted.

In: Progress in Advanced Materials Research
Editor: Nicolas H. Voler, pp. 1-28
ISBN: 978-1-60021-893-4

Chapter 1

Controllable Light Emission and Charge Transport Patterning and Device Fabrications on Multi-Dimensional ZnO

K.C. Hui, R.S. Wang and H.C. Ong*
Department of Physics, The Chinese University of Hong Kong, Hong Kong

Abstract

In this work, the study of modifying the local optical and electrical properties of ZnO down to a sub-micron domain is reported. It is found that an electron beam can be used to change the light emission and charge transport properties of ZnO simultaneously provided an AlO_x cap layer is present. The cap layer, works like a magic layer, gradually enhances the emission while at the same time reduces the conductivity of ZnO when under low electron irradiation. On the contrary, this effect is not observable in bare ZnO. We find the dynamics of electron beam induced effects is strongly dependent on the electron as well as energy doses. By carefully controlling the doses, sub-micron patterns with tailored optical and electrical properties have been fabricated on 2-D and 1-D ZnO nanostructures. Simple functional devices with controllable quantum size effect have been fabricated and are demonstrated. Finally, possible mechanisms that lead to the observed phenomena will be discussed.

Introduction

Miniaturization of functional devices is currently a focus of intense scientific and technological interest [1]. Efforts to date are focused primarily to the fabrication of nanodevices based on nanostructures such as nanowires and nanoribbons [2,3]. Several prototypes such as field-effect transistors [4], laser diodes [5] and light emitting diodes [6] have been demonstrated recently. However, in addition to the size reduction, another aspect

* Email: hcong@phy.cuhk.edu.hk

that has not yet been explored but certainly would go beyond the existing technologies is the development of approaches that enable the nanostructures to perform not only one single task but also multiple tasks. In other words, it is desirable for the nanostructures to exhibit different electrical, optical or chemical properties at different extents under a controllable manner so that their functionalities could be tailored. This implementation will surely extend the application of nanomaterials to suit for various purposes under different circumstances. To achieve such a goal, one way is to control the fundamental properties of materials at a length scale of nanometers so that various characteristics can be exhibited at the same time. This is particularly favorable in materials where a wide range of desirable properties can be found. ZnO is a good candidate because it possesses a combination of optical and electrical properties that are aligned with different electronic and photonic areas [7]. For example, ZnO has a wide band gap (3.3eV) and high excitonic binding energy (60meV) so that it emits strong light at the ultra-violet spectrum [7]. In addition, its visible wavelength emission, spanned from 500-700nm, can be utilized in the phosphor technology [8]. When it is doped with group III elements, ZnO functions as a conductor while at the same time remains optically transparent [9]. Together with its room temperature ferromagnetism [10], photocatalytic effect [11], photoconducting effect [12], photo-induced hydrophilicity [13] and piezoelectric effect [14], ZnO-based semiconductors are considered for making various novel devices such as "nanowire laser" [15], "random laser" [16], "invisible circuit" [17], "nanowire sensor" [18] and "spintronic device" [19], etc.

In this work, we report a simple method to control the local optical and electrical properties in ZnO. We will first show by growing a cap layer (AlO_x) on ZnO thin films, an electron beam can be used to improve the light emission properties of ZnO by "converting" the unwanted green-yellow emitting deep level emission into desirable UV emitting band edge emission. At the same time, the conductivity of the irradiated region is found to decrease significantly. Then, we will show how to use this process to create sub-micron pattern with controlled optical and electrical properties not only on thin films but also on nanoribbons and nanowires. Finally, although no direct evidence, we attribute the modification of light emission and charge transport behaviors of ZnO to the annihilation of detrimental defects "catalyzed" by the presence of the cap layer. Two mechanisms, which are recombination enhanced defect reaction and charging effect, will be discussed at the end.

EXPERIMENTAL

I. Growth ZnO and AlOx Layers and Nanostructures

ZnO thin films were grown on (001) Si and (0001) Al_2O_3 substrates by radio frequency (rf) magnetron sputtering [20]. High purity (5N) 3-inch metallic Zn target was used. During deposition, the working pressure was kept at 6-8 mTorr by mixing ultra-high purity Ar and O_2 and the growth temperature was kept at 550°C. In order to maintain a reasonable deposition rate, the rf power was kept at 150W. As a result, the deposition rate was found to be approximately 80-90nm/hour and the sample thickness of ZnO was kept at ~120nm throughout all the experiments. Si substrates support c-axis textured growth of ZnO while sapphire substrates support epitaxial growth [21]. For AlO_x capping, half of the ZnO film was covered by a shadow mask and then AlO_x was grown directly on the uncovered region by sputtering of an Al target under the same Ar and O_2 ambience. The rf power was kept at 130W and the growth was performed at room temperature. The deposition rate was ~ 10nm/hour and typically 10-20nm thick AlO_x was grown on ZnO. Since two regions, with and without AlO_x capping, were obtained from the same sample at one deposition, all the optical and electrical measurements were carried out on two regions so that direct comparison could always be made. In general, it was found that similar electron beam induced emission enhancement was observed from textured and epitaxial ZnO thin films although the resulting enhancement ratio could be different [20].

Single crystalline, confirmed by transmission electron microscopy (TEM), ZnO nanowires and nanoribbons were grown by hydrothermal method in an autoclave [22]. A solution of 50ml 0.1M $ZnSO_4$, 10ml 0.5M carbamide and 10ml C_2H_5OH was placed in a sealed Teflon autoclave with the pH of the solution adjusted to 14 by NaOH. The solution was then heated at 160°C for 12 hr and the resulting nanostructures were obtained by washing and drying the white precipitates. For electron irradiation experiments, the nanostructures were first diluted by appropriate solvent and then dispersed on substrate by ultrasonic. The samples were then loaded into the sputtering system for AlO_x capping.

For thermal treatment, some of the samples were rapid thermally annealed in an Ulvac-Riko infrared rapid thermal heater at different temperatures in either reducing or oxidizing ambience for 1 to 5mins.

II. Characterizations of ZnO and AlO_x

Various techniques were used to characterize ZnO and AlO_x. X-ray diffraction (XRD) and transmission electron microscopy (TEM) were used for studying the microstructures. The XRD spectrum and cross-section TEM image of AlO_x/ZnO/Si are shown in figure 1. One can see from the θ/2θ scan that the ZnO film is highly c-axis oriented and no peaks can be attributed to AlO_x, which indicates AlO_x is essentially amorphous. Cross-section TEM image further confirms the columnar growth mode of ZnO in c-axis direction and the amorphous nature of AlO_x [21]. Although not shown, electron diffraction pattern performed at different regions shows the layer is highly homogeneous. X-ray photoelectron spectroscopy (XPS) was carried out to identify the chemical composition of ZnO and AlO_x. The measurements were performed in an UHV system equipped with a monochromatized Al Kα source and a hemispherical analyzer. The base pressure of the UHV system was 1×10^{-9} Torr and the XPS energy resolution was calibrated to be ~0.6eV. The binding energy spectra were scaled against the Au 4 f 7/2 line of a sputter-cleaned gold foil at 84.00 eV. Results revealed that ZnO was stoichiometric while AlO_x was slightly oxygen deficient with the O to Al ratio varied from 1.38 to 1.5 [21].

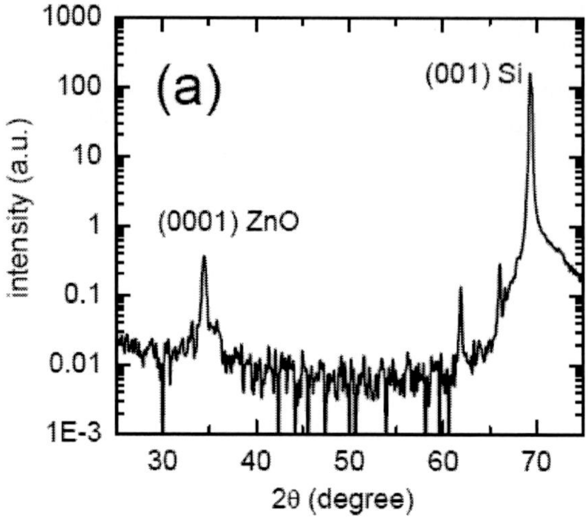

Figure 1. (a) XRD spectrum of AlO_x/ZnO/Si shows the c-axis preferred orientation ZnO peak. No peak is observed from AlO_x indicating the cap layer is amorphous. Peaks at 62° and 66° arise from background. (b) The cross-section TEM shows the columnar growth of ZnO and amorphous AlO_x..

Electron beam irradiation and cathodoluminescence (CL) experiments were carried out in a Cambridge scanning electron microscope (SEM) equipped with an Oxford Instrument MonoCL CL system [23]. Excitation was carried out by irradiating a secondary electron beam on the samples at different accelerating voltages and probe currents ranging from 1 to 10 kV and 100pA to 10 μA, respectively. According to the Kanaya-Okayama model [24], the electron beam with voltage varied from 2 to 10kV would give rise to the penetration depth from 46 Å to 0.75 μm in ZnO. The probe currents were calibrated by using a Faraday cup before and after experiments. For light collection, the signal was firstly dispersed by a 0.3m spectrometer and then was collected by charge coupled device (CCD) detector for spectroscopy or photomultiplier (PMT) detector for luminescence imaging. A 150-grooves/mm grating blazed at 600nm was used. Panchromatic or monochromatic mode could be selected for multiple or single wavelength imaging. Most of the experiments were carried out at room temperature. However, for low temperature experiments, a custom-made cryogenic sample stage was used, which varied the temperature from 77 to 240K by flowing liquid nitrogen.

For local electrical measurement, a conductive atomic force microscope (AFM) was used by modifying a commercial AFM (NanoscopeIII, Digital Instruments, Inc.) with a homemade pre-amplifier and a bias unit [25]. The noise of the pre-amplifier and the bias unit were found to below 61.5 pA and 0.1 mV. Commercial Si_3N_4 cantilevers coated with a 10 nm Cr layer plus 100 nm of gold were used. The loading force was calibrated at about 70 nN. All the measurements were carried out at room temperature.

RESULTS

I. Cathodoluminescence Studies

To illustrate the impact of AlO_x on the CL of ZnO, we first of all show in figure 2(a) the CL spectra of bare epitaxial ZnO film taken at various instants at room temperature as a reference. The accelerating voltage and probe current were kept at 5kV and 30 nA, respectively. The result shows that both the band edge emission at 3.29eV and deep level (DL) emissions at 1.85–2.1eV decrease slightly with irradiation time [20]. The time-dependent integrated band edge emission intensities are then determined by Gaussian fitting the spectra and the results are shown in figure 2(b), which confirm the reduction of intensity under electron irradiation. However, once the film is capped with an AlO_x layer, the scenario changes unexpectedly. As we see from figure 2(c) that while the deep level emissions decrease slowly with electron irradiation, the band edge emission rises almost instantly and reaches the maximum at 600s. The band edge emission from AlO_x/ZnO is plotted in figure 2(b) and it shows the band edge emission increases as a function of irradiation time and finally saturates. Lower currents also induce similar enhancement effect from AlO_x/ZnO but the emission takes a longer time to reach the maximum level. As a result, we find the band-edge emission of the irradiated region is almost 23 times stronger than that of the bare region (figure 2(d)) at saturation. Not only the band edge emission, we also find the deep level (DL) emissions by 5 times. The SEM (left) and panchromatic CL (right) images are illustrated in figure 3. The irradiated region in SEM image appears slightly darker than the unexposed region after irradiation primarily because of the decomposition of hydrocarbon residuals by the electron beam [26], which could explain the slight reduction of emission from bare ZnO after irradiation. Nonetheless, no significant difference in the surface morphology between two regions is found. On the other hand, from the CL image, sharp intensity contrast is clearly seen between two regions. We find the enhancement is highly reproducible in all the AlO_x/ZnO systems regardless of their crystallinity and the emission patterns are very stable with time and are not self-erasable [20]. As we will discuss later, similar phenomenon is observed at lower temperatures down to 77K.

Although no significant difference in the peak position is found between AlO_x/ZnO and ZnO in the course of irradiation, as shown in figure 4, the linewidth of AlO_x/ZnO is slight wider than that of ZnO. Similar results have been observed from the textured samples.

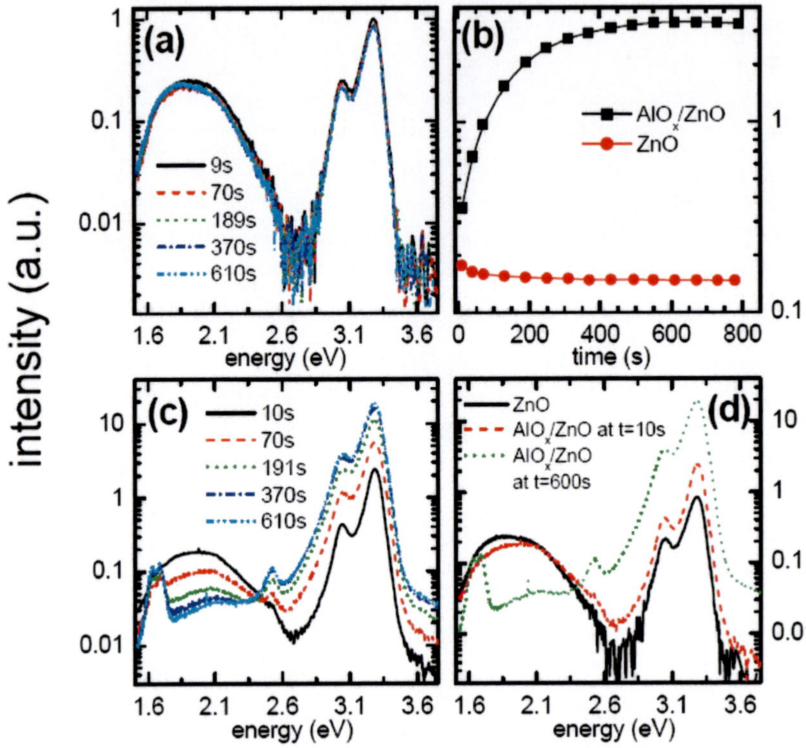

Figure 2. (a) The CL spectra of bare ZnO taken at different irradiation times. Slight reduction of intensity is observed. (b) The integrated band edge emission intensities of bare ZnO (circle) and AlO_x/ZnO are plotted as a function of irradiation time. (c) The CL spectra of AlO_x/ZnO taken at different irradiation times. (d) CL spectra of bare, AlO_x/ZnO before and after irradiation are shown. The emission enhancement is 23 times after irradiation compared to the bare ZnO.

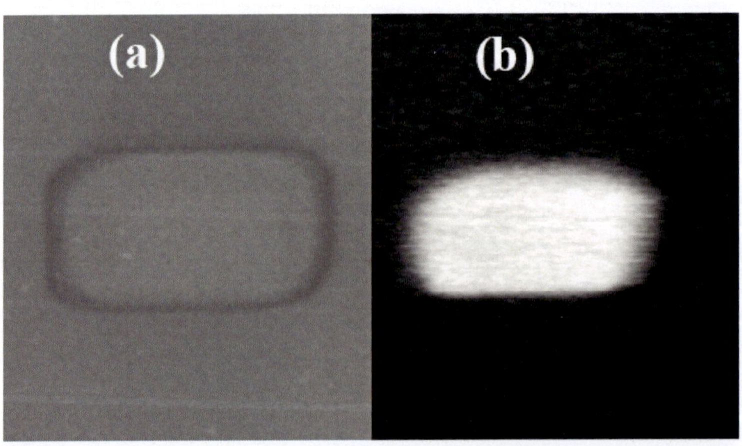

Figure 3. SEM (a) and CL (b) images of AlO_x/ZnO after electron irradiation.

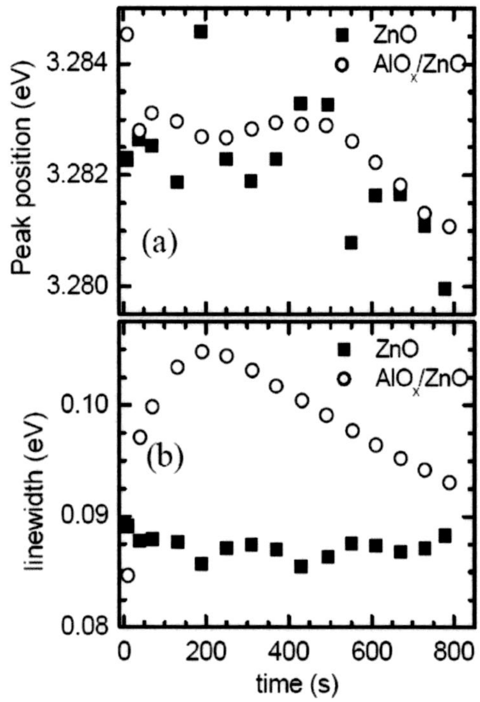

Figure 4. (a) The band-edge emission peak position of bare (solid square) and AlO_x/ZnO (open circle) films as a function of irradiation time. (b) The band-edge emission linewidth of bare (solid square) and AlO_x/ZnO (open circle) films as a function of irradiation time.

In general, our results clearly indicate that electron beam irradiation not only increases the band edge emission from AlO_x/ZnO but also suppresses the DL emissions. In other words, provided an AlO_x cap layer is present, electron irradiation effectively "converts" detrimental DL (green-yellow) emission to desirable band edge (UV) emission. Based on this result, we have developed a simple technique that allows us to create UV emission patterns on green-yellow emitting ZnO films. The basic schematic is outlined in figure 5. First of all, the as-prepared ZnO film is rapid thermally annealed in a reducing ambience to enhance the DL emission (figure 5(a)) [8]. A layer of AlO_x is then deposited on top of ZnO (figure 5(b)). Electron beam is focused on the sample and scanned successively until a desired pattern is produced (figure 5(c)). After the pattern formation, the sample is immersed in a diluted NaOH solution to remove the cap layer (figure 5(d)). Figure 6 illustrates a simple example based on this scheme. Figure 6(a) shows the SEM image of capped ZnO/Si film with the central region being exposed to the electron beam. After the removal of the cap layer, the 2.0 and 3.26 eV monochromatic CL images, which represent the maximum of the DL and band edge emissions, are shown in figures 6(b) and (c). We see that after the treatment, the central irradiated region is converted to UV emitting while the background still remains green-yellow emitting. The electron beam induced effect remains even after the AlO_x is removed. We also find that without the cap layer no further electron beam induced enhancement is observed from the sample, which indicates the presence of AlO_x is important in giving rise to such an effect.

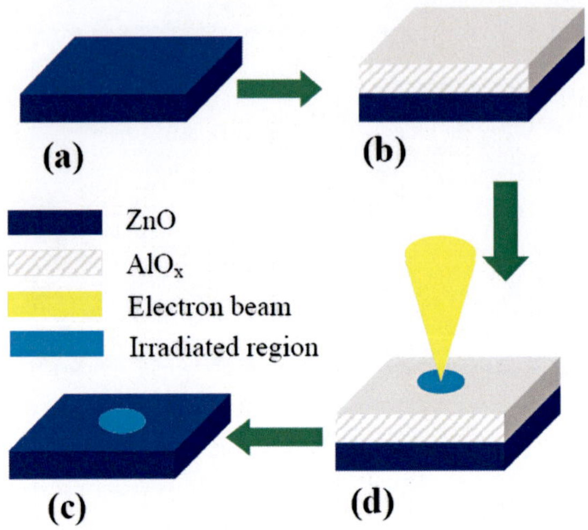

Figure 5. Procedures for fabricating UV emission patterns in green yellow-emitting ZnO films.

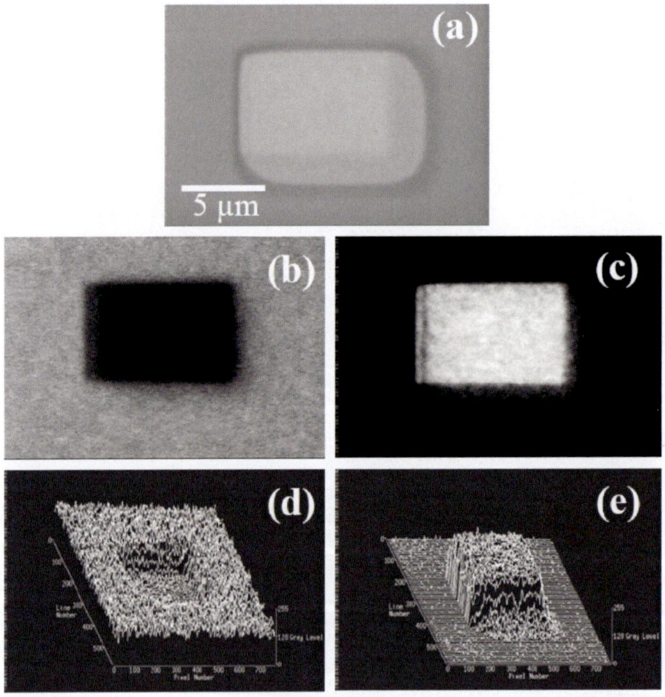

Figure 6. (a) SEM image of the irradiated area of AlO_x/ZnO. (b) and (c) Monochromatic CL images of the pattern taken with the light emission detected at 2.0 and 3.24 eV. (d) and (e) show the 3D plots with reasonable signal to background ratio.

In analogy to electron beam writing, we have fabricated sub-micron light emission pattern on ZnO films. As illustrated in figures 7(a) is the band edge emission pattern created on ZnO/Si with the feature as narrow as 800 nm wide. The contour CL intensity plot shows a reasonably good signal to background ratio (figure 7(b)). The feature size is limited to several hundreds of nanometer due to the limitation of tungsten filament that produces can only produce the beam size on the order of 200-300nm. Finer pattern with size below 200nm, however, is expected by using a field emission type SEM.

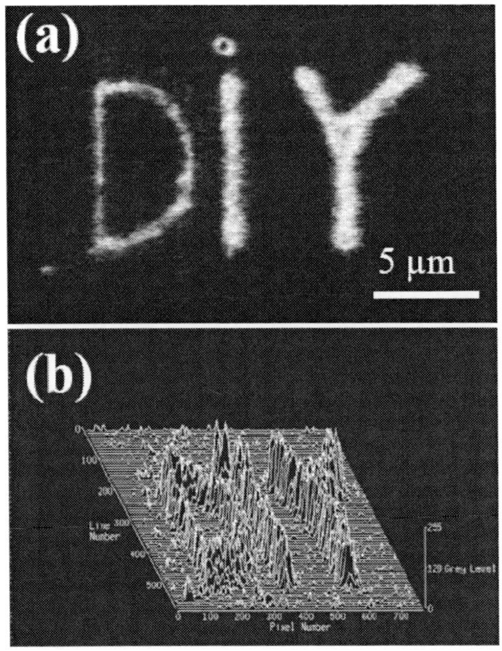

Figure 7. (a) The band edge emission CL pattern fabricated on AlO_x/ZnO. (b) Its corresponding 3-D image plot.

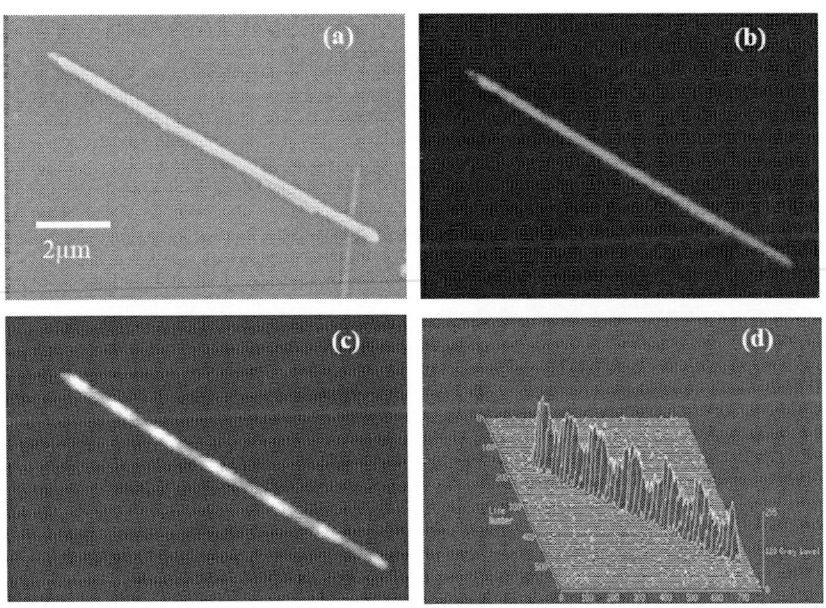

Figure 8. (a) SEM image of a ZnO nanoribbon. (b) CL image of the nanoribbon before irradiation. (c) CL image of the same nanoribbon after electron irradiation at different places. (d) CL emission contour of the ribbon.

Apart from creating emission patterns on a 2-D thin film, this technique also allows us to engineer the emission on nanostructures. ZnO nanowire and nanoribbon powders are dispersed on Si substrates by ultrasonic to reveal individual single nanowire or nanoribbon. Then the nano-materials are capped with AlO_x capping. Figure 8(a) illustrates the morphology of one single nanoribbon on Si. Its corresponding 3.26 eV monochromatic CL image is shown in figure 8(b), which indicates the ribbon has a uniform luminescence across the entire length. A focused electron beam is then irradiated on the ribbon intermittently to create a barcode-like pattern. The resulting CL images are shown in figures 8(c) and (d) and we see a modulation of light emission is displayed along the ribbon demonstrating the success of making an optical nano-tag [20].

II. Local Electrical Measurements

Local conductivity of ZnO subjected to different degrees of electron irradiation has been studied by conductive AFM. An indium tin oxide (ITO) was used as the bottom electrode. Prior to the conductive AFM measurement, the AlO_x cap layer was removed by NaOH. Figures 9(a) and (b) illustrate the CL emission image and conduction patterns of ZnO/ITO/Si taken at different irradiation times. The patterns are denoted as 1, 2, 3 and 4 with spot 1 has the longest irradiation time while spot 4 has the shortest. A positive 5V bias was applied. A quick glance from the images shows that electron irradiation not only affects the optical properties of ZnO but also the electrical conductivity; the irradiated regions are significantly more insulating than their non-irradiated background. In addition, the images also imply there exists a strong dependence between the irradiation time and conductivity. Figure 9(c) illustrates the corresponding CL spectra of ZnO obtained from different spots, which show different emission characteristics. The integrated intensities of the overall emission, band edge emission and DL emissions at different spots are summarized in figure 9(e). We see that after the longest irradiation time, the band edge emission can be 60 times higher than that of the bare ZnO while the DL emissions are reduced by 6 times. The DL/band edge emission ratio decreases from 9.17 to 0.023. Local current-voltage (I-V) profiles of all the regions are shown in figure 9(d) and indeed it is found that the current is lowered by almost 8000 times in the irradiated region even at 0.3V. In addition, different I-V characteristics can be obtained at different intervals. Extraction the conduction threshold in the I-V profile by extrapolating the sharp edge of the current, as indicated by the black dash line in figure 9(d), we find that the threshold is linearly increasing with irradiation time (figure 9(e)). In particular, the threshold is also strongly correlated to the trends of overall emission as well as the band edge emission. Finally, it is noted that the low conducting region is not due to the remaining AlO_x because no trace of Al on the surface is revealed by XPS. We see no distinctive morphological difference between the irradiated and non-irradiated areas. Therefore, our results indicate it is possible to pattern the optical and electrical properties in ZnO simultaneously by carefully varying the irradiation time.

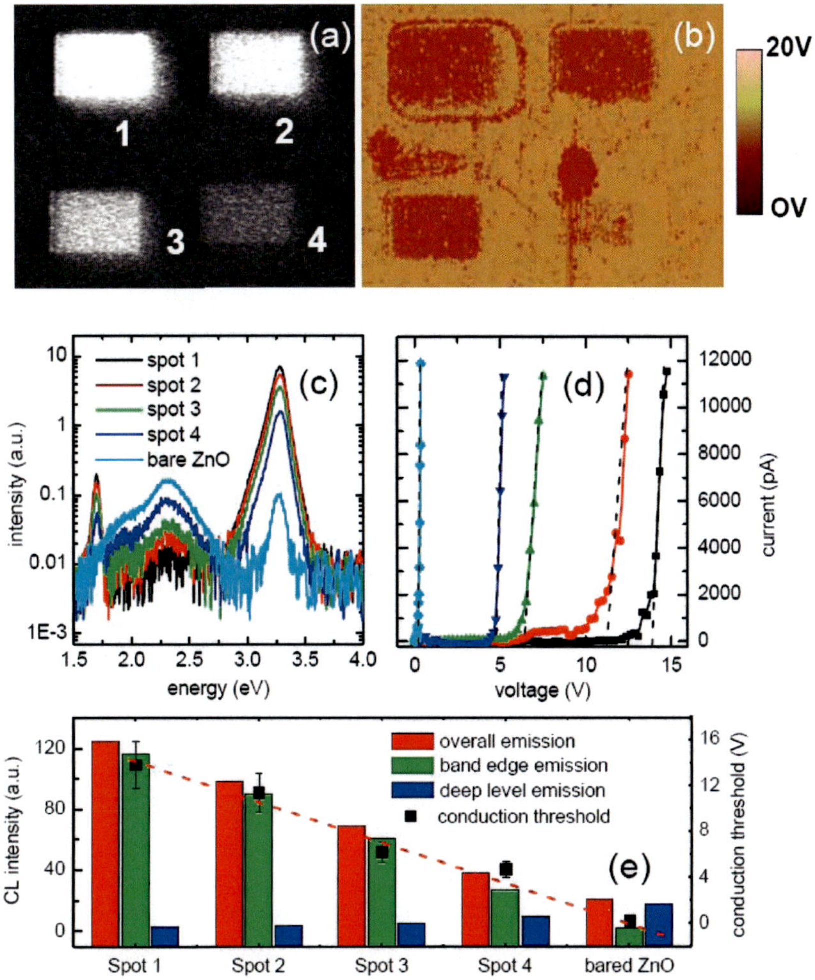

Figure 9. Different patterns (1,2,3 and 4) created by controlling the electron dose. (a) The panchromatic CL image illustrates different emission patterns. (b) The conductive AFM image of the corresponding patterns. (c) The CL spectra of the corresponding patterns including bare ZnO for reference. (d) The local I-V profile of different patterns including bare ZnO for reference. (e) The integrated intensity (red), band edge emission (green) and deep level emissions (blue) taken at different regions. The data shows that after the highest irradiation, the band edge emission increase by 60 times compared with the bare ZnO.

III. Optical Studies

The irradiated patterns are also examined under a 50X optical microscope. Figure 10(b) shows the bright field of an AlO_x/ZnO system after irradiation. One clearly sees distinct reflectance arising from the irradiated and non-irradiated regions. On the other hand, no sharp contrast can be observed from the bare ZnO after the same amount of electron irradiation, as displayed in figure 10(a). Therefore, the optical properties of AlO_x/ZnO, light emission and reflectance, have been altered significantly after electron irradiation.

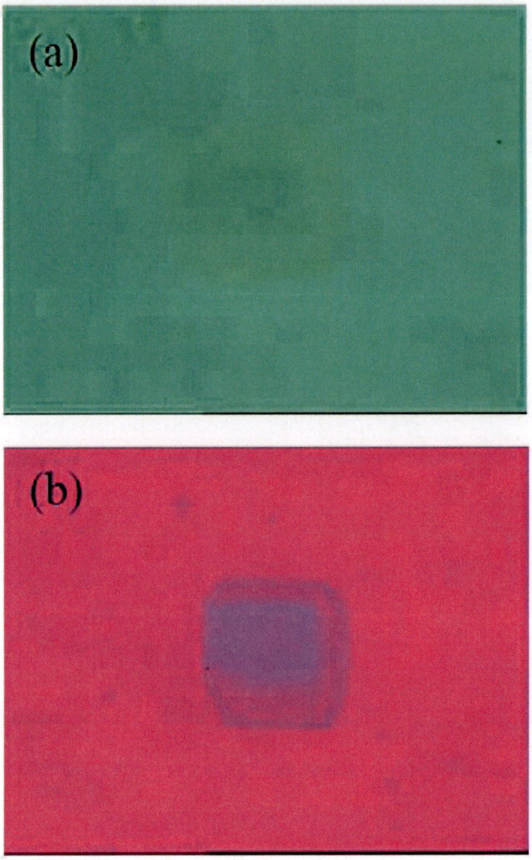

Figure 10. (a) Bright field optical micrograph of bare ZnO after electron irradiation at he center. However, no sharp contrast is observed. (b) The optical micrograph of AlO_x/ZnO after electron irradiation at the center. It is clear that the center region has different reflectance.

IV. Bi-Layer and Multilayer

We have preliminary studied the effects of electron irradiation on capped ZnO bilayers and multilayers (quantum wells) and attempt to gain control on the quantum size effect. We have grown AlO_x/MgZnO/ZnO/Si system by co-sputtering Zn and Mg together to form the intermediate MgZnO layer. The thickness of MgZnO is ~ 80nm while the concentration of Mg to Zn ratio is ~ 0.05-0.1 determined by XPS. Figure 11 shows the time-dependent CL spectra under electron irradiation at room temperature. The emission peak at 3.57 eV can be attributed to the band edge emission from MgZnO while the peak at 3.29 eV is assigned to the emission from ZnO. One can see that, the characteristics of band edge and DL emissions from ZnO change exactly identical to that of single ZnO layer under electron irradiation. However, the emission from MgZnO decreases instead. Although no explanation can be offered to such an effect at this stage, our results show the choice of material is of important in order to obtain the electron beam induced effect. In addition, it also shows that the emission the active ZnO layer can be well controlled even it is buried in a multiplayer structure. Based on this principle, quantum wells with the basic structure of MgZnO/ZnO/MgZnO are then prepared by following the recipes developed in Ref. []. Quantum size effect, which is denoted as the blue shift of ZnO band edge emission, has been obtained from a series of samples. Figures 12 (a) and (b) display CL spectra of 10 layer-$Mg_{0.1}Zn_{0.9}O/ZnO/Mg_{0.1}Zn_{0.9}O$ quantum well

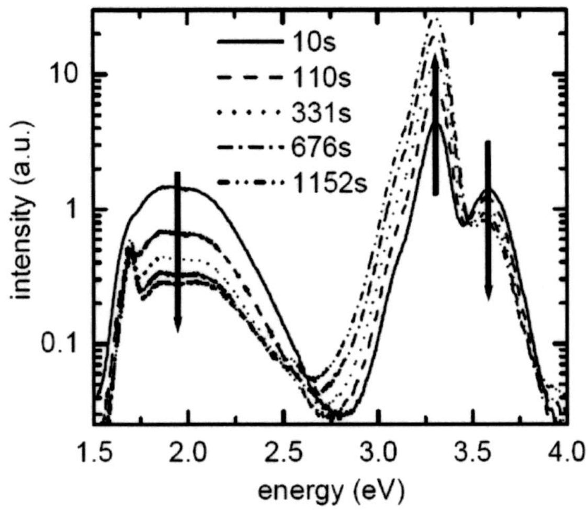

Figure 11. The CL spectra of AlO_x/MgO/ZnO/Si taken at different electron irradiation time. The peaks at 3.29 and 3.57 eV can be attributed to the band edge emission of ZnO and MgZnO, respectively. The accelerating voltage is kept at 5kV and probe current is kept at 30nA.

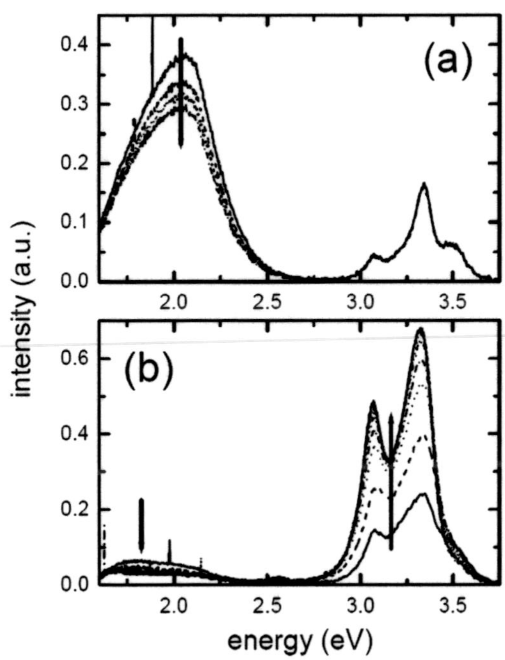

Figure 12. The CL spectra of ZnO-based quantum wells/Si (a) without and (b) AlO_x capping taken at different electron irradiation time. (a) Without capping, the CL spectra remain the same under electron irradiation. (b) However, emission from ZnO quantum wells increases during irradiation if the system is capped.

structure with and without AlO_x capping at different irradiation periods. The beam voltage and current are 5kV and 40nA. The thickness of the active ZnO is estimated as ~6nm according to the deposition rate calibration. From figure 12 (a), the emission peak arises from ZnO quantum wells is found to be blue shifted to 3.34 eV compared to 3.29eV from the bulk counterpart. The shoulder from ~3.5 eV is ascribed to the emission from barrier. It is seen that while no effect is observed from the bare quantum wells under irradiation, gradual increase of emission is obtained from the capped quantum wells.

DISCUSSIONS

Although we have observed a simple AlO_x cap layer can dramatically induce changes both in the optical and electrical properties of ZnO under electron irradiation, the underlying physics remains unclear. In the followings, we attempt to study the characteristics of the electron beam induced emission enhancement effect in detail. Based on our experimental results, we propose the enhancement is related to the deactivation of radiative and non-radiative recombination defect centers under electron bombardment.

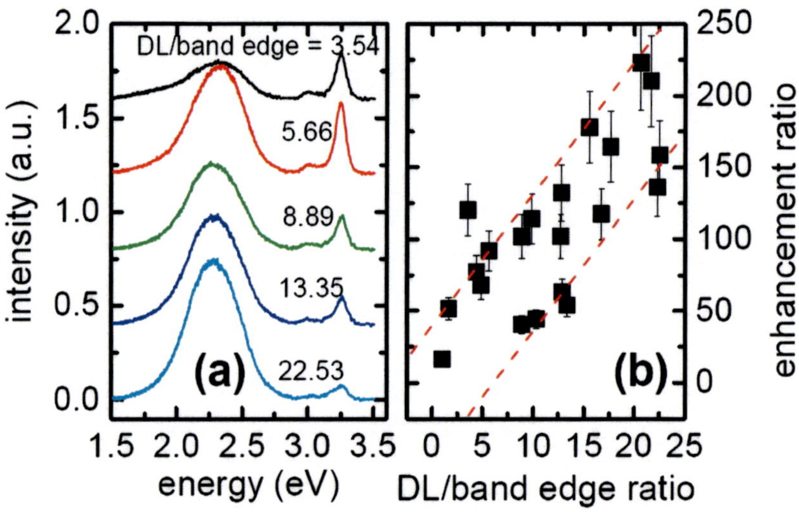

Figure 13. (a) Several CL spectra of ZnO films illustrated with different degrees of DL emissions. The values indicate the DL/band edge intensity ratio. (b) The plot of enhancement ratio of AlO_x/ZnO versus DL/band edge intensity ratio. 15% error bar is added to the data. The dash lines are for guiding.

I. The Dependence of Band Edge Emission Enhancement on the Deep Level/Band Edge Emission Ratio of Bare ZnO

We have found the extent of emission enhancement from AlO_x/ZnO can be correlated to the DL/band edge emission ratio of the ZnO before capping [20]. More than 20 ZnO films with different DL/band edge emission ratios have been examined and some of their CL spectra are illustrated in figure 13(a). After capped with same thickness of AlO_x, the films are then irradiated with 30nA electron beam at 5kV until their band edge emission saturates. By plotting the enhancement ratio, which is defined as the intensity ratio of the band edge emission at maximum to the band edge emission before capping, versus the DL/band edge emission ratio (figure 13(b)), one could see a linear relationship exists between these two ratios; i.e., the higher the original DL/band edge ratio, the stronger will be the enhancement effect. It is also noted that the strongest enhancement achieved can be as high as 200 times if the DL/band edge ratio reaches 20. In addition, it supports the idea that, in the presence of the AlO_x, electron irradiation deactivates the radiative defect centers and at the same time triggers the band edge recombination channel to become more effective.

II. The Dependence of Emission Enhancement on Energy Dose

It is also found that the dynamics of emission enhancement is controlled by energy dose rather than the time of irradiation. To clarify this issue, we have examined the time-dependent band edge emission of AlO_x/ZnO taken at accelerating voltages of 5 and 7 kV under various probe currents and the results are illustrated in figures 14(a) and (b). At each voltage, one can see higher current always leads to stronger emission and shorter time to reach the maximum intensity. In addition, higher voltage also induces stronger emission than that of lower voltage under the same current excitation. In general, stronger emission produced by higher current and voltage can be accounted by the fact that the generation rate in semiconductor is directly proportional to the product of current and voltage [24]. Higher generation rate therefore results in larger excitation and stronger emission [24]. In our examination, in order to eliminate the dependence of the resulting CL on different excitations so that all the emission spectra can be compared directly, we have normalized the intensities with respect to their corresponding probe currents to ensure they are all under the same level of excitation. As shown in figures 14 (c) and (d), the CL spectra now reach the same extent, which implies, when under a given voltage, all the emissions are enhanced by the same factor regardless of the excitation current. More importantly, by normalizing figures 14(a) and (b) with respect to their excitation power defined as ($V_a \times I_p$), where V_a is the accelerating voltage and I_p is the probe current, we again see from figure 15(a) that all the emissions are enhanced by the same factor. However, probe current determines how fast the emission reaches the maximum level, as indicated in figure 14(c) and (d). As a reference, we also have studied the time-dependent CL from bare ZnO taken at 5kV and various currents (figure 16(a)). In contrast to capped ZnO, after the normalization of the intensities with respect to their probe current, all the CL intensities collapse into one single profile (figure 16(b)). No dependence of the speed of reduction can be found on the excitation current. As a result, the distinctive behaviors arising from two samples have led us to believe that, for capped ZnO, excitation power governs emission's speed of increment but not its enhancement factor. In particular, this argument further supports our previous result that the enhancement ratio is strongly dependent on the initial defect density (DL emissions) of the ZnO. For a sample with a given amount of DL defects, emission saturates when the entire defect population is being annihilated.

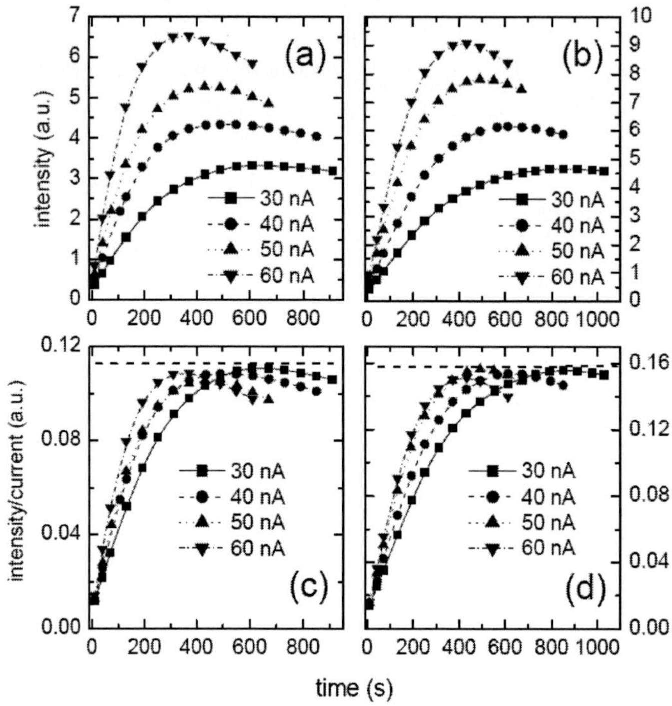

Figure 14. The time dependence of CL emission intensity of AlO_x/ZnO taken at various probe currents at the accelerating voltage of 5 (a) and 7 (b) kV. The time dependence of normalized CL intensity of AlO_x/ZnO with respect to their corresponding probe current taken at 5 (c) and 7 (d) kV.

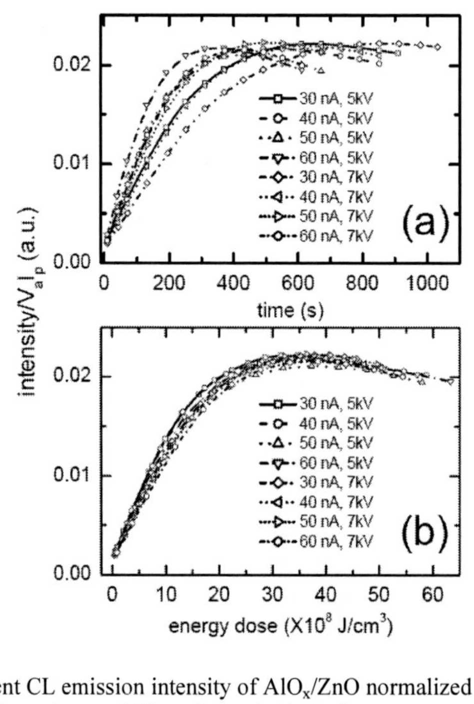

Figure 15. The time dependent CL emission intensity of AlO_x/ZnO normalized with respect to the excitation power. (b)The energy dose dependence of CL emission intensity/power spectra.

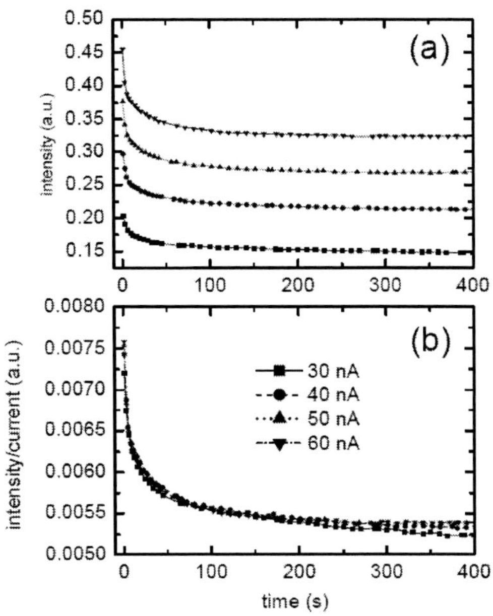

Figure 16. (a) The time dependence of CL emission intensity of ZnO taken at various probe currents at 5 kV. (b) The time dependence of normalized CL intensity of ZnO with respect to their probe current at 5 kV.

Once it is understood that excitation power determines only the rate of emission enhancement but not the enhancement factor, we now turn our attention to reveal the dependence of emission profile on energy dose. By considering the scanning area to be 96 μm^2 obtained from the SEM image after irradiation, we have estimated the energy dose that is deposited into the film according to the following equation given as:

$$\text{Energy dose} = \frac{V_a \times I_p \times t}{A_S \times R_e}, \tag{1}$$

where t is the irradiation time and A_s is the irradiated area. R_e is the penetration depth of electron given by Kanaya-Okayama model [24]. We roughly estimate the penetration depths for AlO_x/ZnO are 190 and 320nm for 5 and 7kV, respectively. Although this is only a rough approximation, figure 15(b) shows all the curves taken at various currents and voltages are merged together, which show there exists a strong dependence of emission profile (emission enhancement rate) on energy dose instead of the irradiation time. More importantly, it further supports the enhancement factor is more or less material dependent with its value can vary significantly from sample to sample.

III. Dependence of Emission Enhancement on Water Desorption

Recently, Xie et al have offered another approach to induce emission enhancement from ZnO under electron irradiation without the use of AlO_x as a facilitator [28]. They claim the use of AlO_x is not necessary and have attributed the observed emission enhancement to the desorption of water molecules under electron bombardment. Thus, we will focus on this issue here and finally conclude two different mechanisms are involved. First of all, a quick comparison between two works indicates several differences [20,28]. For example, our excitation current is ~3 orders of magnitude higher than theirs. In addition, their emission enhancement ratio seems to saturate at ~ 2 shortly after 40s of irradiation while our irradiation usually takes 1000s to reach a ratio from 6 to 200 [20]. These two scenarios imply their findings and ours are different but partially overlap in the final outcome. Secondly, we attempt to reproduce their results by immersing our ZnO films in warm deionized water for over two days in order to absorb water molecules. As shown in figure 17(a), although the CL spectrum shows both the band edge and DL emissions reduce after soaking, no enhancement in band-edge emission is observed from our film even under a wide range of excitation currents (figure 17(b)). However, once the immersed film is capped with AlO_x, a 7-fold emission enhancement is resulted (figure 17(c)). Therefore it shows that either water absorption is not efficient in our procedure or water desorption does not give rise to such an enhancement effect. Knowing that their fabrication methods are primarily aqueous-based in which water absorption is expected during the fabrication process, vacuum-based methods such as sputtering (base pressure routinely reaches 10^{-6} Torr in prior to deposition) should not post a similar problem. To further exclude the possibility of water absorption, we have annealed AlO_x/ZnO in air at 550°C for different durations. Although the treated films show slight reduction in the emissions as a function of annealing time (figure 18(a)), similar or even higher enhancements ratio (~15) are observed from the annealed films after the capping of AlO_x (figure 18(b)). This is completely different from Xie et al's results that no emission enhancement could be observed once the materials have been annealed at high temperature [28]. Finally, AlO_x is grown at a growth temperature of 400°C to eliminate the absorption of water molecules. While no emission enhancement is obtained from bare ZnO films, a 5-fold enhancement is observed after the capping of AlO_x (figure 18(c)). As a consequence, we assume electron-stimulated emission enhancement from water desorption is strongly dependent on the fabrication method and the effect may be significantly weakened in "dry"- or vacuum-based deposited films. In addition, the emission enhancement arises from AlO_x capping is due to another mechanism.

IV. Dependence of Emission Enhancement on other Oxide Systems

We have found not all oxide cap layers induce emission enhancement. For example, although Mg induces moderate enhancement from ZnO, slight enhancement is observed from MgO_x regardless of the excitation power [23]. Figure 19 shows the variation of CL intensity from MgO_x/ZnO as a function of irradiation time. Except a slight increase of intensity at the very beginning of irradiation, emission basically decreases throughout the entire process. Thermal treatment at different temperatures does not provoke any improvement. At this stage, it is still unclear of the role of oxide system on the emission enhancement effect. However, this material selectivity provides a good example for us to elucidate the mechanism in the future. More work is currently underway on MgO/ZnO system.

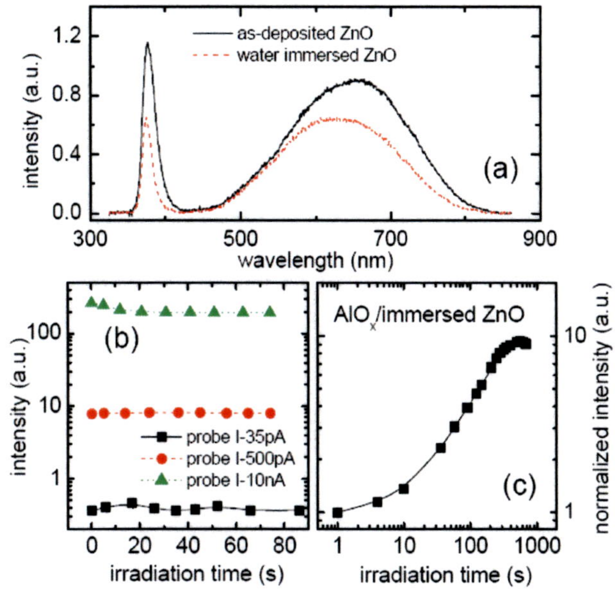

Figure 17. (a) The cathodoluminescence (CL) spectra of as-deposited bare (solid black) and water-immersed ZnO (red dash) film. (b) The integrated band-edge intensity of water-immersed ZnO film irradiated as different probe current. The accelerating voltage is kept at 10kV. (c) The dependence of normalized band-edge intensity of the same water-immersed ZnO film capped with AlO_x on irradiation time. The accelerating voltage and probe current are kept at 5kV and 15nA.

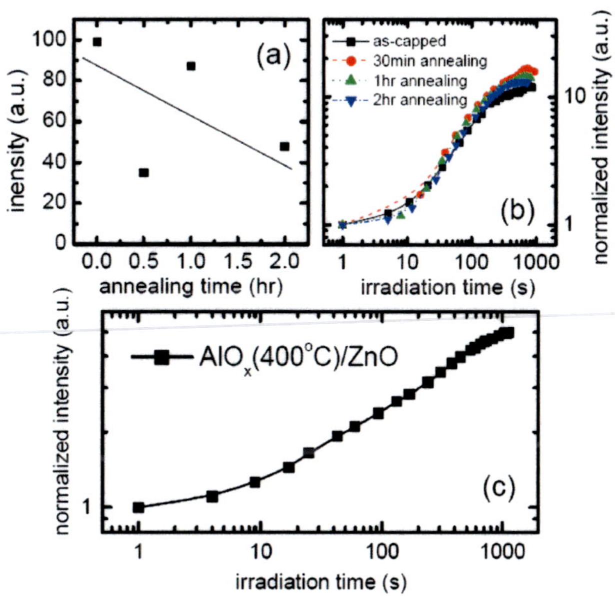

Figure 18. (a) The band-edge CL emission of AlO_x/ZnO films annealed at 550°C in air at different times. (b) The dependence of normalized band-edge CL intensity of the same AlO_x/ZnO films on irradiated time. The accelerating voltage and probe current are kept at 5kV and 20nA. (c) The dependence of normalized band-edge CL intensity of the AlO_x/ZnO film on irradiated time. The AlO_x is grown at 400°C growth temperature. The accelerating voltage and probe current are kept at 5kV and 25nA.

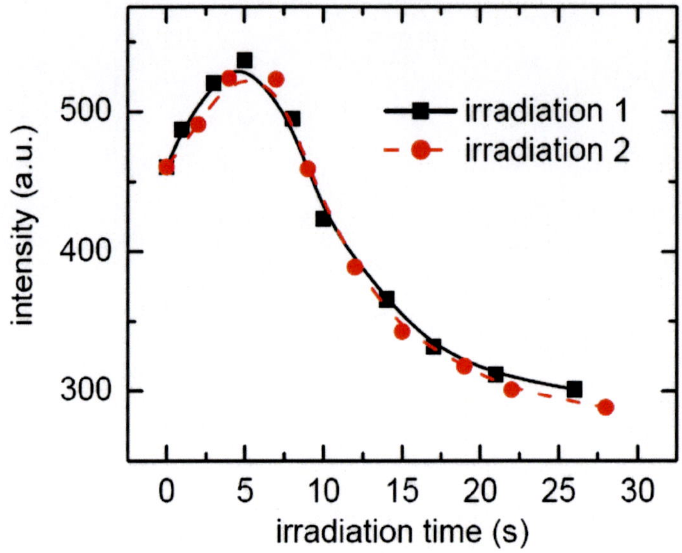

Figure 19. The time-dependences of band-edge CL intensity of MgO_x/ZnO irradiated under an electron beam of 5kV and 15nA at two regions.

Possible Mechanisms

Based on our observations, we tentatively attribute the emission enhancement effect to the defect annihilation facilitated by the presence of an AlO_x cap layer. This attribution arises from three observations. First, the strong correlation between the initial DL/band edge emission ratio and the ultimate enhancement ratio has led us to believe that electron bombardment suppresses the detrimental radiative defect recombination channels, possibly the nonradiative defect channels as well, and facilitates more band to band recombination to occur. Second, it seems that electron irradiation only affects the defect emissions but not the others, such as dopant emissions. For instance, as illustrated in figure 20, rare-earth Er^{3+} dopants have been intentionally introduced into the ZnO matrix during deposition to create additional emission peaks at 532, 586, and 640nm that correspond to the Er^{3+} transitions between 3p and 4f. However, throughout the entire electron irradiation on the capped films, we shall see that all the Er^{3+} emissions remain almost unchanged while the DL emissions decrease with time. The enhancement effect thus mainly arises from the elimination of defect-related emissions. Third, the experiments on energy dose also lead us to conclude that energy is being delivered to the sample for defect passivation. Emission eases to increase when the defects are annihilated.

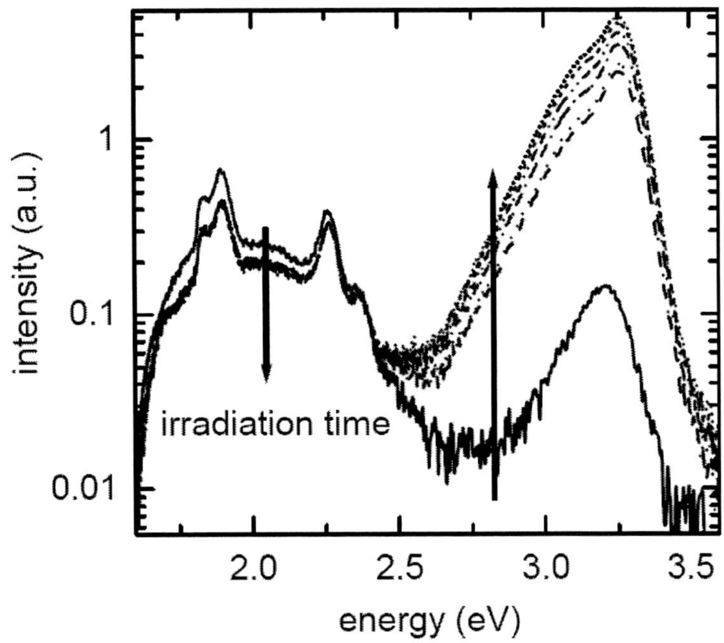

Figure 20. The CL spectra of $AlO_x/ZnO:Er/Si$ taken at different irradiation time.

I. Recombination Enhanced Defect Reaction

In fact, several works have reported that under a nonequilibrium injection of carriers, point defects can be annihilated or generated depending on the treatment conditions [29]. In our case, in the course of electron bombardment, energy released at the nonradiative recombination sites can be used to promote defect reactions such as diffusion and dissociation in the vicinity of the defects, which results in defect annealing [29]. The activation energy required for this process is far less than that of the conventional thermal process. Weeks et al. have found that the enhanced rate constant is proportional to $\exp[-(E^*-\Delta E)/kT]$, where E^* and ΔE are the activation energy of a thermal diffusion process and the energy for the electron-hole pair recombination [30]. This process is named as recombination enhanced defect reaction (REDR) [30].

Since the reaction is found to rely on the recombination energy for electron-hole pair, it implies the reaction will occur easier in a wide gap material such as ZnO. In fact, REDR has been observed to occur in several III-V and II-V semiconductors under a variety of excitation methods [31,32,33]. In II-VI materials, for example, Chuang et al. has proposed a transient REDR model to study the aging effect of Zn-based light emitting diode [31]. By using a simple defect generation rate equation, they find that the normalized optical output goes as 1/t after a long period of time. Likewise, Tang et al. have adopted a similar but revised model to consider the enhancement of emission in their Zn-based quantum well induced by a laser beam [32]. However, their approaches do not consider defect generation and reduction simultaneously [32]. In here, we therefore attempt to explain our experimental results under the framework of REDR. First of all, we combine both models to simulate our transient behavior. To begin with, the continuity equation of carrier density, n, is given as:

$$\frac{dn}{dt} = G - Bn^2 - AN_d n, \tag{2}$$

where G is the generation rate, Bn^2 accounts for the spontaneous recombination rate and AN_dn accounts for the carrier capture at the defect sites. A and B are recombination constants and N_d is defect density. To take into account of the change in defect density being activated by REDR, we modify the rate equation given in Refs [31,32] as:

$$\frac{dN_d}{dt} = -K_1 n^2 N_d \left(\frac{N_d}{N_{do}}\right)^{\alpha} + K_2 n^2 N_d \left(\frac{N_d}{N_{do}}\right)^{\beta}, \tag{3}$$

where $K_1 = K_{o1}\exp[-E_{a1}/(k_BT)]$ and $K_2 = K_{o2}\exp[-E_{a2}/(k_BT)]$ are the defect reduction and generation coefficients, respectively. E_{a1} and E_{a2} are the activation energies for reduction and generation and k_B is the Boltzmann's constant. α and β are constants that governor the speed of reactions. N_{do} is the point defect density at t = 0. Since the carrier density reaches equilibrium much faster than the defect reduction, we therefore consider the steady state by setting $dn/dt \approx 0$. Solving Eqs. (2) and (3), we obtain the differential equation as:

$$\frac{dC}{dt} = \frac{2C^2}{\tau_o}\left[\frac{1-\eta_o C}{1+\eta_o C}\right]\left[\frac{1}{k+1}\right]\left(k\left[\frac{(1-\eta_o C)}{(1-\eta_o)C^{1/2}}\right]^{\alpha} - \left[\frac{(1-\eta_o C)}{(1-\eta_o)C^{1/2}}\right]^{\beta}\right), \tag{4}$$

where C and τ are the normalized CL intensity expressed as C(t)/C(0) and normalized time. τ is defined as t/τ_o where $\tau_o = 1/[(K_1+K_2)n^2(t=0)]$ and η_o is the initial radiative quantum efficiency defined as $Bn(0)^2/G$. The rate coefficient ratio k is K_1/K_2 from which it governs whether the process overall is defect generating or annihilating.

Figure 21(a) illustrates the room temperature time-dependent band edge CL intensity of AlO_x/ZnO taken at various excitation powers (P_{ex}) by keeping the accelerating voltage at 5kV to maintain a fixed penetration depth. Consistent with our previous results, higher excitation leads to higher CL intensity and shorter time to reach for saturation. In figure 21(b) the corresponding normalized CL is plotted against time together with the best fits by using Eq. (4) [31,32]. From the best fits, we have determined η_o, k, α and β to be 0.01393, 2.51746, 3.36176 and 2.5025. Knowing that $k > 1$ and $\alpha > \beta$, they suggest the defect reduction rate is much faster than the generation rate and therefore we expect a net increase of emission with time will result. In addition, under a given temperature, since K_1+K_2 is a constant and τ_o therefore is proportional to $1/n(0)^2$. As the excitation power, P_{ex}, determines $n(0)^2$, we expect a linear relationship appears between τ_o and $1/P_{ex}$ [32]. In fact, as shown in the inset of figure 21(b), a straight line is revealed and the slope is determined to be 3.313, which connects to K_1+K_2.

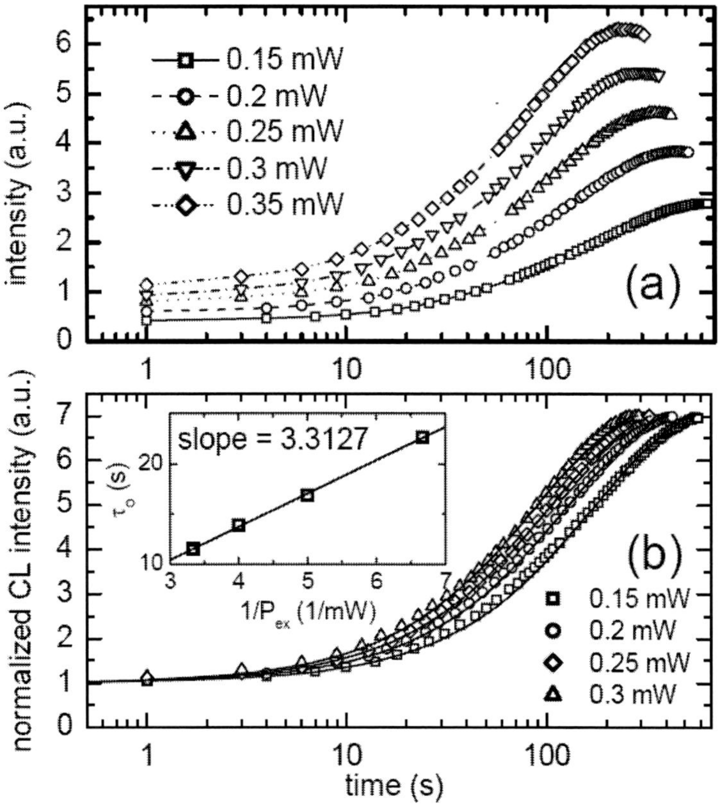

Figure 21. (a) The time-dependent CL emission intensities of AlO_x/ZnO taken at different excitation powers. (b) The corresponding normalized time dependences together with best fits. Inset: The plot of τ_o versus $1/P_{ex}$.

The temperature-dependent CL has also been carried out to elucidate the details of REDR, especially we attempt to explore the interplay between the defect generation and reduction at various temperatures. Figure 22(a) shows the initial emission spectra of AlO_x/ZnO taken from 77 to 290 K. A steady increase of the band edge emission indicates that nonradiative recombination defect centers are progressively suppressed as temperature decreases. Figure 22(b) illustrates the variation of intensity enhancement ratio as a function of temperature. The P_{ex} is kept at 0.05 mW at the same accelerating voltage. It is found that the enhancement effect is strongly dependent on temperature, with the enhancement ratio reduces from 11 to 3.33 as temperature is increased from 77 to 180 K and then increases back to 7 when the temperature reaches 295 K. Our results show that different degrees of interaction exist between the defect generation and reduction processes at various temperatures. As stated in the rate coefficients, K_1 and K_2, of Eq. (3), both the defect reduction and generation rates are expected to lower at lower temperature due to the inactivation of nonradiative recombination defect centers. However, competition among them begins as temperature increases. We note that throughout the entire temperature range being examined, the defect generation rate becomes overwhelming at 180-250K whereas reduction rate becomes dominant when temperature approaching to room temperature.

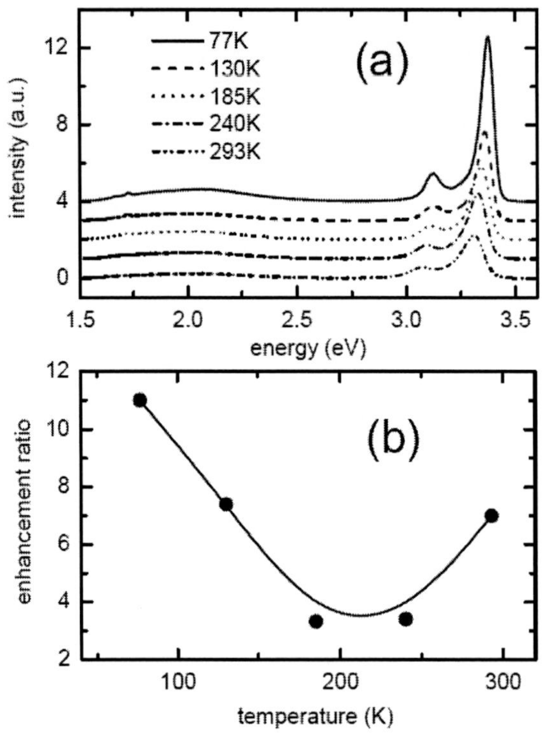

Figure 22. (a) The CL spectra of AlO_x/ZnO taken at different temperature. The spectra are vertically shifted by visualization. (b) The corresponding enhancement ratios after electron irradiation.

Time-dependent CL spectra taken at various temperatures and excitation powers have been examined. As an example, the normalized time-dependent CL taken at 77 K and excitation powers of 0.15, 0.2, 0.25 and 0.3 mW is illustrated in figure 23(a). By using Eq. (4), we obtain reasonably good fits among all the curves. In addition, similar to the case at room temperature, a linear dependence of τ_o versus $1/P_{ex}$ is shown in figure 23(b), which supports the REDR. The plot of α/β at different temperatures is shown in figure 23(c) and we see the rate of defect reduction is suddenly increased over the rate of generation above 250K although its origin remains unclear. We also have plotted Ink against 1/T and a linear relationship is obtained. From the slope, $E_{a2} - E_{a1}$ is determined to be 10meV, which indicates it is easier to activate defect reduction than generation. Finally, it is pointed out that any deviation between the time-dependence profiles and Eqs. (3) and (4) may point out the possibility that the defect generation and reduction rate dependences are more complicated than those as stated in Eq. (3). In addition, more than one type of defect may also participate in the process.

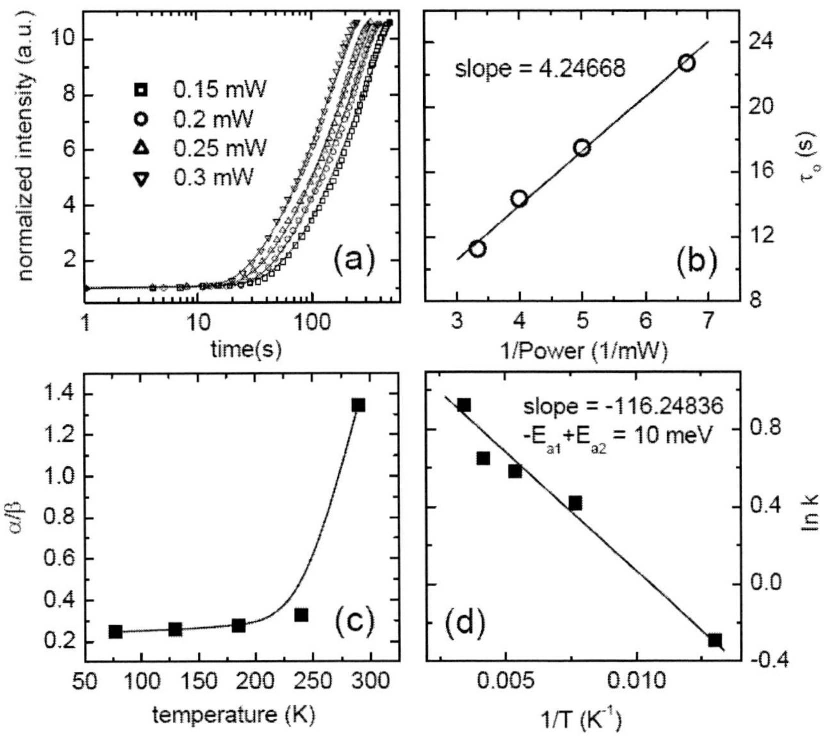

Figure 23. (a) The 77K time-dependent CL emission intensities of AlO_x/ZnO taken at different excitation powers. (b) The corresponding plot of τ_o versus $1/P_{ex}$. (c) The plot of α/β, determined from the best fits and Eq. (4) against temperature. (d) The plot of lnk versus 1/T.

II. Charging Effect

Finally, we would like to remark the recent experiment carried out by Nikiforov et al that should deserve attention. They have observed similar emission enhancement effect from Zn-based quantum wells and have attributed the effect to the presence of internal fields due to charge trapping in the pre-existing defect recombination centers. Because of the charge trapping condition, they note the effect is strongly electron dose dependent. We have determined the electron dose deposited into the film according to the following equation given as:

$$\text{Electron dose} = \frac{I_p \times t}{q \times A_s}, \tag{5}$$

where q is the electron charge. As shown in figure 24, the normalized CL spectra taken at different excitation powers are plotted as a function of electron dose. One can see that all CL spectra are merged into one single profile, which indicates that charging effect may also play an important role in governing the emission profile. More studies are necessary to clarify this issue.

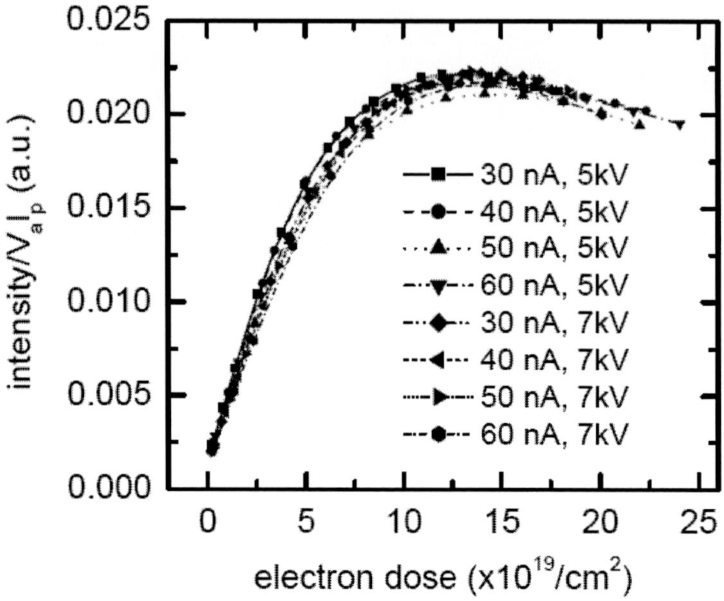

Figure 24. The electron dose dependence of CL emission intensity/power of AlO_x/ZnO taken at various excitation powers.

Conclusion

In summary, we find that by capping the ZnO with a thin layer of AlO_x, its emissions and charge conduction can be modified simultaneously under the electron irradiation. Based on this effect, we have developed a simple technique to pattern UV emission on ZnO films as well as nanostructures. In particular, by carefully controlling the electron and energy doses of the electron beam, well-defined sub-micron patterns can be fabricated on 2-D films and 1-D nanoribbons with controlled optical and electrical properties. In addition, similar effect can be obtained from more complicated ZnO-based multilayer structures and this allows us to gain control on the quantum size effect from quantum wells. Since ZnO is a versatile material that possesses a unique combination of different properties, we believe this technique may be further applied in modifying the local piezoelectric, photoelectrical, photochemical and photo-induced hydrophilic properties of ZnO.

Acknowledgements

The authors would like to thank Prof. J.B. Xu and J. An for conductive atomic force microscopy measurements and Prof. J.Y. Dai and Dr. X.Y. Zhang for providing nanoribbons and TEM measurements. This research was supported by the Chinese University of Hong Kong through the RGC Competitive Earmarked Research Grants (402004 and 402705).

REFERENCES

[1] C.M. Lieber, *MRS Bulletin.* 28, 486 (2003).
[2] N.A. Melosh, A. Boukai, F. Diana, B. Gerardot, A. Badolato, P.M. Petroff, and J.R. Heath, Science 300, 112 (2003); Y. Huang, X.F. Duan, Y. Cui, L.J. Lauhon, K.H. Kim, and C.M. Lieber, *Science.* 294, 1313 (2001).
[3] X.Y. Kong, Y. Ding, R. Yang, and Z.L. Wang, Science 303 1348-1351 (2004); X.F. Duan, C.M. Niu, V. Sahi, J. Chen, J.W. Parce, S. Empedocles, and J.L. Goldman, *Nature.* 425, 274 (2003).
[4] Z.H. Zhong, D.L. Wang, Y. Cui, M.W. Bockrath, and C.M. Lieber, Science 302, 1377 (2003); L.J. Lauhon, M.S. Gudiksen, C.L. Wang, and C.M. Lieber, *Nature.* 420, 57 (2002).
[5] X.F. Duan, Y. Huang, R. Agarwal, and C.M. Lieber, *Nature.* 421, 241 (2003).
[6] W.I. Park and G.C. Yi, *Adv. Mater.* 16, 87 (2004).
[7] S.J. Pearton, D.P. Norton, K. Ip, Y.W. Heo, and T. Steiner, J. Vac. Sci. and Technol. B 22, 932 (2004); D.C. Look, *Mater. Sci. and Eng.* B 30, 383 (2001).
[8] K. Vanheusden, C.H. Seager, W.L. Warren, D.R. Tallant, and J.A. Voigt, *Appl. Phys. Lett.* 68, 403 (1996).
[9] D.S. Ginley and C. Bright, *MRS Bulletin.* (2000), and reference therein.
[10] P. Sharma, A. Gupta, K.V. Rao, F.J. Owens, R. Sharma, R. Ahuja, J.M.O. Guillen, B. Johansson, and G.A. Gehring, *Nature Mater.* 2, 673 (2003).
[11] D.P. Fleischauer, H.K. Alan Kan, and J.R. Shepard, J. Am. Chem. Soc. 94, 283 (1972).
[12] K. Keem, H. Kim, G.T. Kim, J.S. Lee, B. Min, K. Cho, M.Y. Sung, and S. Kim, *Appl. Phys. Lett.* 84, 4376 (2004).
[13] X. Feng, L. Feng, M. Jin, J. Zhai, L. Jiang, and D Zhu, *J. Am. Chem. Soc.* 126, 62 (2004).
[14] M.H. Zhao, Z.L. Wang, and S.X. Mao, *Nano Lett.* 4, 587 (2004).
[15] M.H. Huang, S. Mao, H. Feick, H.Q. Yan, Y.Y. Wu, H. Kind, E. Weber, R. Russo, and P.D. Yang, *Science.* 292, 1897 (2001).
[16] H. Cao, Y.G. Zhao, S.T. Ho, E.W. Seelig, Q.H. Wang, and R.P.H. Chang, *Phys. Rev. Lett.* 82, 2278 (1999).
[17] K. Nomura, H. Ohta, K. Ueda, T. Kamiya, M. Hirano, and H. Hosono, *Science.* 300, 1269 (2003).
[18] Q. Wan, Q.H. Li, Y.J. Chen, T.H. Wang, X.L. He, J.P. Li, and C.L. Lin, *Appl. Phys. Lett.* 84, 3654 (2004).
[19] C.W. Sun, P. Xin, Z.W. Liu, and Q.Y. Zhang, Appl. Phys. Lett. 88, 221914 (2006)
[20] K.C. Hui, J. An, X.Y. Zhang, J.B. Xu, J.Y. Dai and H.C. Ong, *Adv. Mater.* 17, 1960 (2005).
[21] K.C. Hui, H.C. Ong, P.F. Lee, J.Y. Dai, *Appl. Phys. Lett.* 86, 152116 (2005).
[22] X.Y. Zhang, J.Y. Dai, H.C. Ong, N. Wang, H.L.W. Chan, C.L. Choy, *Chem. Phys. Lett.* 393, 17 (2004).
[23] K.C. Hui, C.W. Lai, and H.C. Ong, *Thin Solid Films.* 483, 222 (2005).
[24] B.G. Yacobi and D.B. Holt, Cathodoluminescence Microscopy of Inorganic Solids, Plenum, New York, 1990.
[25] E. Z. Luo, I. H. Wilson, X. Yan, and J. B. Xu, *Phys. Rev.* B 57, R15120 (1998).

[26] C. H. Seager, D. R. Tallant, and W. L. Warren, *J. Appl. Phys.* 82. 4515 (1997).

[27] C.W. Sun, P. Xin, Z.W. Liu, and Q.Y. Zhang. *Appl. Phys. Lett.* 88, 221914 (2006).

[28] R.G. Xie, T. Sekiguchi, T. Ishigaki, N. Ohashi, D.S. Li, D.R. Yang, B.D. Liu, and Y.S. Bando, Appl. Phys. Lett. 88, 134103 (2006); R.G. Xie, D.S. Li, D.R. Yang, T. Sekiguchi, and M.H. Jiang, *Nanotechnology.* 17, 2789 (2006).

[29] D.V. Lang and L.C. Kimerling, Phys. Rev. Lett. 33, 489 (1974); D.V. Lang and L.C. Kimerling, *Appl. Phys. Lett.* 28, 248 (1976).

[30] J.D. Weeks, J.C. Tully, and L.C. Kimerling, *Phys. Rev.* B, 12, 3286 (1975).

[31] S.L. Chuang, N. Nakayama, A. Ishibashi, S. Taniguchi, and K. Nakano, *IEEE J. Quantum Electron.* 33, 970 (1997).

[32] M.X. Tang, K. Shum, L.F. Zeng, and M.C. Tamargo, *Appl. Phys. Lett.* 73, 1541 (1998).

[33] T. Murata, Y. Ohno, S. Kishimoto, and T. Mizutani, *Solid State Electron.* 43, 147 (1999).

[34] A.Y. Nikiforov, G.S. Cargill, S.P. Guo, and M.C. Tamargo *J. Appl. Phys.* 97, 63521 (2005).

In: Progress in Advanced Materials Research
Editor: Nicolas H. Voler, pp. 29-78

ISBN: 978-1-60021-893-4

Chapter 2

IMPROVEMENT OF VORTEX PINNING IN MgB_2 BY DOPING

W. K. Yeoh, J. Horvat, J. H. Kim and S. X. Dou
Institute for Superconducting and Electronic Materials,
University of Wollongong, Wollongong, NSW 2522, Australia

ABSTRACT

The superconductivity of MgB_2 has been hidden for nearly 50 years although it has the highest T_c among the intermetallics superconductors. Beside the high T_c, simple crystal structure, large coherence length, high critical field, transparency of grain boundaries to current flow and low normal state resistivity MgB_2 is a fascinating topic to study for both large scale application and electronic devices. Moreover, the presence of two-gap superconductivity (π and σ band) has been theoretically and experimentally established.

However, critical current density of pristine MgB_2 drops rapidly in the high magnetic field due to the weak pinning centres and low upper critical field. During the past three years, novel techniques and developments for fabrication of useable MgB_2 have been reported, including chemical alloying, irradiation, thermo-mechanical processing techniques and magnetic shielding to improve the critical density, upper critical field and the irreversibility field. Among the studies, atomic substitution, especially using nano-particles, may help in clarification of the superconductivity mechanism thus making it appropriate for practical application. On the other hand, the momentum of enhancing flux pinning using chemical doping is moving to a positive side. Nano-SiC powder is well known for effective improvement of the critical current density.

In this chapter, a brief introduction of superconductivity in MgB_2 will be introduced. Further, we review the basic electronic and magnetic properties of the MgB_2 as well as its crystal structure. We will also discuss the preparation method that has been well developed for experimentation on MgB_2. Along with that, we review the experimental results of the chemical alloying on MgB_2, particularly on the critical current density. In addition, the future prospect of MgB_2 and the development for the application in the current superconductivity market will be given.

1. Introduction

Superconductors are materials that lose their resistance to electrical current flow below certain values of critical temperature (T_c), critical current density (J_c) certain critical field (H_c). Superconductors have generated great interest for power applications including loss-free electric transmission cables, motors, generators, transformers, energy storages, levitation trains, ultra-fast computers etc. Besides that, the superconductivity phenomenon also opens up new challenges to modern physics. Many materials become superconducting when they are cooled to near absolute zero temperature (-273°C) including the C_{60} [1] and the DNA [2].

Critical transition temperature (T_c) is the temperature above which a superconductor loses its superconductivity. The maximum current that a superconductor can carry under superconducting state is called critical current density (J_c). When a superconductor is placed in a weak external magnetic field, the field penetrates the superconductor for only a short distance λ, called the penetration depth, after which it decays rapidly to zero. This is called the Meissner effect, and is a defining characteristic of superconductivity. For most superconductors, the penetration depth is on the order of 100 nm.

Superconductors can be divided into two classes according to their response to magnetic field. All practical superconductors are Type II superconductors. In Type II superconductors, raising the applied field past a critical value H_{c1} leads to a mixed state in which an increasing amount of magnetic vortices penetrates the material, but there remains no resistance to the flow of electrical current as long as the current is not too large. At the second critical field H_{c2}, superconductivity is destroyed.

It is very important that these vortices do not move in response to magnetic fields if superconductors are to carry loss-less currents. Vortex movement results in resistivity. Vortices can be effectively pinned at sites of atomic defects, such as inclusions, impurities, and grain boundaries. All these critical parameters determine the performance of a superconductor and their value in the applied field.

The phenomenon of superconductivity was observed for the first time in mercury by Dutch physicist, H. K. Onnes in 1911. At the temperature of 4.2 K, he observed electrical resistivity of mercury abruptly dropping to zero. In the following years, other metals such as tin and lead were also found to exhibit superconductivity at very low temperatures. However, the majority of superconductors are not pure elements, but alloys and compounds. Intermetallics compounds, such as niobium-tin, vanadium-silicon and niobium-titanium were developed from the 1930s onwards and considered as the low temperature (LT) superconductors. And, in 1962 scientists at Westinghouse developed the first commercial superconducting wire, an alloy of niobium and titanium (NbTi). The highest critical temperature observed before the discovery of high temperature superconducting materials was 23.2K, for Nb_3Ge.

A complete microscopic theory of superconductivity was developed by Bardeen, Cooper, and Schrieffer proposed in 1957, so called BCS theory. Under this theory, the superconducting charge carriers are pairs of electrons (Cooper pairs) interacting through the exchange of phonons. Before 1986, physicists believed that BCS theory prohibited superconductivity at temperatures above 30 K. However in 1986, researchers at the IBM Research Laboratory, Bednorz and Müller, discovered superconductivity in a lanthanum-based cuprate perovskite material, which had a transition temperature of 35 K and high temperature (HT) superconductivity was born.

The major break through was shortly found by Paul C. W. Chu et al. where they discovered the YBCO superconductor by replacing lanthanum with yttrium in the lanthanum based superconductor. YBCO superconductor which has the critical temperature of 92 K superconducts at temperatures warmer than liquid nitrogen and cheap liquid nitrogen could be used as a coolant. Since then, other classes of high temperature superconductors have been discovered, such as bismuth, thallium and mercury based cuprate superconductors. Figure 1 shows the history of superconductor discovery. Up to recent date the highest superconducting temperature (T_c) is still held by the Hg-based superconductor (153K), that was found in 1993. There are a total of 13 Nobel laureates in superconductivity, from H.K. Onnes (1913) to the recent one P. C. Lauterbur and Sir P. Mansfield (2003) on their contributions to theory, experimental results and application of superconductivity.

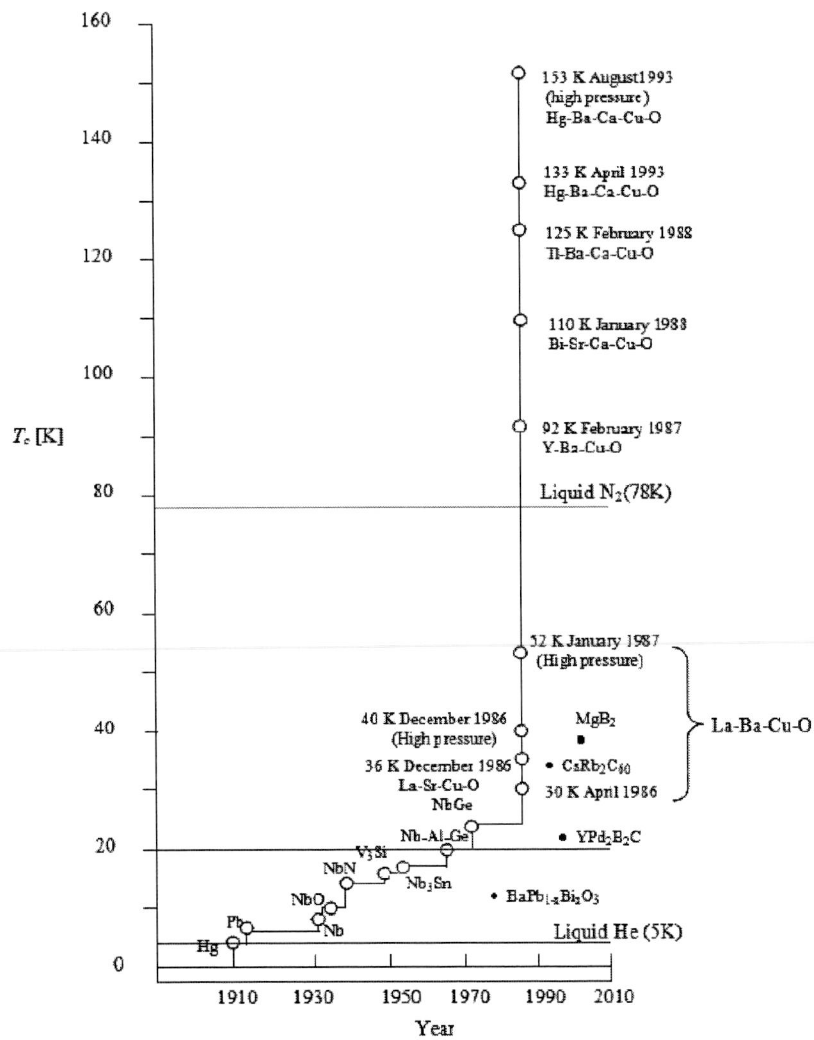

Figure 1. Evolution of T_c with time.

The superconductivity of MgB_2 has been unknown for nearly 50 years after its first synthesis [3] although it has the highest T_c among the intermetallic superconductors. The discovery of superconductivity in MgB_2 [4] has revolutionized our approaches to both theory and to experiment. The limit of T_c in metallic superconductors had been believed to be ~ 30K, as predicted by the BCS theory, yet T_c of MgB2 is 39K. The recent discovery also demonstrated that we still have not come close to discovering all the materials that exhibit superconductivity and it is quite likely that new superconducting materials will be found in the future. At the time this review is written, the first paper on MgB_2 by J. Akimitsu in Nature has been cited for nearly 1950 times and this is just a conservative estimate.

The discovery of MgB_2 also prompted a new wave of investigation on high T_c materials in similar systems. Large number of experimental and theoretical works has been carried out in a short time [5-9] because of its unexpectedly high T_c. Due to the high T_c, simple crystal structure, large coherence length, transparency of grain boundaries to super currents and low normal state resistivity, MgB2 is a fascinating topic to study for both large scale application and electronic devices. Moreover, the presence of two-gap superconductivity (π and σ band) has been theoretically and experimentally established. Beside this, MgB2 shows a significant isotope effect like a low temperature superconductor.

However, critical current density of pristine MgB_2 drops rapidly in the high magnetic field due to the weak vortex pinning and relatively low upper critical field. During the past three years, novel techniques for fabrication of useable MgB_2 have been reported including chemical alloying, irradiation, thermo-mechanical processing techniques and magnetic shielding to improve the critical density, upper critical field and the irreversibility field. Although many difficulties need to be overcome, rapid progress in the fabrication of various forms of MgB_2 in single crystals, bulk samples tapes, wires and thin films is quite promising for the realization of practical applications in the near future.

Today, the well-established Nb-based LT superconductor has become the dominant commercial superconductor and represents 99.99% of world production of superconducting materials because it can be economically manufactured in a ductile form with the prerequisite nano-structure needed for high critical current. Similarly, Nb_3Sn based strand, although based on a brittle A15 superconducting phase, can be manufactured into strong composites in km lengths and microstructures that promote high critical current densities. These superconductors are often termed "technical superconductors" because of their applicability to engineering tasks. All these conductors require cooling to 4 K (liquid He is the most common coolant).

Although the HT superconductors have a higher critical temperature compared to the LT superconductor and can be made superconducting by immersion in liquid nitrogen with a temperature of 77K (–196°C), most of them are unsuitable for practical applications because the weak link effect or the toxicity of the starting materials, like thallium and mercury.. The most promising in HT superconductors in application are the bismuth-based and the yttrium based superconductors, where development of Bi2223/Ag tapes and YBCO coated conductors were major focus of interest due to promise of high J_c and T_c of the superconductors.

Today's MRI machines use expensive niobium-alloy wires and require costly liquid helium refrigeration to maintain superconducting properties. The prices of pre-reacted MgB_2 cost about US$750/kg to US$300/kg, depending on the volume demand. This results in the cost of production (CP) for MgB_2 wire in the range of US$0.16 to US$0.88/kA per m, at 25

K, 1 T. When we compare this result with CP of niobium–titanium (NbTi) and niobium–tin (Nb_3Sn) which is roughly \$1/kA/m at 4.2 K and 2 T, and \$10/kA·m at 4.2 K and 10 T, respectively, we could predict that MgB_2 wire could potentially replace NbTi for power devices such as transformers and MRI. With a smaller conductor, the cost of wire used in MRI magnets could be reduced from US \$3-10 per kiloampere of current per meter of wire to only US \$1-2 [10]. The development of less expensive wires by the private sector could put an MRI machine in every doctor's office and even in veterinary hospitals.

In this chapter, we review the basic electronic and magnetic properties of the MgB_2 as well as its crystal structure. We will also discuss the preparation method that has been well developed for experimentation on MgB_2. Along with that, we review the experimental results of the chemical alloying on MgB_2 particularly on the critical current density. In addition, the future prospect of MgB_2 and the development for the application in the market of superconductor today will be given.

2. The Discovery of MgB_2

In January 2001, Prof. J. Akimitsu announced the discovery of superconductivity in magnesium diboride with feasible superconductivity observed up to 39K (-234°C). The discovery has generated enormous interest and excitement in the superconductivity community and the world. Many newspapers and news channels such as the New York Times, the Washington Post, the Chicago Tribune, CNN, etc., have reported on it. The UK based journal Nature published Akimitsu group's paper in 2001 and revived the interest of superconductivity in non-oxides and boron related compounds [4].

Table 1. Al-B_2 type borides superconductor with the transition temperature

Formula	Transition temperature (K)	Reference
NbB_2	0.62	[14]
$NbB_{2.5}$	6.4	[15]
$Nb_{0.76}B_2$	9.2	[16]
$MoB_{2.5}$	8.1	[15]
BeB_2	0.72-0.79	[17]
ZrB_2	5.5	[18]
TaB_2	9.5	[19]

Boron atoms have a suitable size and electronic structure for forming direct B-B bonds than can form various kinds of boron network. Higher dimensionality of network is formed with increasing the boron content [11]. There are more than 50 boride compounds with different structures reported to be superconductors. The details can be found in the review paper [12]. However, the most outstanding discovery is that of B, which under pressure of 160 Gap will exhibit superconductivity with the transition temperature of 11.2K [13]. All the Al-B_2 type borides are listed in the table 1 with their transition temperature. Among all of these compounds some were discovered before the discovery of the superconductivity in MgB_2 and some were reported after the discovery of superconductivity in the MgB_2. A more systematic investigations and studies of the relationship between stoichiometry and crystal structure is need to be carried out since there were some contradictory results in some compounds especially in compounds like TaB_2 and ZrB_2. The fact that some borides have the controversial results on the critical temperature, suggests that non-stociometry may be an important factor in the superconductivity of this family.

3. Crystal and Electronic Structure of MgB_2

MgB_2 posses an AlB_2-type hexagonal structure (space group P6/mmm) with alternating boron honeycomb planes and magnesium triangular planes as shown in the figure2. The Mg atoms are located at centres of hexagons formed by boron and they donate their electron to the boron planes; hence the B-B bonding is strongly anisotropic. With the structure of MgB_2 similar to the graphite, the distance between the boron planes is significantly longer than the distance between the boron atoms in the planes. He et al. analysed the crystal structure of MgB_2 using the Rietveld method and found that the compound has the unit cell parameters: a = 0.3081361(14) nm and c = 0.351782(17) nm at room temperature [20]. The same conclusions on the crystal structure have been made on the basis of studies of high resolution transmission electron microscopy (HRTEM) [21], high resolution powder diffraction [22] and electron energy loss spectroscopy [23]. This value of lattice parameters for MgB_2 is in the middle of the values of lattice parameters of AlB_2-type compounds. There is not sign of structural transition for MgB_2 down to 2K or under the high pressure of 40GPa.

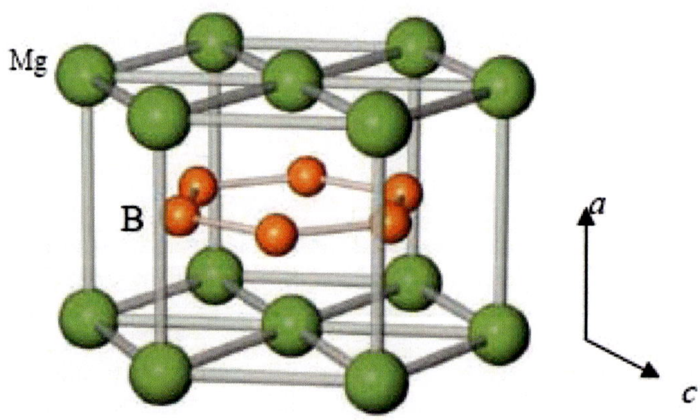

Figure 2. Crystal structure of MgB_2 [24].

The band structure of MgB_2 has been long reported before and after the discovery of the superconductivity of MgB_2 [25-31] and it is now known in detail. Despite its crystal structure being similar to graphite intercalated compounds, the band structures are more similar to those of graphite.

In MgB_2, the sp^2 boron orbitals overlap, creating σ bonds within the neighbouring atoms in the plane while the reaming p orbitals extend above and below the plane and create the π

bands. As shown in the figure 3a the golden hexagonal network is associated with the σ bands, whereas the six green lobes above and below the boron plate are associated with the π bands. These 2D and 3D metallic-type states contribute almost equally to the state of the Fermi level and the electrons are delocalized throughout the honeycomb. Unlike conventional metallic bands, the σ band has charge concentrated along the B-B axes rather spread through the unit cell (figure 3b). In the Fermi surface of the MgB_2 superconductor, the green and blue surface comes from the bonding $p_{x,y}$ bands, the blue tubular network from the bonding p_z bands, and the red tubular network from the antibonding p_z band. However, the complexity of the electronic structure of MgB_2 and especially coexistence of both quasi-2D covalent cylindrical network and a 3-D metallic type interlayer conducting network raise the possibility of having two distinct superconducting energy gaps.

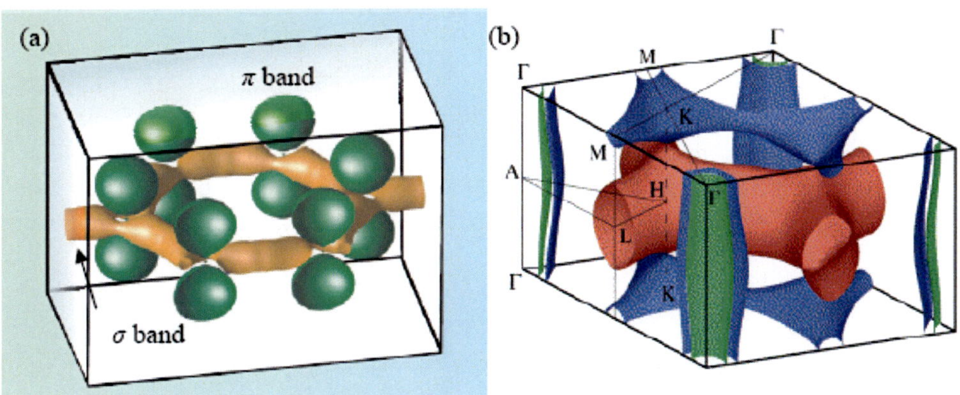

Figure 3. Electronic structure of MgB_2 (a) the 2-D network of σ bands and 3-D network of π bands [32] (b) Fermi surface of MgB_2. The vertical sections of cylinders at the corners are associated with the σ bands; the more 3 D network of tunnels and caves in the centres of the zone is associated with π bands. [28].

Historically, two-band superconductivity was proposed right after the formulation of the BCS theory as the extension of the BCS theory by Suhl et al. [33]. In the early 1960s, experimental claims of two-bands superconductor in some transition metals like V, Nb and Ta [34, 35] and later in SrTiO3 [36] and borocarbide [37] were observed.

MgB_2 appears to be the first system for which multiband superconductivity has been identified by several experimental techniques: heat capacity, tunnelling spectroscopy, Raman spectroscopy, penetration depth measurements and angle-resolved photoemission spectroscopy using polycrystalline samples or single crystals. The imperfectly filled of 2-D σ bands and the weak coupling of 3-D π bands lead to existence of two energy gaps in this material. The two bands interact with each other and manifest the unique properties of MgB_2 with different degree of interaction. There are three different impurity scattering channels in MgB_2 alloys: intraband scattering within each σ and π band and interband scattering between them. As a result, by tuning the rate of interactions, the upper critical field can be significantly enhanced [38, 39].

In most of the experimental reports, it has been observed that the σ gap decreases with the decreasing of T_c while the π gap remains constant and independent of the T_c. The reports from Gonnelli et al. [40, 41] demonstrated a different behaviour of the superconducting gaps depending of the type of dopant. E. Ohimichi suggested that the two-gap superconductivity is still valid in carbon-substituted samples, and that the smaller π-band gap is robust against carbon substitution in contrast to an appreciable decrease in the larger σ-band gap [42]. The interband impurity scattering between σ and π bands is exceptionally small, due to the particular electronic structure of MgB_2, so that in the superconducting state the two gaps in the σ and π bands are preserved even in ''dirty limit'' samples with a considerably reduced T_c and a broad range of normal state resistivities. However, a recent point contact study on MgB_2 crystals containing high C substitution levels showed a merging of the σ and π gaps for the first time as predicted by the theoretical prediction that the σ and π band gaps merge into a single gap giving T_c =25 K [43].

4. Electron Phonon Coupling in MgB_2

MgB_2 with the reported T_c of 39K seems to be the upper limit that can be explained by the phonon-mediated BCS superconductivity. BCS predicted that high frequency phonons in the low atomic masses produce a high T_c. From the fundamental point of view, explanation of superconductivity in MgB_2 will show whether the new systems can be understood within the conventional electron-phonon (EP) mechanism or, the high transition temperature might imply exotic mechanism coupling.

According to initial findings, MgB_2 seems to be a low- T_c superconductor with its properties resembling those of conventional superconductors. The measurements of the isotope effect revealed the importance of the electron-phonon interaction for the Cooper pairing mechanism in MgB_2: T_c shifted by 1K when ^{11}B was substituted for the ^{10}B [44, 45], as shown in figure 4.

Besides that, most of the experimental results like linear T-dependence of the H_{c2} with a positive curvature near T_c [47], shift of the T_c and $T_{c\text{-}onset}$ to a lower temperature with the increasing of magnetic field in the resistivity measurements [48, 49] were reported to be similar to the low temperature superconductors.

On the other hand, contradicting reports indicated unconventional superconductivity similar to cuprates, like quadratic T-dependence of the penetration depth λ(T) [50-52] and reversal of the sign of the Hall coefficient near T_c [53]. More recently, optical studies in *c*-axis oriented superconducting MgB_2 films indicate that the electron-phonon coupling is very weak and insufficient to produce the high T_c in the MgB_2 [54]. The value of the isotope-effect coefficient of Mg, 0.32, h is much lower than the expected value of 0.5 for BSC superconductor [44].

Majority of the theorists support the BSC mechanism, *i.e.*, an electron-phonon driven *s*-wave mechanism of the superconductivity in MgB_2. However, the theory of hole superconductivity explains the paring mechanism in the MgB_2 in another way. Superconductivity in this theory occurs through lowering of kinetic energy of the holes as they form Cooper pairs, through decrease of the effective mass of the charge carriers with creation of Cooper pairs. This is opposed to the BCS theory, where Cooper pairs are created through lowering of potential energy of charge carriers. In the case of theory of hole superconductivity, the planes of B atoms are proposed to be akin to the Cu-O planes in the cuprates and the high temperature superconductivity in MgB_2 arises from the boron $p_{x,y}$

orbitals in the negatively charges B-planes [55, 56]. Large values of T_c, observed isotope effect, reversal of the sign of Hall coefficient and many other observed phenomena seem to be a natural consequence of the creation of Cooper pairs in the hole theory of superconductivity. At this stage, though, the theory is not widely accepted.

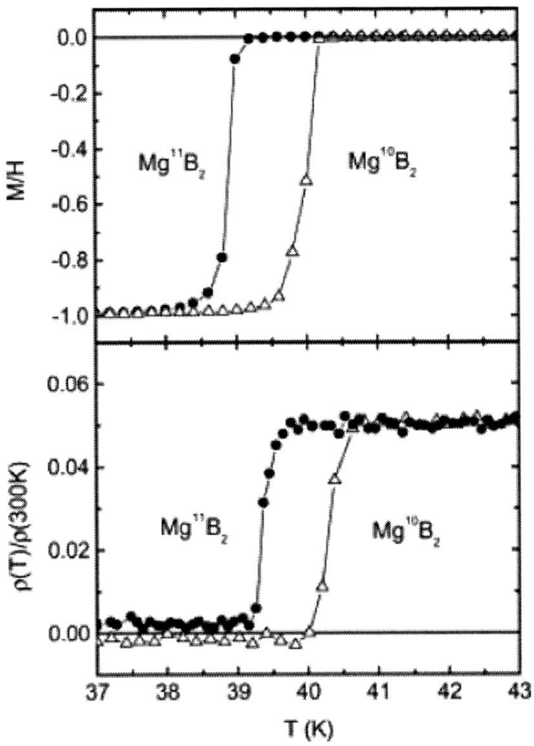

Figure 4. Boron Isotope Effect in Superconducting MgB_2 [46].

5. Preparation Method

Low cost and simple preparation methods of MgB_2 promise the important large-scale and electronic device application. In this section we will concentrate on the fabrication on bulk pellet and wires of MgB_2 and some of the important synthesis issue of MgB_2. Although magnesium diboride is available from chemical supplier as commercial powder, the critical properties like critical current density and upper critical field of this commercial powder are not so desirable and need further improvement.

5.1. Bulk MgB_2 Superconductors

5.1.1. In-Situ Reaction

MgB_2 can be synthesized by a simple route: reaction of B with Mg vapor, generally at temperature above 650°C which is the melting point of the Mg. Up to date this method has been used to form MgB_2 powder, sintered pellets, wires and films. The binary diagram for Mg-B is shown in the figure 5. MgB_2 is shown as a line compound that decomposes above 1500°C when it is in equilibrium with Mg vapor. As from the phase diagram, the liquid phase of MgB_2 varies from the around 650°C to 1000°C which also indicated the big variation of sintering temperature to form MgB_2. Prolonged liquid phase is necessary to yield a homogenous MgB_2 with a better superconducting properties. Yamamoto et. al reported a formation of MgB_2 below the melting point of magnesium is possible under many hours of sintering, which is also known as solid-solid reaction of magnesium and boron[57]. Kumakura et al. succeeded in synthesis of high J_c in situ MgB_2 tapes by low temperature reaction for 1 h below 650°C, using ultrafine magnesium or MgH_2 and high purity boron powders as starting materials [58, 59]. GaN, which has low melting point, has been used to promote the liquid phase sintering in MgB_2 and thus the J_c was enhanced by a factor of 2 and 10, for temperatures of, 5 K and 20 K, respectively [60].

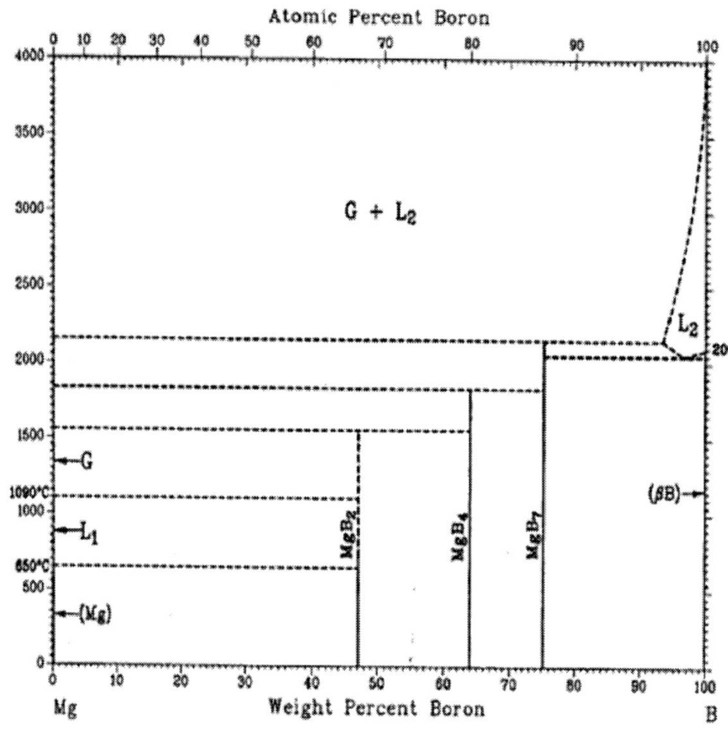

Figure 5. Phase Diagram of Mg+B [61].

DTA traces of *in situ* MgB_2 wire showed two exothermal peaks were observed at around 457°C and 619°C respectively. According to the Goldacker *et al.*, the first peak is corresponding to the formation of MgO and melting of B_2O_3 due to Mg react with B_2O_3 to form MgO + B [62]. The second exothermal peak was consequence of the formation of MgB_2 between the melting of Mg and the amorphous B. However, the both the exothermal peaks were slightly shifted to the higher temperature for both nano C and CNT doping which due to the presence of additional impurities like nano C and CNT in the mixture of B and Mg that delayed the reaction Mg and B powder [63].

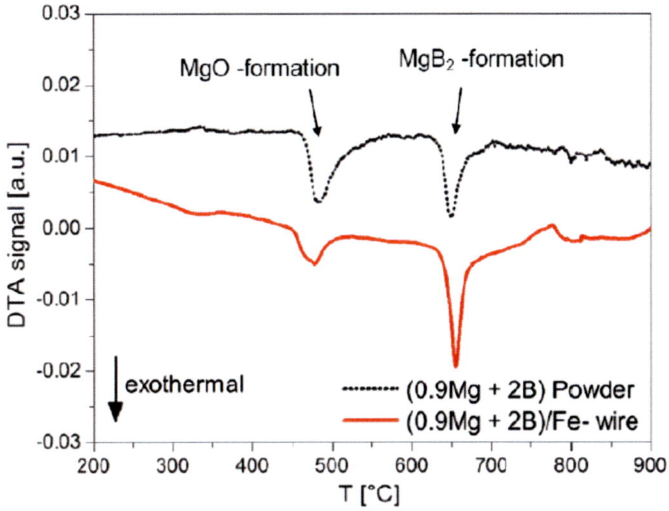

Figure 5A. DTA characterization of a MgB2/Fe wire and the corresponding Mg + B precursor powder mixture [62].

5.1.2. HOT ISOSTATIC PRESSING (HIP)

HIP process for MgB_2 is promising as it can produce highly-dense, low-porosity, crack-free samples which allow to improve mechanical properties of MgB_2. HIP has been reported to improve superconducting properties in bulks and wires samples [64-67].

In particular, a hot isostatic pressed (HIPed) MgB_2 pellet was found to have significantly increased J_c at high fields as compared to its un-HIPed counterpart as shown in the figure 6. The improvement is attributed to elimination of porosity present in the MgB_2 pellet resulting in an improved intergrain connectivity (figure 7). The HIP process was also found to disperse the MgO that existed at the grain boundaries of the un-HIPed MgB_2 pellet and to generate more dislocations in the pellets [67].

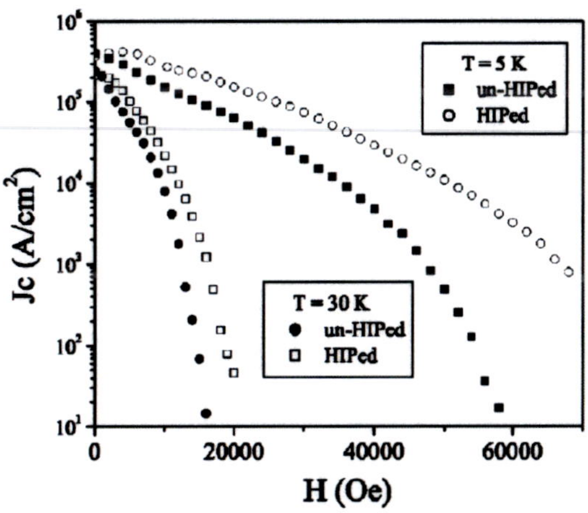

Figure 6. Magnetization critical current density J_c as a function of magnetic field H for the un-HIPed and HIPed samples at 5 and 30 K. The J_c at 0 field is nearly the same for both samples, but the differences between the samples increases with field, and the drop in J_c at higher fields is remarkably faster in the un-HIPed sample than in the HIPed one [66].

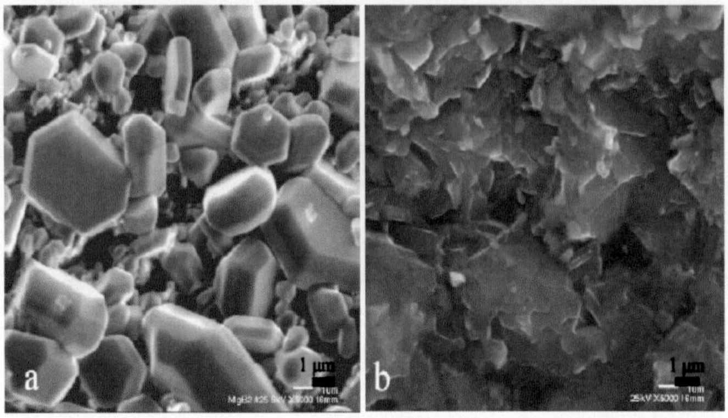

Figure 7. SEM micrographs of MgB_2 samples (a) surface of the un-HIPed and (b) surface of HIPed sample [66].

5.1.3. MECHANICAL ALLOYING

It is a well-known fact that, mechanical alloying is a non-equilibrium processing technique to prepare metastable amorphous, quasi crystalline, and nanocrystalline materials [68]. The use of mechanical alloying (MA) for the preparation of nanocrystalline MgB_2 superconductors was shown to be an appropriate technique that yields a highly pure, homogeneous, nanocrystalline microstructure and refined particle size, thus, enhancing flux pinning ability of the grain boundaries [68-76].

Starting materials and milling balls are placed together in a milling vial, which is exposed to a strong vibration or rotational acceleration causing impacts of milling balls leading to repeated fractioning and cold welding of powder particles. The milling conditions can be adjusted so the desired phase forms via solid-state reaction at moderate temperatures. Normally the milling process will be performed in the Ar atmosphere in the sealed tungsten carbide vial on a planetary on SPEX ball milling. However, long time of sintering will expose the final product to impurities contamination from the vial. Using mechanically alloyed elemental Mg and B with subsequent hot pressing; J_c and the irreversibility line have been substantially improved [69]. The increment in J_c was ascribed to the large number of grain boundary pinning contributed by nanoparticles. Meanwhile, mechanical milling on the commercially available MgB_2 powder has led to the deterioration of T_c and superconductivity was totally lost after 20 h of milling. However, T_c was subsequently restored upon annealing [77]. More recently, it was demonstrated for MgB_2 with carbon doping under high-energy mechanical milling combined with HIP can substantially enhance the critical current density and the upper critical field [78].

5.2. MgB_2 Superconductor Tape and Wire

5.2.1. POWDER IN TUBE (PIT)

It has been mentioned that MgB_2 is not significantly affected by weak links, and thus high transport critical current densities of the order of 10^6 A/cm^2 have been reported in the temperature range between 4.2 and 25 K in tape and wire samples [79-81]. This has important consequences for the application of MgB_2 in wires and tapes for cable and magnet industries to be competitive with low T_c superconductors like NbTi and Nb_3Sn for operation at higher temperature at around 20K.

The general process of tape and wire fabrication is the powder-in-tube (PIT) process, [82-87] which is characterized by filling metallic tubes with powder and then drawing and rolling into tapes (figure 8). Two methods have been reported in the literature so far for wire application. One is the in situ PIT technique, which consists of filling of the metallic tubes with elemental Mg and B powders and subsequent deformation and heat treatment [82, 83, 88].

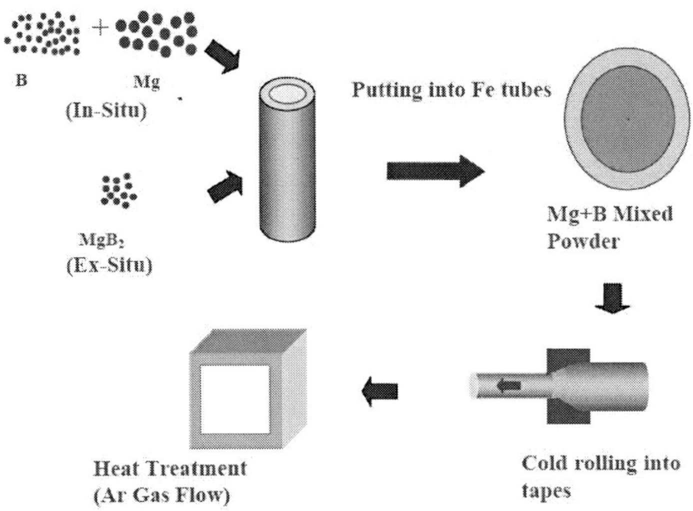

Figure 8. Process of Powder In Tube (PIT) for in-situ and ex-site method.

Another is the ex situ PIT technique, which consists of direct filling of the metallic tubes with reacted MgB_2 powder, followed by deformation [85, 86]. Heat treatment at 600-1000 °C is usually applied to the cold-worked tape or wire to obtain superconductivity. Although the fabrication of ex-situ method is cheaper and simple, the in-situ method still holds advantages that make it worth further exploring. For example, it is easy to carry out impurity doping [89-91]. Already, long mono core wires over 100 m in length have been successfully developed by the PIT method [92].

Generally, normal metal cladding provides parallel electrical conduction, thermal stabilization, and mechanical protection for the superconductor cores. In particular, MgB_2 is mechanically hard and brittle and therefore impossible to draw into the desired wire geometry. Sheath materials which are applied to MgB_2 core were stainless Steel (SS) [86, 93-96], Cu [82, 85, 97], Ag [82, 85], Ni [85, 88], Cu–Ni [96], and Fe [83, 88, 98-100]. The selection of sheath materials is limited due to the reactivity of Mg for the *in situ* process. Mg tends to react and combine with many metals such as Cu or Ag to form solid solutions or intermetallics with lowered melting points [61] which make the metal cladding not suitable during sintering of MgB_2 at around 900 to 1,000°C.

The sheath material's toughness and chemical compatibility with MgB_2 are crucial for fabricating long-length MgB_2 tapes or wires with high critical current. Sheath materials with high mechanical strength, such as stainless steel (SS) and carbon steel (CS), can achieve high

J_c values for tapes prepared by the ex situ process [86, 101]. However, sheath materials with high mechanical strength usually show poor ductility. Others, such as Nb and Ta, increase the fabrication cost and weight and reduce the merits of using MgB_2. Among these metals, Fe is one of the most promising materials due to its ductility, low cost and light weight as well as the suitability of this technique for large-scale industrial production.

Besides that, J. Horvat et al. report that field dependence of I_c is strongly improved by the presence of the iron sheath, which suggests that interaction between ferromagnet and superconductor can be used to substantially improve the field performance of MgB_2/Fe wires at relatively low fields [102, 103]. The improvement of J_c(H) by the iron sheath was expected because the ferromagnetic sheath would shield the superconducting core from the external field. Therefore, up to a certain value of external field, the field inside the sheath would be zero. This would provide an almost constant value of J_c up to that particular field, which depends on the permeability and thickness of the iron sheath. This would be followed by a decrease of J_c with H for higher fields in the same manner as if there was no iron sheath. This can be very effectively employed for decoupling the superconducting filaments in multifilamentary MgB_2/Fe wires to substantially lower the AC loss in the wires [104]. An overview of the current state of development of MgB_2 wires and tapes can be found in the review paper [105].

5.3. Alternative Preparation Routes

The very first superconducting MgB_2 wires were prepared by the diffusion of Mg into commercially available B whiskers of 200 μm diameters or exposure of boron filaments to Mg vapour [106]. The wires obtained by this method exhibit very low residual resistivity values, ρ_o of the order of 0.6 μΩ cm and have better than 80% the theoretical density of MgB_2.

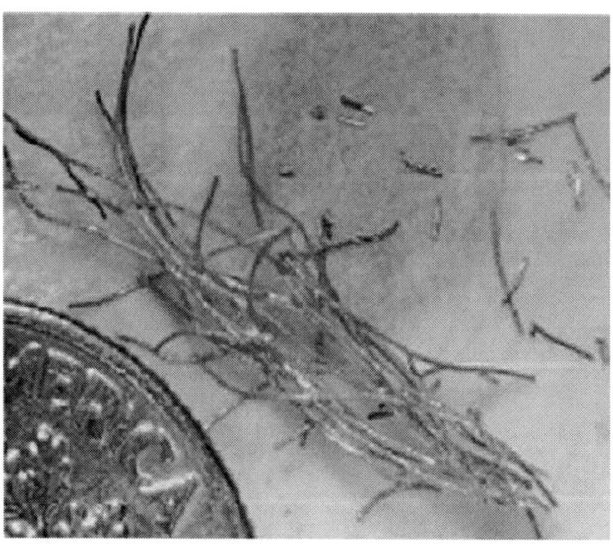

Figure 9. MgB_2 wire segments made from 100 μm diameter boron filaments [106].

Another technique was so-called continuous tube filling/forming (CTFF) process [107] (see figure 10). In the CTFF process, powder is dispensed onto a strip of metal as it is being continuously formed into a tube. High purity long Fe strips (23 mm wide, 0.20 mm thick) are formed continuously into a tube with an overlap-closed tube process and powder being enclosed in a sheath. After that, the closed tube can be inserted into a seamless Monel or Cu–30Ni tube and deformed to the round wire. In fact, this technique has been used to prepare commercial MgB_2 wires and tapes by the Hyper Tech Research, Incorporated, Columbus, USA.

A variation of the PIT preparation route was reactive liquid infiltration [108]. This technique has been successfully applied to the manufacture of both monofilamentary and multifilamentary wires of several tens of meters in length. The process starts with a steel clad thin niobium (Nb) tube filled with cylindrical magnesium rod and fine boron powder. The MgB_2 composites formed and disturbed inside the Nd tube after annealing, resulting a very compact, dense and finely grained MgB_2.

More recently, Togano et al. proposed a new approach for the fabrication of MgB_2 tape utilizing an interface diffusion reaction between a Fe–Mg alloy substrate and a boron layer [109]. In this process, the Fe-Mg alloy will be formed into a tape by a lamination method. The Mg in the Fe–Mg alloy diffuses towards the interface during the heat treatment, forming a thin Mg rich layer along the interface. Then, the Mg rich thin layer acts as a source of Mg for MgB_2 formation, which proceeds to the inside of the B layer. A dense structure of MgB_2 with much less MgO impurities can be produced by such formation mechanism, with a J_c value of 3×10^4 A cm^{-2} at 5 K and 7 T.

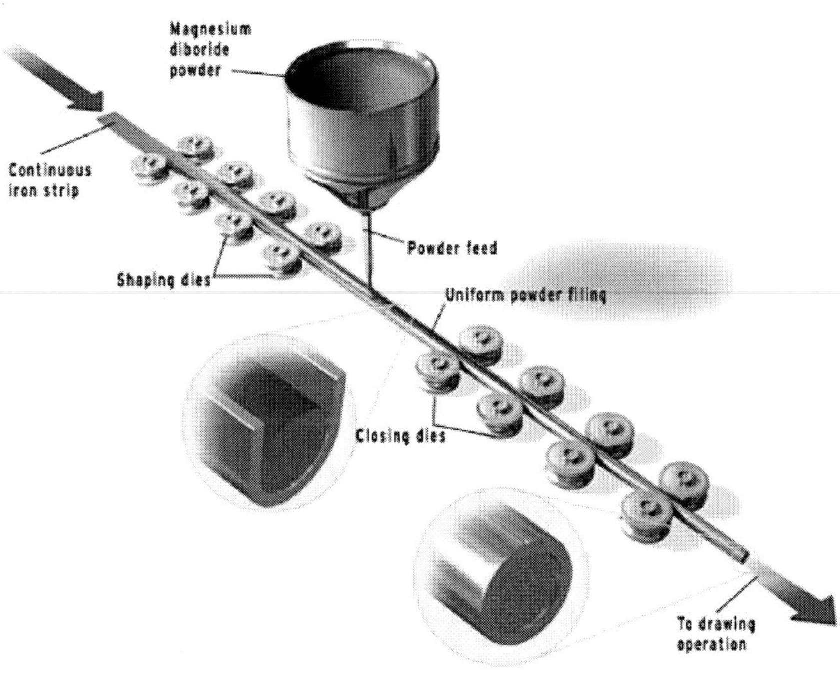

Figure 10. Schematic of the continuous tube filling/forming (CTFF) process.

6. Basic Properties of MgB_2

6.1. Critical Temperature, T_c

A primary difference between pure MgB_2 and its alloyed forms is that the former is a line compound and, once formed, has the same composition everywhere, whereas the latter is a solid solution and requires diffusion to move alloying elements. Alloying elements might not be distributed homogeneously since defect energies are high, which could have important consequences for the observed superconducting properties

Unfortunately, there has been no report stating that chemical doping may enhance T_c of MgB_2. One of the difficulties associated with doping of MgB_2 may be the fact that the MgB_2 structure is robust as mixing of the dopant with the doped element before or in the process of synthesis is not likely to achieve the substitution. Chemical substitutions change the physical quantities of the system such as carrier density, lattice constants and the crystallinity. The critical temperature of MgB_2 was reported to decrease at various rates for different substitutions.

However, $T_{c\text{-}onset}$ of pure MgB_2 increased, from 20 K to 38 K after the sintering time increased from 1 h to 24 h, and sutured at 38.7 K for samples sintered longer than 60 h has been observed [57]. The improvement of the T_c can be explained by improvements in superconducting transitions with heating time coinciding well with the improvement of crystallinity. As pointed out by Ribeiro et. al, T_c and the residual resistivity ratio (RRR) of MgB_2 bulks are very sensitive to boron purity [110]. The small decrease of T_c is believed to be due to both slightly poor crystallinity and low quality of the starting boron powder with 99% purity containing silicon, iron and carbon as a major impurities.

6.2. Critical Current Density, J_c, and Flux Pining

Although MgB_2 has a relatively high critical current density, J_c, the value of J_c for pristine MgB_2 drops rapidly with the increasing magnetic field due to its weak vortex pinning strength in magnetic field and low value of H_{c2}. The critical current density in MgB_2 is influenced by various factors, e.g., the quality and size of the powder particles and the annealing temperature and time. However it is clear that there are two main obstacles that have to be overcome to enhance the $J_c(H)$: vortex pinning and H_{c2}.

Superconducting properties of MgB_2 wires can be affected not only by a magnetic field, but also by an electric current. In the mixed state of type II superconductors, the field penetrates the superconductor in the form of thin filaments, so called magnetic vortices. Movement of the vortices is associated with a change of magnetic flux in and around the vortices [111]. A current driven by external source will force these vortices to move and this is described on a macroscopic level by a Lorentz Force, F_L, which can be defined as

$$F_L = J \times B$$

where J and B are spatially averaged transport current density and external magnetic field, respectively.

The moving of the vortices will lead to the dissipative losses in the superconductor and a voltage drop is observed. Pinning centres are defective sites in the superconductor where vortices tend to stick, provided that their size is comparable to the coherence length. They tend to stick there because it requires less energy for the vortices to stay there. When the transport current flows through the superconductor, the vortices will start moving only if the Lorentz force becomes strong enough to overcome pining and tear a vortex away, generating resistance [111]. The current at which the resistance just starts occurring is called the critical current, I_c. Thermal excitations also promote this de-pinning of vortices, resulting in a weaker effective pinning at higher temperatures.

In general, local sharp change in the condensation energy of superconducting charge carriers is desirable for flux pinning. In other words, if the transition from superconducting to local weak superconducting or nonsuperconducting regions, such as defects in the crystal lattice, dislocations, twin boundaries, grain boundaries, small precipitates, compositional fluctuations and artificial defects, is sharp, these regions can act as effective pinning sites if their size is comparable to the diameter of the vortices, i.e.ξ. In the case of MgB_2, candidates for effective pinning centres are different form that for high temperature superconductors because of relative long superconducting coherence length and much more isotropic nature of MgB_2.

Attempts to enhance J_c and flux pinning have been made by using several techniques, including impurity chemical doping, irradiation and thermo-mechanical processing. However, irradiation and thermo-mechanical processing techniques are not suitable for the preparation of MgB_2 wires and tapes because of their complicated nature and the practical difficulty in the use on very large scale. Among these, the doping of chemicals (mostly nano-particles) into MgB_2, which is a low cost fabrication technique to improve the pinning centres, is considered to be the most practical for application. S. Soltanian et al. have showed that grain size of the precursor SiC has a strong effect on the critical current density and its field dependence of MgB_2 superconductor. The smaller the SiC grains are, the better the field performance J_c and H_{irr} is as shown in figure 11[112]. Improving the pinning by introducing nano-size particles into precursor powders has several advantages over the use of larger particles. Firstly, a stronger pinning force will be introduced due to the size of nanoparticles being comparable to the coherent length (ξ). Secondly, nano size doping would not lower the volume of superconductivity or affect the formation of superconducting phase substantially, enabling introduction of large number of pinning centers. If reaction between the introduced nanoparticles and constituents of precursor powder occurs, the resulting products of this reaction will be on the nanometer scale. Further, if chemical reaction is desired, it will be more efficient with nano-size particles.

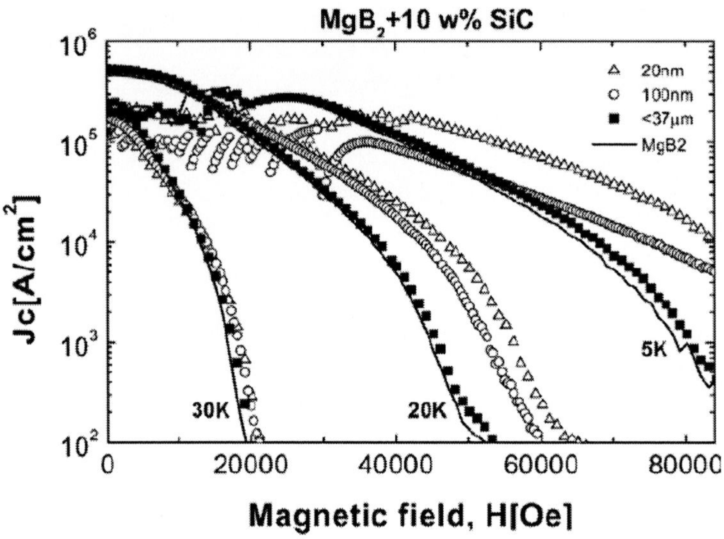

Figure 11. The J_c field dependence of MgB_2 samples doped by 10 wt% of different SiC powders as well as the reference sample at different temperatures of 5, 20 and 30 K [112].

6.3. Crystalline and Grain Size Effect on Critical Current Density

It has been reported that grain boundaries are believed to act as pinning centres [113-116]. Yamamoto et al. has systematically studied the crystallinity of MgB_2 by using full width at half maximum (FWHM) value derived from powder x-ray diffraction (XRD) peaks [117]. FWHM of the <110> peak reflects the in-plane disorder while FWHM of <002> peak reflects the out plane disorder. The FWHM is apparently larger at the low sintering temperature, as compared to the high sintering temperature, suggesting that poor crystallinity occurred in the low temperature sintering reaction. MgB_2 sintered at low temperature was composed of small grains while significant grain growth occurs upon high sintering temperature. The degradation of crystallinity originates from disordered crystal lattice caused by various types of lattices defects or intragranular precipitates [118]. These are believed to increase the intraband scattering, shorten the electron mean free path l and ξ according to the equation $1/\xi = 1/l + 1/\xi_0$ [38], where ξ and l are the coherence length and mean free path, respectively. Since pinning force at grain boundaries f_{GB} is proportional to the inverse of grain size and lattice strain, both small crystal size and lattice distortion are considered to enhance the grain boundary pinning [117].

6.4. Effect of Mg Precursor Powder on MgB_2 Superconductivity

It is well known that Mg is highly volatile at high temperatures. Therefore, it is inevitable that Mg is easily lost during the preparing process of the sample. That is to say, the prepared samples usually have Mg deficiency. Nevertheless, Mg, as the other constitutional element, is a very reactive element. Mg powder has a large surface area, and it oxidizes easily during powder production, storage and handling. The effects of MgO on MgB_2 superconducting properties should be studied to facilitate the production of MgB_2 on an industrial scale.

All of the studies on Mg precursor have only focused on the Mg stoichiometry effect while there is no detail report in the purity of Mg on the critical current density and transition temperature [110, 119-122]. However, all the reports on the effect of the critical temperature so far vary significantly, which is most probably due to different preparing condition and the starting precursor. In general the Mg deficiency degrades the critical temperature. According to Lezza et al., addition of 5% Mg to MgB_2 powder increased T_c and did not affect neither J_c nor B_{c2} [123]. However, recent results indicated that preparing MgB_2 samples with a slight excess of Mg might not be advantageous when developing devices for high-field applications [124].

Kumakura's group has systematically studied the effect of the Mg on the critical current density. They achieved twice as large J_c by using MgH_2 as Mg source as for samples prepared using the mixture of Mg + B powders [125]. Replacement of commercial Mg powder with nano Mg powder is effective to enhance the reaction between Mg and B and hence improve grain connections. This is because the surface of commercial Mg powder is mostly oxidized, while the nano Mg powder is very active and promotes the reaction between Mg and B.

Yamada et al. has proved that transport J_c values of the non-doped and 10 mol% SiC-doped tapes prepared with nanometer-size Mg powder reached 90 and 250 A/mm^2 at 4.2 K and 10 T, respectively. These values were about five times higher than those of the tapes prepared with commercial Mg powder [59, 126]. The use of ultrafine Mg and B powder mixture prepared by ball milling also effectively improved the grain connections and enhance the J_c values [75, 127].

6.5. Effect of Boron Precursor Powder on Superconductivity

The influence of the nature of the boron precursor on the superconducting properties of polycrystalline MgB_2 was studied, mostly its effect on critical current density, resistivity and phase formation [110, 121, 122, 124-131]. Chen et al. and Zhou et al. showed that samples made from different purity amorphous boron exhibited different J_c, and also that the highest purity boron (99.99%) yielded the highest J_c. In addition, samples made from crystalline boron powders have about an order of magnitude lower J_c compared to those made from amorphous precursors which is due to different reactivity rates and as a result of the presence of oxide impurities in the MgB_2. Xu et al. have studied the effect of boron powders on the superconducting properties of MgB_2 [129]. The 92% and 96% purity boron powders produce lower surface reactivity and larger particle size of MgB_2 than the 99% boron powder, indicating that the low purity powders cannot be used to achieve the same superconducting properties as those of samples made from pure 99% boron powder. However, the purity of 92% and 96% boron powders can be improved by using a simple chemical process, leading to enhanced magnetic critical current densities J_c. More recently, W. Haßler et al. discovered that T_c and the microstructure of the MgB_2 tapes are correlated to oxygen impurities and the grain size of the B precursor powder. They also suggested that the nominal purity of boron specified by the suppliers was not sufficient for the characterization of the boron precursor powder as the information considers only metallic impurities [130]. Jiang et al. demonstrated that MgO phase can be significantly reduced in the MgB_2 by removing the B_2O_3 through boron purification [131].

6.6. Porosity and Density Effect

On the other hand, MgB_2 bulks synthesized from the powder mixture of magnesium (1.74 g/cm^3) and boron (2.2 g/cm^3) have porous microstructure, having low bulk density, approximately 50% of the theoretical density (2.62 g/cm^3). With larger particle size of Mg, large voids will be formed in MgB_2 matrix when Mg diffuses into B to form MgB_2 as shown in the figure 12. This leads to the artifact of the sample size-dependent critical current density J_c and irreversibility field obtained from magnetic measurements, due to the different length scale of superconducting screening in the magnetic measurement [132]. Besides that, grain size of MgB_2 is much dependent on the B powder and smaller grain size is preferable for enhanced J_c where grain boundary is believed to contribute to the flux pinning. If the highly dense MgB_2 bulks or tapes are easily obtained, higher J_c will be expected because of an increase of effective cross-section area for superconducting current.

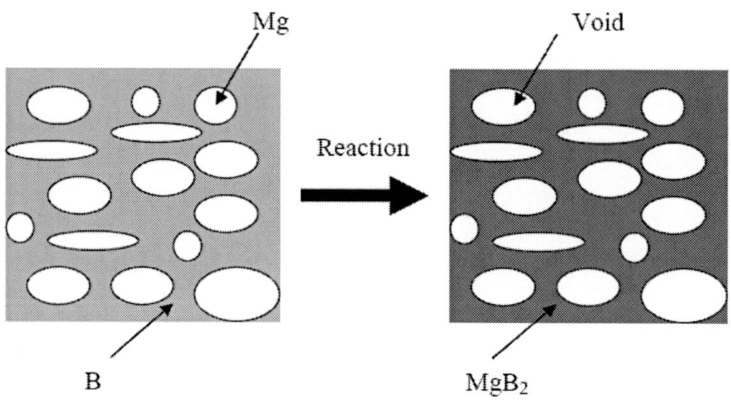

Figure 12. Theoretical concept of "in-situ" reaction of Mg and B [11].

7. Effect of Chemical Doping on Superconductivity of MgB_2

Although numerous studies were reported on enhancing the critical current density by using mechanical alloying and irradiation treatment, chemical doping is considered the most viable in term of practical application. Further more, MgB_2 has a relatively large coherence length and small anisotropy, which opens the window to the success of chemical doping in this material. Despite that nano impurities are expected to act as pinning centers, moderate substitutions of elements will yield changes of electronic state, lattice distortions and crystallinity which will drive MgB_2 to the "dirty limit" and hence enhance the H_{c2} and H_{irr} [38].

7.1. Carbon Doping

Aimed at substitution for boron in MgB_2, various carbon sources, like SiC, TiC, B_4C, C, Na_2CO_3, diamond, graphite, nanohorn and CNT, have been studied aggressively by several groups as shown in the table 2. However, among all the reports on C doping, most of the studies only focused on the effect on superconductivity, especially on the two-band study and the solubility of C [110, 133-168]. In this section, the topic of C solubility and the flux pinning of most of the carbon sources doped into MgB_2 will be reviewed.

Due to the difference between the nominal and actual C substitution level, the correlation between the C substitution and superconducting properties of MgB_2 is obscured. The inconsistencies originate from the precursor materials, fabrication techniques and processing conditions used. Early study on the carbon solubility ranging from 1.25% to 30% was reported when elemental magnesium, boron, and carbon were used as starting materials [134, 140]. However, another study shows that a higher substitution of carbon for boron can be obtained, as revealed by the magnitude of the measured decrease of T_c and the change of a-lattice parameter for C doped MgB_2 samples [155]. By using the neutron diffraction study to estimate C concentration, Avdeev et al. found a linear relation between the unit cell parameter-*a* and the C concentration. The authors also showed that earlier reports overestimated the carbon content in MgB_2 because polycrystalline-carbon-substituted samples may contain significant amounts of impurity phases and the nominal content was assumed most often to be equal to the actual one [146]. S. Jemima et al. have evaluated the actual C fraction substituted into the polycrystalline MgB_2 matrix by conventional solid–vapour reaction route and discovered that the actual composition is in disagreement with the nominal composition as plotted in the figure 13 [155].

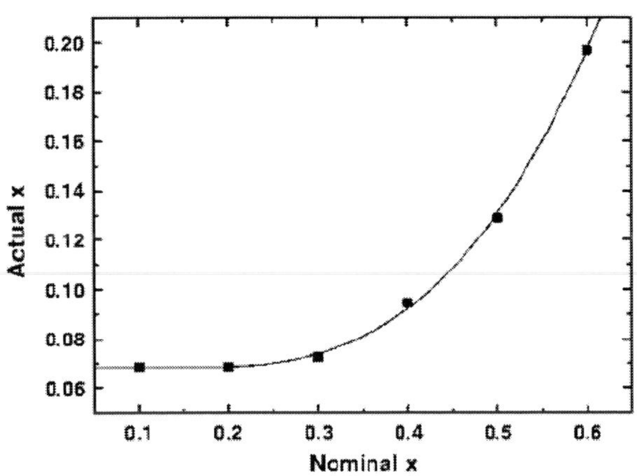

Figure 13. Variation of the actual carbon plotted against the nominal carbon in the $MgB_{2-x}C_x$ [155].

Another approach for carbon substitution of MgB_2 was suggested by Mickelson *et al.* to overcome the diffusion problem by using C as starting material [138]. Using boron carbide B_4C as a source of carbon they achieved better mixing of B and C atoms. The samples had an estimated composition $Mg(B_{0.9}C_{0.1})_2$, and T_c was decreased by 7 K. However, the synthesis using B_4C yields multiphase samples containing another borocarbide MgB_2C_2. Ribeiro *et al.*

followed this approach and reported the optimization of synthetic conditions of carbon doped samples leading to nearly single-phase $Mg(B_{1-x}C_x)_2$ with T_c=22 K [142].

Ueda et al. and Yamamoto et al. showed that C can substitute at the B site with low temperature of 850°C using B_4C as starting materials [169-171]. Substantially enhanced J_c properties under high magnetic fields were observed in the B_4C doped samples due to low temperature heating and carbon substitution effects. On the other hand, the values of the Full Width Half-Maximum (FWHM) of the MgB_2 (110) peak for the B_4C doped samples increased with increase of its doping levels, suggesting the introduction of lattice defects and/or decrease of the crystal size of MgB_2 occurring upon B_4C doping. More recently, Lezza et al. has successfully obtained J_c values of 1×10^4 A cm^{-2} at 4.2 K and 9 T for 10 wt% B_4C powders added to MgB_2/Fe wires at reaction temperature at 800 °C [172].

Wilke et al. used chemical vapor deposition, CVD, method to co-deposit B together with a doping element to form long lengths of carbon doped boron fibre [152]. More recently, bulk samples of milled carbon and MgB_2 with hot isostatic pressing and Mg vapour annealing have achieved upper critical fields in excess of 32 T and critical current density approaching 10^6 A/cm^2 for H=0T, which was due to lattice disorder introduced in the milling process [78].

The effect of C doping on the flux pinning and critical current density in MgB_2 using amorphous carbon, diamond, Na_2CO_3, carbon nanohorn and graphite have been studied, all showing improvement of J_c at elevated field. The effective pinning improvement of all the samples is mainly attributed to partial C substitution despite finely dispersed small particles like MgO and MgB_2C_2 in MgB_2 matrix. All kinds of carbon sources with the magnetic field and transport J_c(H) at 5K and 20K are shown in the table 2. Recently, Yamanda et al. have fabricated *in situ* PIT MgB_2/Fe tapes by using aromatic hydrocarbons benzene, naphthalene, and thiophene as additives [173]. They found that aromatic hydrocarbons were very effective for increasing the J_c values. The J_c values of 20 mol% benzene-added tapes reached 13000 A cm^{-2} at 4.2 K and 10 T. This value was almost comparable to that of 10 mol% SiC-added tapes and about four times higher than that of tapes with no additions. Microstructural analyses suggest that J_c enhancement is due to both the substitution of carbon for boron in MgB_2 and the smaller MgB_2 grain size.

More recently, Dou's group has demonstrated the advantages of carbohydrate doping over other carbon dopants, resulting in an increase of in-field J_c by more than one order of magnitude without any degradation of self-field J_c. In addition to providing homogenous mixture with the Mg and B powder, hydrocarbon doping offers fresh atomic C to substitute into B site at low sintering temperature [174, 175].

Among various carbon precursors, CNT [176] may have the potential to improve the mechanical and thermal properties of MgB_2 wires since CNTs have been used as a component for a number of composites to improve their mechanical properties [177, 178]. It has been demonstrated that carbon nanotubes can be embedded in Bi-2212, thus offering a promising method for introduction of extended defects in superconductors [179]. The magneto-optical image investigations have revealed that CNT's are functioning like columnar defects, enhancing J_c in the Bi-based superconductor [180]. However, the flux-pinning properties were improved only at low temperatures, below 52 K. Similar method has been reported by Yang et al for a significant enhancement in $J_c(H)$ for high temperature superconductor by introducing nanorods as columnar pinning centers in to the composites [181-183]. Recently, Wei et al. have studied the superconductivity of MgB_2-carbon nanotube composites [184]. However, the effect of carbon nanotube doping on critical current density and flux pinning has not been reported. Besides the carbon substitution effect, CNT is believed to provide better electrical, mechanical and thermal properties for MgB_2 wire.

The advantages of CNT doping over other additives can be described as follows.

1. Multiwalled CNTs can carry current densities up to 10^9 – 10^{10} A/cm^2 and remain stable for extended period of time [185]. The CNT doping can improve the current path and connectivity between grains in MgB_2.
2. The thermal conductivity for isolated multiwall CNTs is estimated to be over 3000W/K m at room temperature, which is higher than that of graphite [186] and hence could benefit the heat dissipation and thermal stability of the MgB_2 wires.
3. Bundled CNTs have very high axial strength and stiffness, approaching that for ideal carbon fibres [187]. CNT is typically one-dimensional with high aspect ratio of several orders of magnitude and can act as good pinning centres.

Dou et al. have studied the effects of CNT doping on the transition temperature, lattice parameters, critical current density and flux pinning [188-190]. Figure 14 shows the $J_c(H)$ curves at 5K and 20K for the samples of $MgB_{2-x}C_x$, where x = 0, 0.05, 0.1, 0.2 and 0.3 are the nominal values for C content, with all the samples sintered at 800°C for 30min. All the $J_c(H)$ curves for doped samples have a higher J_c than the undoped sample at high fields. The sample doped with 10% of carbon nanotubes (x = 0.2) gives the best J_c at high fields: J_c was more than 10,000A/cm^2 at 20K and 4T, and 5K and 8.5T, respectively. At higher doping level (x = 0.3), although the J_c in low field regime was depressed, the rate of J_c drop is much slower than for all other samples, clearly indicating strong flux pinning induced by the C nanotube doping. Besides that, H_{c2} line for the doped sample has a much steeper slope than the undoped sample due to the disorder on B site of carbon substitution that leads to impurity scattering and enhancement of H_{c2}. A similar enhancement has also been observed in the single wall-carbon nanotube (SWCNT) doped MgB_2, where transport J_c for SWCNT doped wire was estimated to be 10200 Acm^{-2} at 10 T and 4.2 K [191].

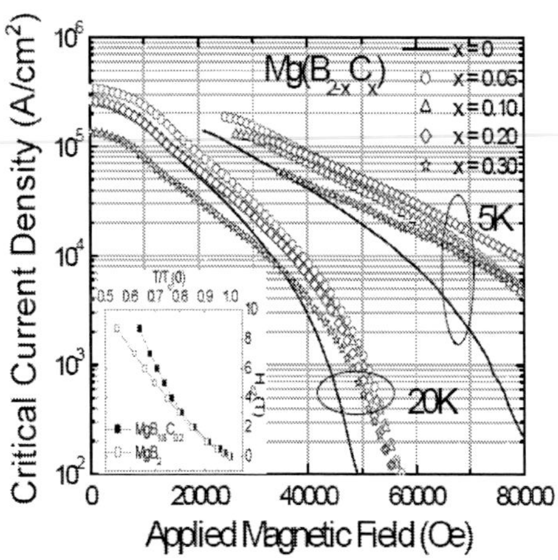

Figure 14. Critical current density as a function of magnetic field at 5K and 20K for $MgB_{1.8}C_{0.2}$, sintered at different temperatures for 30 minutes. The carbon was in the form of multi-walled carbon nanotubes. Inset : Normalized temperature $(T/T_c(0))$ dependence of the upper critical field, H_{c2}, for pure MgB_2 and $MgB_{1.8}C_{0.2}$ [188].

However, longer CNTs tend to entangle and agglomerate, that makes their mixing with precursor powder inhomogeneous [192]. This also suggests that the unreacted CNT's will block the transport of the current density and suppresses the J_c. Galvan et al. faced the same problem when they embedded nanotubes in the Bi-based superconductor [193]. This suggests that shorter length nanotubes would be the solution for more homogenous mixing and substitution. As a result, magnetization measurements of CNT doped MgB_2 indicated a change in the critical current density with the length of nanotube and not with its outside diameter [190]. However, by using sonication method, Yeoh et al. have successfully improved homogenous mixing of CNTs with the MgB_2 matrix [194]. Besides that, ultrasonication of CNT doped MgB_2 resulted in a significant enhancement in the field dependence of critical current density, while avoiding the side-effects that would occur at higher sintering temperatures. More recently, Chen et al. and Dou et al. have showed that CNT doped MgB_2 wires exhibit anisotropy in J_c in relation to the direction of applied field and significantly improve heat transfer and dissipation during materials processing [195, 196].

In order to understand the role of CNT in MgB_2, Yeoh et al. have systematically studied the carbon substitution effect on the MgB_2 [197]. Under the optimum conditions, magnetic J_c was enhanced by two orders of magnitude at 5 K for a field of 8 T, and by a factor of 33 at 20 K for a field of 5 T for carbon doped bulk samples while enhancement by a factor of 5.7 was observed in the transport I_c at 12 T and 4.2 K for a wire sample. Samples sintered at high temperature (900°C and 1000°C) exhibited excellent J_c, approximately 10000 A/cm^2 in fields up to 8 T at 5 K. The result is in agreement with the Ma et al. report of carbon substitution effect on MgB_2 tape, where J_c for all the C-doped samples was enhanced by more than an order of magnitude in magnetic fields above 9 T, as compared to the undoped tapes [198]. At 4.2 K, the transport J_c for the 5 at. % doped tapes reached 1.85 x10^4 A/cm^2 at 10 T and 2.8 x 10^3 A/cm^2 at 14 T, respectively. Comparative study on the carbon substitution effect of CNT and nano C on MgB_2 superconductor demonstrated that the lattice distortion and optimum doping level is different in the CNT and nano C samples [199]. The reactivity of the carbon source plays an important role to determine the carbon substitution in the MgB_2. Carbon substitution is more direct process in the nano C precursor while carbon substitution only occurred after break out of carbon atoms from CNT. Although both dopants showed significant enhancement in critical current density at high sintering temperatures, like 900°C, nano C doping showed a better J_c performance in high field compared to the CNT samples (see figure 15). However, the prospect of the CNT for improving the thermal stability, heat dissipation, and mechanical support of MgB_2 superconductor wire should not be ignored.

The improvement of critical current of MgB_2 has been confirmed to be the most effective by doping with SiC nano-powder. Although their density was only half of the theoretical density, a 10wt% SiC-doped MgB_2 bulk samples showed H_{irr}~8T and J_c~10^5 Acm^{-2} under 3T at 20K. In fact, the SiC-doped MgB_2 achieved the highest J_c in fields at 20K ever reported for MgB_2 wires and bulks [89, 200]. At 20 K and 2 T, the J_c for the doped sample was 2.4x10^5 A/cm^2, which is comparable to J_c values for the best Ag/Bi-2223 tapes. At 20 K and 4 T, J_c was twice as high as for the best MgB_2 thin films and an order of magnitude higher than for the best Fe/MgB_2 tapes, as shown in figure 16.

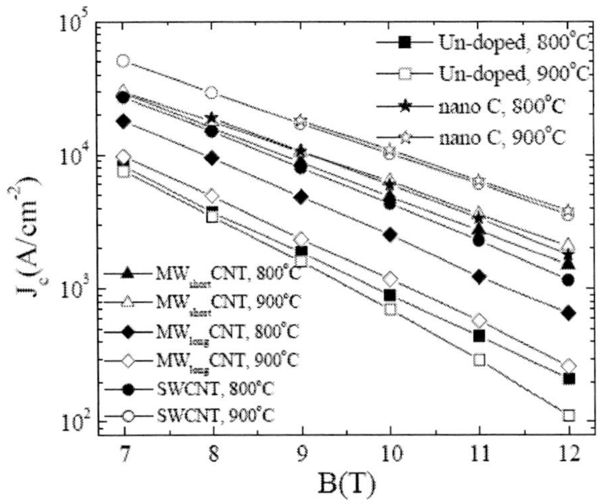

Figure 15. Transport critical current at 4.2 K at fields up to 12 T for nanoC and various of CNT wire produced at 800 and 900°C sintering temperature. The pure MgB_2 results were included for comparison [199].

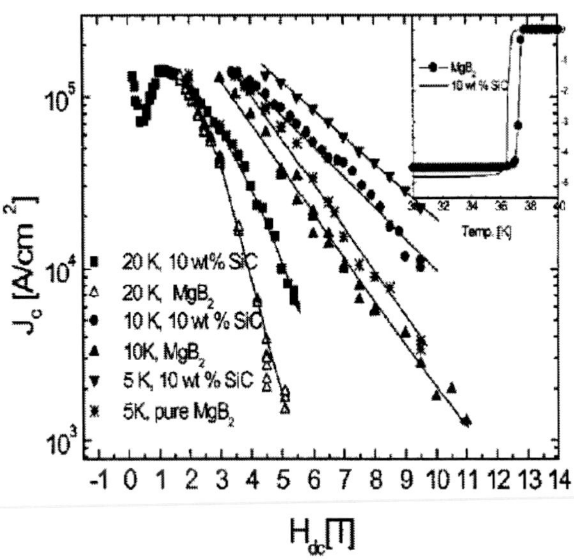

Figure 16. Significant enhancement of transport J_c(H) by SiC doping [89].

There are a few reasons nano-SiC can be considered as a good substitution in the MgB_2:

1. Carbon is the only element confirmed to be able to partially substitute at the B site and enhance the J_c(H) in all forms and sources.
2. SiC was found to provide moderate amounts of carbon substitution, largely independent of heating temperature, 600 and 850 °C
3. The enhancement of $J_c(H)$ extends to all temperatures, even at the temperatures close to T_c, unlike certain doping elements that show effective pinning only at low temperature

4. Besides the partial substitution effect, highly dispersed fine particles(< 10nm) occurring due to nano-SiC doping, like $MgSi_2$, BC, BO_x and $SiBO_x$, match the coherence length of MgB_2, making them good pinning centers
5. T_c reduction is not pronounced even in the heavy SiC doped sample
6. Partial C substitution for B causes disorder on the B sites which results in the impurity scattering and contributes to the enhancement of H_{c2}.
7. SiC can provide suitable amount of carbon into MgB_2 even through the reaction below the melting point of magnesium. This advantageous point is desirable for low temperature synthesis of MgB_2, which shows high critical current performance due to smaller grains with poor crystallinity.

Several leading groups all around the world like NISM and Hyper Tech Research Center have confirmed the significant enhancement of J_c and H_{c2} for SiC doped MgB_2. Nowadays, most of the commercial MgB_2 wires are prepared with the nano SiC doping. Serquis et al. demonstrate that the high critical current of SiC doped MgB_2 can be retained in longer wire lengths. They made a 6 layer coil using 25 meters of wire [201]. All of these results demonstrated that the enhancement effect on the J_c(H) performance from SiC doping is highly reproducible.

Table 2. Comparison of effects of all carbon source doping of MgB_2 on magnetic and transport J_c(H) at 4 and 20K. All the J_c(H) were determined by the magnetization unless stated otherwise

Carbon source	Type	J_c at 8T (A/cm^2)	J_c at 6T (A/cm^2)	J_c at 2T (A/cm2)	Reference
4.25/5K					
Graphite	Wire/SS	Transport $2x10^4$	$8x10^3$		[202]
Carbon nanohorn	Wire (suspension spinning method)	Transport $4x10^3$	$4x10^3$	$4x10^3$	[203]
Carbon	Tape/Fe	Transport $4.4x10^4$			[198]
	Bulk 4x3x0.5 mm^3	$3x10^3$	$2x10^4$		[204]
	Bulk 2.87x2.70x3.12 mm^3			$3x10^5$	[205]
	Bulk	$7x10^4$	10^5		[78]
	CVD with Mg vapor			$2x10^3$	[163]
	Transport thin film	$6x10^4$	$2x10^5$	10^6	[206]
	Bulk (2.5 x 1.5 x 1 mm^3)	$1.6x10^4$	$4.3x10^4$		[197]
	Wire	Transport $3.2 x10^4$			[197]
MWCNT	Bulk (2.5 x 1.5 x 1 mm^3)	$9.4x10^3$	$2.9x10^4$		[189]
	Wire	Transport $1.8 x10^4$			[190, 199]
SWCNT	Wire	Transport $2.7 x10^4$			[191]
B_4C	Bulk			$>8x10^5$	[169]
	Bulk			$2x10^4$	[170, 171]

Table 2. (Continued).

Carbon source	Type	J_c at 8T (A/cm^2)	J_c at 6T (A/cm^2)	J_c at 2T (A/cm2)	Reference
	Wire	Transport > 10^4			[172]
0.5%Ti+1.1 %C	Bulk			$2x10^5$	[158]
SiC	Bulk			>$8x10^5$	[171]
	Bulk 4x2.8x0.7 mm^3			$2.4x10^4$	[200]
	Bulk 0.56x2.17x3.73 mm^3	$2.5x10^4$	$9x10^4$		[89]
	Wire	Transport $4.1x10^4$	$8.4x10^4$		[89]
	Tape/Fe	Transport $3x10^4$			[95]
	Tape/Fe	$2x10^4$			[207]
	Tape/Fe	Transport $3x10^4$			[58]
	Tape/Fe	Transport $6x10^4$			[59]
	Strand (1m)	Transport $2x10^5$	$3x10^5$	$4x10^5$	[107]
Benzene	Tape	Transport 3.2 $x10^4$			[173]
Sucrose	Bulk	10^4	6 $x10^4$	2 $x10^5$	[174]
Maelic Acid	Bulk	2.6 $x10^4$	10^5		[175]
		J_c at 4T (A/cm^2)	J_c at 2T (A/cm^2)	J_c at 0T (A/cm^2)	
20K					
Graphite	Wire(SS)	Transport 10^3	$8x10^3$	>10^4	[202]
Diamond	Bulk 0.7x2.1x2.7 mm^3	$8x10^3$	$7x10^4$	$5x10^5$	[208]
Carbon	Bulk 4x3x0.5 mm^3	$5x10^3$	10^5	$3x10^5$	[204]
	Bulk 2.87x2.70x3.12 mm^3		$2x10^4$	$3x10^5$	[205]
	Bulk	10^4	$7x10^4$		[78]
	CVD with Mg vapor		$5x10^2$	>10^5	[163]
	Thin film	$7x10^4$	$2x10^5$	$4x10^6$	[206]
	Bulk (2.5 x 1.5 x 1 mm^3)	$1.2x10^4$	$8.7x10^4$	$2.9x10^5$	[197]
MWCNT	Bulk (2.5 x 1.5 x 1 mm^3)	$6.8x10^3$	$6.1x10^4$	$2.5x10^5$	[189]
B_4C	Bulk	10^4	10^5	$7x10^5$	[169]
	Bulk	10^4	$8x10^4$	$4x10^4$	[170, 171]
Na2CO3	Bulk	$8x10^3$	$9x10^4$	$5x10^5$	[209]
SiC	Bulk	$6x10^3$	$2x10^5$	$7x10^5$	[169]
	Bulk 0.56x2.17x3.73 mm^3	$5x10^4$	$2x10^5$	$7x10^5$	[89]
	Wire/Fe	Transport 3.2 $x10^4$	>10^5		[89]
Sucrose	Bulk	$8x10^4$	$6x10^4$	$2x10^4$	[174]
Maelic Acid	Bulk	3 $x10^4$	10^5	4x 10^5	[175]

7.2. SINGLE ELEMENT DOPING

The effects of metals doping in MgB_2 have been reported due to that many metals can form metal borides having AlB_2 structure. Some metals were found to be partially substituted in the Mg site while some do not substitute into MgB_2 phase and form other impurities with B. However, there are some element like Si that were expected to substitute at the B site instead of the Mg site.

7.2.1. AL DOPING

Although most of the reports on the Al substitution at the Mg site reveal dramatic decrease of J_c [207-212] due to change of carrier density and density of states at the Fermi surface, Berenov et al. have successfully improved the J_c , particularly at lower fields at 20 K. For example at 5 T, J_c was enhanced by a factor of 20 for a doping level of 1 at.% Al. The critical temperature (T_c) remained largely unchanged, which may be the effect of low doping levels of Al on the crystal structure and superconducting properties of MgB_2 [213]. On the other hand double doping with Al and Zn or Li show improvement of the J_c compared to the single Al doping but the value is still well below the pure MgB_2 [214].

7.2.2. ZN DOPING

However, measurements by Martínez et al. demonstrated that single Zn doping does not improve, nor deteriorate the field dependence of the superconducting wires and powders [215]. The results indicate that the solubility limit of Zn in the MgB_2 structure is very small and a very small amount of Zn enters the Mg site thus, it does not alter significantly the superconducting properties.

7.2.3. AU DOPING

The effect of a gold coating on MgB_2 thin film has been studied by Choi et al. where they believe that the good thermal conduction of the gold film helps the thermal stability of the MgB_2 thin film [216]. When the thickness of the thin gold film was increased, the suppression of the J_c due to flux jump was drastically reduced and J_c is fully recovered.

7.2.4. CD DOPING

Similarly to the Zn doping, Cd does not occupy the atomic Mg sites in the MgB_2 crystal structure, but merely reacts with Mg and forms $MgCd_3$ phase. It is striking to note that although the nonsuperconducting phase $MgCd_3$ is as high as about 67 vol.% in $Mg_{0.5}Cd_{0.5}B_2$ (nominal composition), the J_c of 5.0×10^5 A/cm^2 (5 K, 0 T) has been achieved. It is suggested that the improvement of J_c in $Mg_{0.5}Cd_{0.5}B_2$ is primarily due to pinning effects induced by $MgCd_3$ and better grain linkage provided by the $MgCd_3$ [217].

7.2.5. TI DOPING

Several groups have studied the effect of Ti doped into MgB_2 in the form of bulk, wire and tape and all the reports showed positive results [218-228]. Zhao *et al* have observed a 1 nm TiB_2 layer at grain boundaries in *in-situ* synthesized MgB_2 bulk material with titanium additions [228]. This, they argue, contributes significantly to flux pinning and also prevents coarsening of the MgB_2, producing a grain size of just 20 nm. Similar substantial increases in performance have been reported for Ti additions to *in situ* high pressure synthesized MgB_2

bulk material [226]. In the magnetic fields up to 2 T, high-pressure synthesized MgB_2 (with 10% of Ti) at 20 K has a critical current density comparable to that of Nb_3Sn at 4.2 K. Finnemore et al. and Anderson et al. have introduced a new processing method for the doping of magnesium diboride wire by using the chemical vapour deposition (CVD) to produce long lengths of suitably doped starting boron fibre [224, 225]. It was found CVD methods are an effective way to distribute the Ti impurities and can be used to produce conductors with J_c values in the range of 10,000 A/cm^2 at 25 K and 1.3 T. The samples show a fine dispersion of Ti through out the grains and no conspicuous precipitation of TiB_2 on the MgB_2 grain boundaries. This is to be contrasted with the precipitation of TiB_2 on MgB_2 grain boundaries observed for samples prepared by solid state reaction of Ti, Mg, and B powders.

7.2.6. FE DOPING

Jin *et al* have studied the effect of Fe addition on T_c and the J_c–H characteristics of MgB_2 [83]. They found that 5 wt% Fe additions had no effect on T_c and had a negative effect on J_c. However, it had the least negative effect on the J_c–H behaviour of all the elements that they studied, including Cu, Ag, Y, Ti and Mo. Dou et al. discovered that the $J_c(H)$ performance was severely depressed at above 2% doping level of Fe in the MgB_2 sample [229]. The detrimental effect of nanoscale Fe doping on $J_c(H)$ is attributable to both the Fe substitution for B in the lattice structure and the inclusions of Fe and FeB which act as weak links at grain boundaries.

7.2.7. LA DOPING

High density of LaB_6 nanoparticle inclusions were found in MgB_2 grains in the La doped MgB_2 sample, as a major impurity [230, 231]. In fact, the production of nano-particles of LaB_6 impurity phase within the bulk MgB_2 is believed to be responsible for the enhancement of J_c. The optimum result on J_c is obtained for $Mg_{0.97}La_{0.03}B_2$ at 5 K; J_c reaches ~1.4×10^7 A cm^{-2} in self-field, ~2.1×10^6 A cm^{-2} at 1 T, ~2.5×10^5 A cm^{-2} at 2.5 T and ~1.8×10^4 A cm^{-2} at 4.5 T.

7.2.8. ZR DOPING

Similar to Ti doping, the majority of Zr atoms form ZrB_2 phase while a small amount of Zr atoms may be introduced into the lattice of MgB_2 [219, 232-234]. The combination of good grain connection, the reduction of grain size and small ZrB_2 particles in the sample may be responsible for the significant enhancement of J_c in Zr-doped samples

7.2.9. TA DOPING

Prikhna et al. have observed a positive influence of Ta during high pressure (HP) synthesis of MgB_2 on the critical current density in the fields up to 3 T and on the irreversibility fields of the material [235]. At 20 K, HP-synthesized MgB_2 has critical current density of 68 kA/cm^2 in the 1 T field and of 160 kA/cm^2 in the zero field, its irreversibility field at 20 K being 5.7T.

7.2.10. OTHER SINGLE ELEMENT DOPING

Tachikawa *et al.* explored doping *ex situ* wires in nickel sheaths by different kinds of metal powder, like Ag, Cu and Ni and also low melting point metal powders of Sn and In [236, 237]. They have shown that 10 wt% indium addition increased J_c values by a factor of

6–7, which was attributed to the improvement of weak inter-grain links that would consist mostly of unsintered MgB_2 powder particles. But this J_c increase is still one order of magnitude lower than that achieved by high temperature annealing which causes the whole MgB_2 core to recrystallize.

Cimberle et al. attempt to introduce defects in the MgB_2 structure by different chemical treatments like substitution of lithium on the magnesium site and doping of the precursor boron powders with Si [211]. J_c always increased in doped or substituted samples (up to a factor 3) and this fact is meaningful, in particular in the light of the small level of substitution. The best result in terms of J_c is achieved by Si doping that, moreover, does not significantly decrease the transition temperature. However, neutron diffraction and Rietveld analysis showed no evidence of Si substitution into the lattice in the Si doped MgB_2 samples [238]. Nano-particle inclusions, such as Mg_2Si, precipitates of MgO, unreacted nano particles observed using TEM are proposed to be responsible for the enhancement of flux pinning in high fields.

A wide variety of metal powders, including Nb, Hf and W have all been used at the 10 wt% in the *ex situ* wires by Kovac's et al. [232]. Transport current measurements have shown that metal particle addition leads to an especially significant improvement in current carrying capacity. The presence of normal particles influences both the resistivity and also the thermal conductivity of the MgB_2 core, and the improved internal stability may be responsible for the increased critical current values. Goto et al. have proved that J_c values were also improved to $4.8x10^5$ A/cm^2 with 1% Hf doping [218]. On the other hand, the J_c value of MgB_2 tapes was increased from 1500 A/cm^2 to 4500 A/cm^2 at 4.2 K and 5 T with Cu addition [239].

7.3. OXIDE AND OTHER COMPOUND DOPING

7.3.1. DOPING

Delfany et al. have reported that J_c values decrease by more than one order of magnitude in the AlO_2 doped MgB_2 whereby the amount of MgO was found to increase with increasing Al_2O_3 level [240]. However, smaller decrease of J_c is observed by Rui et al. when they used a nano alumina dopant [241]. It is believed that the nanocomposite superconductor consists of MgB_2 grains with nano-alumina (10–50 nm) inside the grains and at grain boundaries could be strong pinning centers, unlike in the case of doping with the coarse alumina.

7.3.2. BI-2212 DOPING

Shen et al. [242] found that J_c of bulk polycrystalline MgB_2 is significantly enhanced by the Bi-2212 addition. Microstructural analysis indicates that a small amount of Bi-2212 is decomposed into Cu_2O and other impurity phases while a significant amount of not decomposed Bi-2212 particles remains in MgB_2 matrix, and these act as effective pinning centres for vortices. The enhanced pinning force is mainly attributable to these highly dispersed inclusions inserted in the MgB_2 grains.

7.3.3. FE_2O_3 DOPING

Prozorov et al. have performed a study on magnesium diboride slurries with magnetic Fe_2O_3 nanoparticles. By using ultrasonic method, considerable enhancement of magnetic hysteresis of such MgB_2 was obtained, which implies an increase of vortex pinning strength due to embedded magnetic nanoparticles [243].

7.3.4. MGO DOPING

Besides the main phase of MgB_2, MgO is always present as the major impurity phase in MgB_2 superconductor. It was shown that an MgO layer on the surface of MgB_2 grains would cause a strong depression of I_c [74]. However, Kovac et al. have concluded that by controlling the amount of MgO, pinning in *ex situ* MgB_2 wires can be increased. The level of MgO should be optimized as a compromise between inter-grain connectivity and introducing pinning centres in ex-situ MgB_2 wire [244]. It was found that light doping of nano-MgO (62.5 wt.%) could increase the J_c property of MgB_2 tapes through enhancing the grain connectivity and core density. A 50% enhancement in J_c (10 T, 4.2 K) was obtained when adding 1.25 wt % of nano-MgO into MgB_2 [245]. In the mean time, Perner et al. discovered that doping with MgO, which improves J_c (i.e. 1.6 $\times 10^6$ A cm^{-2} at 7.5 K in self field), is more effective than doping with SiO_2 [76, 246].

7.3.5. SILICIDES DOPING

Besides SiC, several silicon compounds, such as WSi_2, $ZrSi_2$, Mg_2Si, Si_3N_4 except for SiO_2 have also been added into MgB_2, and almost all of them showed a positive effect on enhancing the J_c values of MgB_2 superconductors [247-251]. Among these additives, the best result in terms of J_c was achieved for $ZrSi_2$ doping, by up to a factor of 3.4 at 4.2 K in magnetic fields up to 12 T. Scanning electron microscopy investigation revealed that the WSi_2 and $ZrSi_2$ doping enhanced intergranular connectivity, thus the critical temperature for the doped tapes decreased slightly ~less than 0.7 K. Under the EDX, small precipitates and homogenous segregation of each elements were observed in the SiC doped MgB_2 which explained why SiC doped MgB_2 is much better than other silicides impurities [248].

7.3.6. SIO_2 DOPING

Microstructural analysis reveals that a high density of $MgSi_2$ nanoparticles (10–50 nm) exists inside the SiO_2 nanoparticles added MgB_2 grains, leading to the formation of a nanocomposite superconductor [252]. The lattice constant in the *c* direction increases with additive content due to a small amount of Si being doped into the lattice of the MgB_2. The addition of SiO_2 nanoparticles also improves the J_c–H and H_{irr}–T characteristics of MgB_2 when the additive content is lower than 7%. However, the degradation of J_c values was obtained by using MgH_2 as a precursor instead of Mg in SiO_2 doping [95]. It was speculated that the H_2 released from the decomposition of MgH_2 deoxidizes SiO_2 to form H_2O and this H_2O degrades the MgB_2 core.

7.3.7. NB_XB_2 DOPING

The crystal structure of Nb_xB_2 is similar to MgB_2 in both AlB_2-type structure and in values of the lattice parameters [253, 254]. So, utilizing Nb_xB_2 as the additive for strong pinning centre for the MgB_2 tape is expected to have some advantages with respect to matching the MgB_2 grains, densification of the tape core and alignment of the grains. The best $J_c(H)$ value was more than 1000 A/cm^2 for 5 mol % of Nb_xB_2 at 4.2 K and 12T compared to the undoped one.

7.3.8. MG$_2$CU DOPING

Kikuchi's group have studied the Mg_2Cu (568 °C) doping that has a lower melting point than pure MgB_2 (600°C), thus promoting a diffusion reaction for synthesizing MgB_2 through their liquid states [255]. Both J_c and H_{irr} at 4.2 K improved with decreasing the size of Mg_2Cu powders as a starting material. J_c of 10^4 A cm $^{-2}$ was obtained at 4.2 K and 4 T for the Fe-sheathed MgB_2 wire made by using Mg_2Cu powder with a diameter of a few microns. These improvements may be due to the enhancement of homogeneity of the powder mixture with small $MgCu_2$ particles with a size of about 10 nm, which were dispersed in the MgB_2 matrix.

7.3.9. O$_2$ DOPING

Haruta et al. has showed that effective pinning centres can be introduced by the deposition of the MgB_2 thin film in an O_2 atmosphere [256]. The values of J_c of the O_2-doped film have exceeded those of the non-doped film in magnetic fields applied perpendicular and also parallel to the *c*-axis. The peak of J_c caused by the *c*-axis correlated pinning centres has been observed in the angular dependence of J_c. It is found that B_{c2} increased by the introduction of O_2 in both the directions of the applied field. For example at 15T, both B_{c2}, for ($B \perp c$) and (B//c), of O_2-doped film were enhanced by a factor of 1.1 compared to the pure MgB_2 thin film at 15.5 K and 13 K respectively.

7.3.10. TIO$_2$ DOPING

Xu et al. have achieved an increase of H_{irr} and J_c performance of TiO_2 doping [257]. The H_{irr} at 20 K increases from 4.3 to 4.9 T as the TiO_2 content increases from 0 to 10 wt%, and the J_c increases from 450 to 4250 A/cm^2 at 4.2 T. Besides TiO_2, the Mg-deficiency and existence of MgO are suggested to be an important factor in increasing the H_{irr} and J_c.

7.3.11. ZRB2 DOPING

According to Bhatia et al. an increase in B_{c2} from 20.5 T to 28.6 T and enhancement of H_{irr} from 16 T to 24 T were observed in the ZrB_2 doped sample as compared to the binary MgB_2 sample at 4.2 K [258]. In addition, the critical field increases similarly as for SiC doped MgB2 at 4.2 K.

7.3.12. ZRO$_2$ DOPING

The *in situ* and *ex situ* ZrO_2 doped MgB_2 was studied by Chen et al. in order to understand the contamination effect by mechanical alloying [259]. Short hours of milling ZrO_2 and MgB_2 powders (> 4 hours) results in no new phases, while 400 h milled powder was partially amorphized and showed the formation of a minority ZrB_2 phase. The result from the magnetic measurement shows that J_c of MgB_2 is deteriorated by *in situ* ZrO_2 addition while *ex situ* ZrO_2 addition with annealing did not degrade the T_c of MgB_2. However, there is not report on the J_c of the *ex situ* ZrO_2 addition doped MgB_2.

7.3.13. ZRH$_2$ DOPING

Xu et al. used ZrH_2 as another form of Zr for doping effect in the MgB_2 as ZrH_2 decomposes to be Zr and H_2 at high temperature [260]. Some J_c values of the undoped wires are higher than those of the doped samples, due to the microstructure caused by the ZrH_2 doping and the sintering temperature. J_c values of all the doped MgB_2 wires decrease with the increase of sintering temperature, but the optimum doping ratio goes up correspondingly.

Table 3. Comparison of other addition source doping for magnetic and transport $J_c(H)$ at 4 and 20K. All the $J_c(H)$ were determined by the magnetization unless stated

Dopant	Type	J_c at 8T (A/cm^2)	J_c at 6T (A/cm^2)	J_c at 2T (A/cm2)	Reference
4.25/5K					
Al	Bulk		2x10^3	2x10^4	[210]
	Bulk 2x2x3 mm^3			1x10^4	[211]
	Bulk	7x10^2	1.5x10^5		[212]
Al_2O_3	Bulk 0.5×1.7×2.9mm^3			4x10^3	[240]
Au	Thin film			4x10^5	[216]
BN	Wire (*ex-situ*)	Transport	5x10^2		[232]
Cd	Bulk			2x10^5	[217]
Co_3O_4	Bulk		10^6	5x10^7	[263]
Cu	Tape (*in-situ*) magnetic	5x10^2	5x10^3		[239]
Dy_2O_3	Bulk (6K)			2.9 x10^5	[262]
Fe	Bulk			2x10^5	[229]
Hf	Wire (*ex-situ*)	Transport 10^3	8x10^3		[232]
In	Wire (*ex-situ*)	Transport		2x10^4	[236, 237]
La	Bulk			2x10^5	[230]
Li	Bulk 2x2x3 mm^3			7x10^3	[211]
MgO	Wire (*in-situ*)	Transport		3x10^5	[245]
MgSi	Tape (*in-situ*)	Transport 0.5x10^4	4x10^4		[248]
Na	Bulk	7x10^3	2x10^5		[212]
Nb	Wire (*ex-situ*)	Transport 8x10^2	6x10^3		[232]
N_xB_2	Tape (*in-situ*)	Transport 1.5x10^4	5x10^4		[254]
Nb_2O_5	Wire (*ex-situ*)	Transport 6x10^2	5x10^3		[232]
Si	Bulk 2x2x3 mm^3			1.5x10^4	[211]
Si_3N_4	Tape	Transport 1.6x10^4	5x10^4		[250]
SiO	Tape (*in-situ*)	Transport 4x10^3	10^4		[248, 95]
Sn	Wire (*ex-situ*)	Transport		1x10^4	[236, 237]
Ti	Bulk		2x10^4	6x10^5	[219, 221]
	Bulk		4x10^3		[223]
	Chemical Vapor Deposition (CVD) conductor			2x10^6	[224, 225]
TiO_2	Wire (*ex-situ*)	Transport 6x10	6x10^2		[232]
	Bulk 1.35x2.83x4.45 mm^3			8x10^5	[257]
W	Wire (*ex-situ*)	Transport 10^3	6x10^3		[232]
WB	Wire (*in-situ*)	Transport 2x10^4	7x10^4	7x10^5	[261]
WSi_2	Tape (*in-situ*)	Transport 8x10^3	2x10^4		[247-249]
Y_2O_3	Bulk	2x10^3	10^4	2x10^5	[90]
Zn	Bulk	2x10^4	2.5x10^5		[186]
Zn	Wire magnetic			10^4	[188]

Table 3. (Continued).

Dopant	Type	J_c at 8T (A/cm^2)	J_c at 6T (A/cm^2)	J_c at 2T (A/cm2)	Reference
Zr	Bulk		2x10^3	2x10^5	[219]
	Bulk 0.7x0.9x1.4 mm^3		10^4	8x10^5	[233]
	Wire (*ex-situ*)	6x10^2	5x10^3		[232]
ZrB_2	In-situ Tape	Transport 8x10^3	2x10^4		[247]
$ZrSi_2$	In-situ Tape	Transport 8x10^3	2x10^4		[247-250]
		J_c at 4T (A/cm^2)	J_c at 2T (A/cm^2)	J_c at 0T (A/cm^2)	
20K					
Al	Bulk	8x10	7x10^3	7x10^4	[210]
			2x10^5	4x10^5	[213]
Li and Al	Bulk		7.8x10^4	10^6	[214]
Zn and Al	Bulk		7x10^5	2x10^6	[214]
Al_2O_3	Bulk 20x10x2mm^3		2x10^2	10^5	[241]
	Bulk 0.5×1.7×2.9mm^3		2x10^2	4x10^4	[240]
Bi-2212	Bulk 1.4 × 1.5 ×3.0 mm^3		5x10^4	10^6	[242]
Cd	Bulk	10^2	10^4	2x10^5	[217]
Dy_2O_3	Bulk	1.9x10^3	9x10^4		[262]
Fe	Bulk		4x10^4	4x10^5	[229]
Hf	Bulk	7x10^2	3x10^4	3x10^5	[218]
La	Bulk		2x10^3	10^5	[230]
MgO	Wire (*in situ*)	Transport 4x10^3	8x10^4	3x10^5	[245]
O	Thin film (B⊥c)	5x10^4	10^5		[256]
Si	Bulk 0.56x2.7 x 3.73 mm^3	2x10^4	2x10^5	9x10^5	[238]
SiO	Bulk (0.21x0.21x 0.21 mm^3)	2x10^3	10^5	7x10^5	[252]
Ta	Bulk	10^3	4x10^4	1.5x10^5	[239]
Ti	Bulk	10	10^5	10^6	[219, 222]
	In-situ Ta/Cu tape	Magnetic		9x10^5	[220, 222]
	Bulk	6x10^2	3x10^4	3x10^5	[191]
	Bulk	10^4	2x10^5	5x10^5	[222]
	Bulk	3x10^4	2x10^5	5x10^5	[225]
	Bulk	4x10^4	2x10^5	4x10^5	[227]
TiO	Bulk 1.35x2.83x 4.45 mm^3	6x10^3	2x10^5		[256]
Y_2O_3	Bulk	10^4	10^5	3.5x10^5	[90]
Zr	Bulk	10	5x10^4	8x10^5	[219]
	Bulk 0.7x0.9x1.4 mm^3	2x10^2	10^5	7x10^6	[233, 234]
	Bulk	2x10^3	3x10^4	3x10^5	[218]
ZrB_2	Bulk			3x10^5	[258]

7.3.14. WB Doping

The effect of WB addition on the microstructure and superconducting properties of MgB_2 tapes has been investigated by Fuji et al. [261]. The microstructure was not influenced appreciably by the addition, whereas the superconducting properties changed. However, the transport critical current density at 4.2 K increased about twice by the 5 mol% addition and reached 15 and 5 kA cm^{-2} at 8 and 10 T, respectively, suggesting that pinning centres effective in a high-field region were introduced.

7.3.15. Y_2O_3 Doping

Wang et al. doped MgB_2 with 5, 10, and 15 wt% Y_2O_3 nanoparticles, using a simple solid-state reaction route [90]. Transmission electron microscopy showed a fine nanostructure consisting of 3–5 nm YB_4 nanoparticles embedded within MgB_2 grains of 400 nm size. Compared to an undoped control sample, an improvement in the in-field critical current density J_c and irreversibility fields were observed with a value of 8 x 10^4 Acm^{-2} at 2T and 5.5 T for J_c and H_{irr}, respectively, at 20K.

7.3.16. Dy_2O_3 Doping

Recently, Chen et al. studied the rare earth oxide, Dy_2O_3 nano particles by in-situ method [262]. They found that the T_c remained largely unchanged, as J_c was enhanced by a factor of 4, as compared to the pure sample at 20K and 1T.

7.3.17. Co3O4 Doping

Study showed that magnetic critical current density (J_c) of nano Co_3O_4 doped MgB_2 increased by nearly an order of magnitude for 2% and 4% but decreased sharply for the 6% sample at nearly all temperatures and applied fields [263]. In addition, all the samples exhibited clear and sharp diamagnetic transitions at around 38 K, in zero-field-cooled (ZFC) magnetic susceptibility measurements with a sizeable signal.

8. Practical Applications and Challenges of MgB_2

In the past three years, rapid R and D progress of MgB_2 towards its development as a technical conductor has reached remarkable milestones. With the knowledge of the basic properties of the MgB_2 like the H_{c2}, two gaps superconductivity, flux-pinning mechanisms and weak link free grain, the problems of fabricating MgB_2 conductors commercially are being addressed by Hyper Tech Research in Columbus, Ohio [264-266], Columbus Superconductors based in Genova, Italy [267], Japan based Hitachi which has collaborated with National Institute of Material Science (NIMS), Japan [248], Diboride Conductors in Cambridge and many other laboratories worldwide. All these groups have emerged to exploit MgB_2 in applications ranging from fault current limiters for electrical power transmission systems to low field magnetic resonance imaging (MRI). Besides that, it is believed that MgB_2 will eventually find its way into generators and motors in ships and marine systems, as such superconducting components would be smaller and lighter than conventional conductor version. Because the promised HTS superconductors such as $YBa_2Cu_3O_7$ coated conductor are still in a very infant stage of scaling up and also suffers from low engineering critical current density, J_{ceng}, this opens up a possibility of MgB_2 to compete with the well established

Nb-based low temperature superconductors, LTS, such as NbTi and Nb_3Sn that represent 99.99% of world production of superconducting materials. The prospects for an operation of MgB_2 around 20-25K and fields up to 5T in the cryocooler cooled magnet (~20K) (MRI with field 2-5T), high field magnet (~4.2K), liquid hydrogen cooled electrical transmission system and superconducting magnetic-energy storage, SMES, will give a very competitive challenge to the LTS. The demonstration of coils with magnetic field exceeding 1 T at 25K under self-field and 4K under 1.25 T has demonstrated the advanced MgB_2 competitiveness with LTS for superconducting transformer and low field MRI [268, 269]. Figure 17 shows the critical current density of MgB_2 in comparison to other commercial superconductor materials. One can note that the SiC doped MgB_2 stand very strongly at 20K low field and it is comparable to the value of J_c for Nb-Ti at 4.2K, which is very useful for application of MRI.

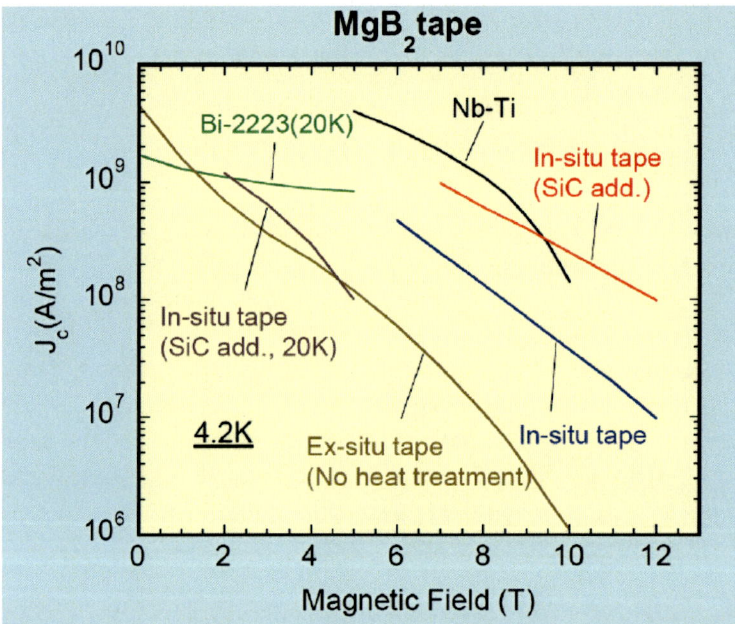

Figure 17. Comparison of J_c of MgB_2 with other commercial superconducting wire [270]. All results were measured at 4.2 K unless labeled.

On the economic side, dramatic savings could be achieved if the superconductors can be put to practical use. Advantage of MgB_2 is that the raw material costs of both B and Mg are low; reasonable estimates suggest that even for appropriately purified material, they will be several times less than those of Nb-based superconductors. And instead of requiring a silver cladding, like Bi2223/Ag, MgB_2 works just fine with a sheath of cheap metal, such as iron. Even though low-temperature superconductors, such as niobium-titanium (NbTi), cost only $1 per kA/m to produce, they must be cooled to 4 K with expensive liquid helium. Operating at 77 K, HTS wire allows the use of cheap liquid nitrogen to cool the material to its operating temperature. However, the first generation of HTS wires cost around US$200 per kA/m although large-scale manufacturing might drive the cost down to US$50 per kA/m. MgB_2 becomes superconducting below 39 K and it will operate at temperatures between 20 and 30 K. In fact, 20 K is good news for many applications, because it allows the use of inexpensive cryocoolers, a type of no hassle refrigerator that can be plugged into the wall power socket.

After factoring in all the costs for wire, the cooling equipment and electrical losses MgB_2 wire will eventually be better than HTS and standard copper wire as in figure 18 [271]. Predictions that the annual market for HTS-based wire and machinery will approach $2 billion by 2025 suggest that there's plenty of room for sales of MgB_2. According to Hyper Tech Research, they will drive down the prices to around US$ 5 to 7/kA/m at 20K in a 2T field, with a 200A of current by 2009. If successful, this price will match the US$ 13/kA/m price tag on the copper wire used in electric motors.

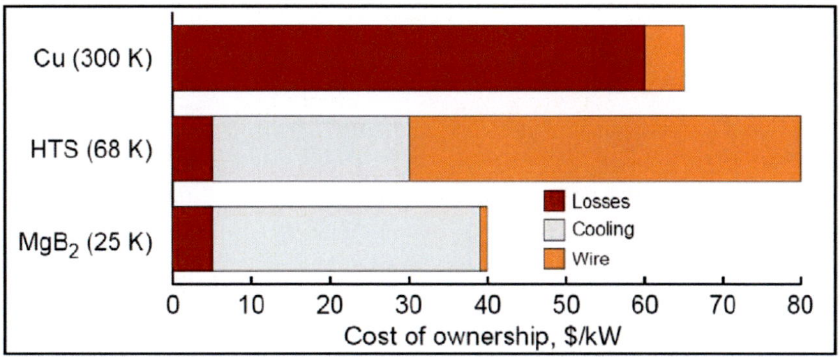

Figure 18. MgB_2 wire cost less than those based on copper or high-temperature superconductors [271].

Despite all the optimistic prospects, many problems have yet to be tackled in the development of MgB_2 conductors. Iron can be used as a protective layer between MgB_2 and a stabilizing outer copper layer while providing the magnetic shield of the filaments as well. The Fe barrier, which provides a poor thermal stability to MgB_2 due to the electrical resistivity, and does not have the same flow stress as the high-purity Cu or Ag needed for electrical stabilization. This leads to drawing instabilities, and wire failure. The mechanical reinforcement of conductor is important to withstand handling during coil winding, thermal stress and Lorentz forces in magnetic field. Study on V(I) transition are commonly described by means of a power law equation, $V=I^n$ which defines a characteristic exponent, so called n-values. In fact, n-factor of superconducting wire or tape has to exceed value of 30 to be used for magnets in persistent mode [272, 273].

Although high n values above 100 were observed for MgB_2 conductors at low fields [123, 274-278], consideration needed to be addressed for variety of fundamentals properties that influenced the n values like sheath materials, filament size and geometry. Poor thermal stabilization of high current-carrying MgB_2 wires and tapes leads to burn-out even for currents below the true I_c. In AC-applications, superconductor exhibits AC-losses, which should be minimized since they translate into cooling costs during normal operation. Besides that, the design of the conductor composites and the choice of sheath materials have to meet several important technical conductor specifications, which cannot be fulfilled in the single sheath core material. As a result, development of high quality multi filament wire geometries is now becoming important for reasons of strain tolerance, flux jump stability, transport current stability and AC loss reduction.

REFERENCES

[1] Schon, J. H.; Kloc, C.; Batlogg, B. *Nature.* 2000, *408*, 549-552.

[2] Kasumov, A. Y.; Kociak, M.; Guĕron, S.; Reulet, B.; Volkov, V. T.; Klinov, D. V.; and Bouchiat, H. *Science* 2001, *291*, 280-282.

[3] Jones, M. and Marsh, R. *J. Am. Chem. Soc.* 1965, *76*, 1434-1436.

[4] Nagamatsu, J.; Nakagawa, N.; Muranaka, T.; Zenitani, Y and Akimitsu, J. *Nature.* 2001, *410*, 63-64.

[5] Kortus, J.; Mazin, I. I., Belashchenko, K. D.; Antropov, V. P. and Boyer, L. L. *Phys. Rev. lett.* 2001, *86*, 4656-4659.

[6] An, J. M. and Pickett, W. E. *Phys. Rev. Lett.* 2001, *86*, 4366-4369.

[7] Bud'ko, S. L.; Lapertot, G.; Petrovic, C.; Cunningham, C. E.; Anderson, N. and Canfield, P. C. *Phys. Rev. Lett.* 2001 *86*, 1877-1880.

[8] Sharoni, A.; Felner, I.; Millo, O. *Phys. Rev. B* 2001, *63*, 2205081-2205084.

[9] Kotegawa, H.; Ishida, K.; Kitaoka, Y.; Muranaka, T.; Akimitsu, J. *Phys. Rev. Lett.* 2001, *87*, 127001-127004.

[10] Grant, P. M. *Industrial Physicist.* 2001, *7*, 22-23.

[11] Muranaka, T.; Zenitani, Y.; Shimoyama, J. and Akimitsu, *J. Superconductivity in MgB_2 Frontier in Superconducting Materials,* 2005 Springer-Verlag Berlin Heidelberg.

[12] Buzea, C. and Yamashita, T. *Supercond. Sci. Technol.* 2001, *14*, R115-R146.

[13] Eremets, M. I.; Struzhkin, V. V.; Mao, H. K. and Hemley, R. J. *Science.* 2001, *293*, 272-274.

[14] Leyarovska, L. and Leyarovski, E. *J. Less Common Met.* 1979, *67*, 249-255.

[15] Copper, A. S.; Corenzwit, E.; Longinotti, L. D.; Matthias, B. T. and Zachariasen, W. H. *Proc. Natl. Acad. Sci. U.S.A.* 1970, *67*, 313-319.

[16] Yamamoto, K. A.; Takao, C.; Masui, T.; Izumi, M. and Tajima, S. *Physica C.* 2002, *383*, 197-206.

[17] Young, D. P.; Goodrich, R. G.; Adams, P. W.; Chan, J. Y.; Fronczek, F. R.; Drymiotis, F. and Henry, L. L. *Phys. Rev. B* 2002, *65*, 1805181-1805183.

[18] Gasparov, V. A.; Sidorov, N. S.; Zver'kova, I. I. and Kulakov, M. P. *JETP Letters.* 2001, *73*, 532-535.

[19] Kaczorowski, D.; Zaleski, A. J.; Zogal, O. J. and Klamut, J. (2001) "Incipient superconductivity in TaB_2" *Preprint cond-mat/0103571*

[20] He, L. H.; Hu, L. G. Q.; Zhang, P. L. and Yan, Q. W. *Chin. Phys*. 2001, *10*, 343-344.

[21] Sung, G. Y.; Kim, S. H.; Kim, J. H.; Yoo, D. C.; Lee, J. W.; Lee, J. Y.; Jung, C. U.; Park, M. S.; Kang, W.N.; Du, Z. L. and Lee, S. I. *Supercond. Sci. Tech.* 2001, *14*, 880-883.

[22] Margadonna, S.; Muranaka, T.; Prassides, K.; Maurin, I.; Brigatti, K.; Ibberson, R. M.; Arai, M.; Takata, M. and Akimitsu, J. *J. Phys-:Condens. Matter*. 2001, *13* L795-L802.

[23] Yu, R. C.; Li, S. C.; Wang, Y. Q.; Kong, X.; Zhu, J. L.; Li, F. Y.; Liu, Z. X.; Duan, X. F.; Zhang, Z. and Jin, C. Q. *Physica C.* 2001, *363*, 184-188.

[24] Larbalestier, D.; Gurevich, A.; Feldmann, D. M. and Polyanskii, A. *Nature.* 2001, *414*, 368 -377.

[25] Tupitsyn, I. I. *Sov. Phys. Solid State.* 1976, *18*, 1688-1690.

[26] Armstrong, D. R. and Perkins, P. G. *J. of the Chem. Soc., Faraday Transactions. 2* 1979, *75*, 12-16.
[27] Medvedeva, N. I.; Ivanovskii, A. L.; Medvedeva, J. E. and Freeman, A. J. *Phys. Rev. B* 2001, *64*, 020502-020505.
[28] Kortus, J.; Mazin, I. I.; Belashchenko, K. D.; Antropov, V. P. and Boyer, L. L. *Phys. Rev. Lett.* 2001, *86*, 4656-4659.
[29] An, J. M. and Pickett, W. E. *Phys. Rev. Lett.* 2001, *86*, 4366-4369.
[30] Satta, G.; Profeta, G.; Bernardini, F.; Continenza, A. and Massidda, S. *Phys. Rev. B* 2001, *64*, 104507-104515.
[31] Belashchenko, K .D.; Schilfgaarde, M. V. and Antropov, V. P. *Phys. Rev. B* 2001, *64*, 092503-092506.
[32] Canfield, P.C. and Grabtree, G. W. *Physics Today.* 2003, *56*, 34-40.
[33] Suhl, H.; Matthias, B. T. and Walker, L. R. *Phys. Rev. Lett.* 1959, *3*, 552-554.
[34] Shen, L. Y. L.; Senozan, N. M. and Phillips, N. E. *Phys. Rev. Lett.* 1965, *14*, 1025-1027.
[35] Radebaugh, R. and Keesom, P.H. *Phys. Rev.* 1966, *149*, 209-216.
[36] Binnig, G.; Baratoff, A.; Hoenig, H. E. and Bednorz, J. G. *Phys. Rev. Lett.* 1980, *45*, 1352-1355.
[37] Shulga, S. V.; Drechsler, S. L.; Fuchs, G.; Müller, K. H.; Winzer, K.; Heinecke, M. and Krug, K. *Phys. Rev. Lett.*1998, *80*, 1730-1733.
[38] Gurevich, A. *Phys. Rev. B* 2003, *67*, 184515-184527.
[39] Braccini, V.; Gurevich, A.; Giencke, J. E.; Jewell, M. C.; Eom, C. B.; Larbalestier, D. C., Pogrebnyakov, A.; Cui, Y.; Liu, B. T.; Hu, Y. F.; Redwing, J. M.; Li, Q.; Xi, X. X.; Singh, R. K.; Gandikota, R.; Kim, J.; Wilkens, B.; Newman, N.; Rowell, J.; Moeckly, B.; Ferrando, V.; Tarantini, C.; Marré, D.; Putti, M.; Ferdeghini, C.; Vaglio, R. and Haanappel, E. *Phys. Rev. B* 2005, *71*, 012504-1-012504-4.
[40] Gonnelli, R. S.; Daghero, D.; Calzolari, A.; Ummarino, G. A.; Dellarocca, V.; Stepanov, V. A.; Kazakov, S. M.; Zhigadlo, N. and Karpinski, J. *Phys. Rev. B* 2005, *71*, 060503-1-060503-4.
[41] Gonnelli, R. S.; Daghero, D.; Ummarino, G. A.; Calzolari, A.; Dellarocca, V.; Stepanov, V. A.; Kazakov, S. M.; Jun, J. and Karpinski, J. (2004) *"A Point-Contact Study of the Superconducting Gaps in Al-Substituted and C-Substituted MgB_2 Single Crystals"* Prepint Cond-matt 0407267
[42] Ohmichi, E.; Masui, T.; Lee, S.; Tajima, S. and Osada, T. *J. Phys. Soc. of Jap.* 2004, *73*, 2065-2068.
[43] Liu, A. Y.; Mazin, I. I. and Kortus, J. *Phys. Rev. Lett.* 2001, *87*, 870051-870054.
[44] Hinks, D. G.; Claus, H. and Jorgensen, J. D. *Nature.* 2001, *411*, 457-460.
[45] Bud'ko, S. L.; Lapertot, G.; Petrovic, C.; Cunningham, C. E; Anderson, N. and Canfield, P. C. *Phys. Rev. Lett.* 2001, *86* 1877-1880.
[46] Canfield, P. C.; Bud'ko, S. L. and Finnemore, D. K. *Physica C.* 2003, *385*, 1-7.
[47] Bud'ko, S. L.; Kogan, V. G. and Canfield, P. C. (2001) "Determination of superconducting anisotropy from magnetization data on random powders as applied to $LuNi_2B_2C$, YNi_2B_2C and MgB_2" *Preprint* cond-mat/0106577
[48] Lee, S.; Mori, H.; Masui, T.; Eltsev, Y.; Yamamoto, A. and Tajima, S. *J. Phys. Soc. Jpn.* 2001, *70*, 2255-2258.

[49] Xu, M.; Kitazawa, H.; Takano, Y.; Ye J.; Nishida, K.; Abe, H.; Matsushita, A. and Kido, G. *Appl. Phys. Lett.* 2001, *79*, 2779-2781.

[50] Panagopoulos, C.; Rainford, D.; Xiang, T.; Scott, C. A.; Kambara, M. and Inoue, I. H. *Phys. Rev. B* 2001, *64*, 945141-945143.

[51] Pronin, A. V.; Pimenov, A.; Loidl, A. and Krasnosvobodtsev, S. I. *Phys. Rev. Lett.* 2001, *87*, 970031-970034.

[52] Klein, N.; Jin, B.B.; Schubert, J.; Schuster, M.; Yi, H. R.; Pimenov, A.; Loidl, A. and Krasnosvobodtsev, S. I. (2001) "Energy gap and London penetration depth of MgB_2 films determined by microwave resonator measurement" *Preprint* cond-mat/0107259

[53] Jin, R.; Paranthaman, M.; Zhai, H. Y; Christen, H. M.; Christen, D. K. and Mandrus, D. *Phys. Rev. B* 2001, *64*, 2205061-2205064.

[54] Marsiglio, F. *Phys. Rev. Lett.* 2001, *87*, 2470011-2470014.

[55] Hirsch, J. E. *Phys. Lett. A* 2001, *282*, 392-398.

[56] Hirsch, J. E. and Marsiglio, F. *Phys. Rev. B* 2001, *64*, 1445231-1445238.

[57] Yamamoto, A.; Shimoyama, J. I.; Ueda, S.; Katsura, Y.; Horii, S. and Kishio, K. *Supercond. Sci. Tech.* 2005, *18*, 116-121.

[58] Kumakura, H.; Kitaguchi, H.; Matsumoto, A. and Hatakeyama, H. *Appl. Phys. Lett.* 2004, *84*, 3669-3671.

[59] Yamada, H.; Hirakawa, M.; Kumakura, H.; Matsumoto, A. and Kitaguchi, H. *Appl. Phys. Lett.* 2004, *84*, 1728-1730.

[60] Chen, S. K.; Lockman, Z.; Wei, M.; Glowacki, B. A. and MacManus-Driscoll, J. L. *Appl. Phys. Lett.* 2005, *86*, 242501-242503.

[61] Massalski, T. B. In *Binary Alloy Phase Diagrams*; Okamoto H.; Subramanian P.R. and Kacprzak L.; Ed.; 0871704064; ASM International: Materials Park, OH (2nd), 1990.

[62] Goldacker, W. and Schlachter, S. I. In *Preparation and properties of advancedMgB2 wires and tapes, Frontiers in Superconducting Materials*; Narlikar A.V.; Ed.; 354024512X; Springer: Berlin, 2005 pp 937-981.

[63] Yeoh, W. K.; Kim, J. H.; Horvat, J.; Xu, X. and Dou, S. X. (2006) *Control of Carbon Substitution for Enhancing the Superconductivity of MgB_2* manuscript submitted to Applied Superconductivity Conference 2006.

[64] Frederick, N. A.; Li, S.; Maple, M. B.; Nesterenko, V. F. and Indrakanti, S. S. *Physica C.* 2001, *363*, 1-5.

[65] Indrakanti, S. S.; Nesterenko, V, F.; Maple, M. B.; Frederick, N. A.; Yuhasz, W. M. and Li, S. *Philos. Mag. Lett.* 2001, *81*, 849-857.

[66] Serquis, A.; Liao, X. Z.; Zhu, Y. T.; Coulter, J. Y.; Huang, J. Y.; Willis, J. O.; Peterson, D. E.; Mueller, F. M.; Moreno, N. O.; Thompson, J. D.; Nesterenko, V. F. and Indrakanti, S. S. *J. Appl. Phys.* 2002, *92*, 351-356.

[67] Serquis, A.; Civale, L.; Hammon, D. L.; Liao, X. Z.; Coulter, J. Y.; Zhu, Y. T.; Jaime, M.; Peterson, D. E.; Mueller, F. M.; Nesterenko, V. F. and Gu, Y. *Appl. Phys. Lett.* 2003, *82*, 2847-2849.

[68] Schultz, L. and Eckert, J. In Glassy Metals III: Amorphization Techniques, Catalysis, Electronic and Ionic Structure (Topics in Applied Physics), Beck, H.; Ed.; 0387574409; Springer, Berlin 1994; Vol. 72, pp 69.

[69] Gűmbel, A.; Eckert, J.; Fuchs G.; Nenkov, K.; Műller, K. H. and Schultz, L. *Appl. Phys. Lett.* 2002 *80*, 2725-2727.

[70] Gűmbel, A.; Perner, O.; Eckert, J.; Fuchs, G.; Nenkov, K.; Műller, K. H. and Schultz, L. *IEEE Trans. Appl. Supercond.* 2003, *13*, 3064-3067.
[71] Eckert, J.; Perner, O.; Fuchs, G.; Nenkov, K.; Műller, K. H.; Häßler, W.; Fischer, C., Holzapfel, B. and Schultz, L. In *Advances in Solid State Physics*; Kramer, B.; Ed.; 3540401504; Springer: Berlin 2003; Vol. 43, pp. 703.
[72] Abe, H.; Naito, M.; Nogi, K.; Matsuda, M.; Miyake, M.; Ohara, S.; Kondo, A. and Fukui, T. *Physica C.* 2003, 391, 211-216.
[73] Chen, S. K.; Glowacki, B. A.; MacManus-Driscoll, J. L.; Majoros, M.; Vickers, M. and Pawar, S. H. (2003) *EUCAS* 2003 (Poster presentation)
[74] Hassler, W.; Rodig, C.; Fischer, C.; Holzapfel, B.; Perner, O.; Eckert, J.; Nenkov, K. and Fuchs, G. *Supercond. Sci. Technol.* 2003, *16*, 281-284.
[75] Fischer, C.; Rodig, C.; Hassler, W.; Perner, O.; Eckert, J.; Nenkov, K.; Fuchs, G.; Wendrock, H.; Holzapfel, B. and Schultz, L. *Appl. Phys. Lett.* 2003, *83*, 1803-1805.
[76] Perner, O.; Hassler, W.; Fischer, C.; Fuchs, G.; Holzapfel, B.; Schultz, L. and Eckert, J. *IEEE Trans. Appl. Supercond* 2005, *15*, 3192-3195.
[77] Gao, Y. D.; Ding, J.; Rao, G. V. S.; Chowdari, B. V. R.; Sun, W. X. and Shen, Z. X. *Phys. Status Solid A* 2002, *191*, 548-554.
[78] Senkowicz, B. J.; Giencke, J. E.; Patnaik, S.; Eom, C. B.; Hellstrom, E. E. and Larbalestier, D. C. *Appl. Phys. Lett.* 2005, 86, 202502-202504.
[79] Kambara, M.; Babu, N. H.; Sadki, E. S.; Cooper, J. R.; Minami, H.; Cardwell, D. A.; Campbell, A. M. and Inoue, I. H. *Supercond. Sci. Technol.* 2001, 14, L5-L7.
[80] Takano, Y.; Takeya, H.; Fujii, H.; Kumakura, H.; Hatano, T. and Togano, K. *Appl. Phys. Lett.* 2001, 78, 2914-2916.
[81] Dhalle, M.; Toulemonde, P.; Beneduce, C.; Musolino, N.; Decroux, M. and Flukiger, R. *Physica C* 2001, 363, 155-165.
[82] Glowacki, B. A.; Majoros, M.; Vikers, M.; Evetts, J. E.; Shi, Y. and Mcdougall, I. *Supercond. Sci. Technol.* 2001, *14*, 193-199.
[83] Jin, S.; Mavoori, H.; Bower, C. and van Dover, R. B. *Nature* 2001, *411*, 563-565.
[84] Soltanian, S.; Wang, X. L.; Kusevic, I.; Babic, E.; Li, A. H.; Liu, H. K.; Collings, E. W. and Dou, S. X. *Physica C* 2001, *361*, 84-90.
[85] Grasso, G.; Malagoni, A.; Ferdeghini, C.; Roncallo, S.; Braccini, V.; Cimberle, M. R.; and Siri, A. S. *Appl. Phys. Lett.* 2001, *79*, 230-232.
[86] Kumakura, H.; Matsumoto, A.; Fujii, H. and Togano, K. *Appl. Phys. Lett.* 2001, *79*, 2435-2437.
[87] Matsumoto, A.; Hatakeyama, H.; Kitaguchi, H.; Fujii, H.; Togano, K. and Kumakura, K. *Physica C*, 2002, *382*, 207-212.
[88] Suo, H. L.; Beneduce, C.; Dhalle, M.; Musolino, N.; Genoud, J. Y. and Flukiger, R. *Appl. Phys. Lett.* 2001, *79*, 3116-3118.
[89] Dou, S. X.; Soltanian, S.; Horvat, J.; Wang, X. L.; Zhou, S. H.; Ionescu, M.; Liu, H. K.; Munroe, P. and Tomsic, M. *Appl. Phys. Lett.* 2002, 81, 3419-3421.
[90] Wang, J.; Bugoslavsky, Y.; Berenov, A.; Cowey, L.; Caplin, A. D.; Cohen, L .F.; MacManus-Driscoll, J. L.; Cooley, L .D.; Song, X. and Larbalestier, D. C. *Appl. Phys. Lett.* 2002, 81, 2026-2028.
[91] Dou, S. X.; Pan, A. V.; Zhou, S.; Ionescu, M.; Liu, H. K. and Munroe, P. R. *Supercond. Sci. Technol.* 2002, 15, 1587-1591.

[92] Tanaka, K.; Kitaguchi, H.; Kumakura, H.; Hirakawa, M.; Yamada, H. and Okada, M. *IEEE Trans. Appl. Supercond.* 2005, *15*, 3180-3183.
[93] Serquis, A.; Civale, L.; Hammon, D. L.; Coulter, J. Y.; Liao, X. Z.; Zhu, Y. T.; Peterson, D. E. and Mueller, F. M. *Appl. Phys. Lett.* 2003, *82*, 1754-1756.
[94] Song, K. J.; Lee, N. J.; Jang, H. M.; Ha, H. S.; Ha, D. W.; Oh, S. S.; Sohn, M. H.; Kwon, Y. K. and Ryu, K. S. *Physica C* 2002, *370*, 21-26.
[95] Matsumoto, A.; Kumakura, H.; Kitaguchi, H. and Hatakeyama, H. *Supercond. Sci. Technol.* 2003 *16*, 926-930.
[96] Kumakura, H.; Matsumoto, A.; Kitaguchi, H. and Hatakeyama, H. *Physica C* 2002, *182*, 93-97.
[97] Pradhan, A. K.; Feng, Y.; Zhao, Y.; Koshizuka, N.; Zhou, L.; Zhang, P. X.; Liu, X, H.; Ji, P.; Du, S. J. and Liu, C. F. *Appl. Phys. Lett.* 2001, *79*, 1649-1651.
[98] Goldacker, W.; Schlachter, S. I.; Obst, B.; Liu, B.; Reiner, J. and Zimmer, S. *Supercond. Sci. Technol.* 2004, *17*, S363-S368.
[99] Dou, S. X.; Horvat, J.; Soltanian, S.; Wang, X. L.; Qin, M. J.; Zhou, S.H.; Liu, H. K. and Munroe, P. R. *IEEE Trans. Appl. Supercond.* 2003, *13*, 3199-3202.
[100] Kovac, P.; Husek, I.; Pachla, W.; Melisek, T.; Diduszko, R.; Frohlich, K.; Morawski, A.; Presz, A. and Machajdik, D. *Supercond. Sci. Technol.* 2002, *15*, 1127-1132.
[101] Fujii, H.; Kumakura, H. and Togano, K. *J. Mater. Res.* 2002, *17*, 2339-2345.
[102] Horvat, J.; Wang, X. L.; Soltanian, S. and Dou, S. X. *Appl. Phys. Lett.* 2002, *80*, 829-831.
[103] Horvat, J.; Soltanian, S.; Wang, X. L. and Dou, S. X. *IEEE Trans. Appl. Supercond* 2003, *13*, 3324-3327.
[104] Majoros, M.; Glowacki, B. A. and Campbell, A. M. *Physica C.* 2001, *334*, 129-140.
[105] Flukiger, R.; Suo, H. L.; Musolino, N.; Beneduce, C.; Toulemonde, P. and Lezza, P. *Physica C* 2003, *385*, 286-305.
[106] Canfield, P. C.; Finnemore, D. K.; Budko, S. L.; Ostenson, J. E.; Lapertot, G.; Cunningham, C. E. and Petrovic, C. *Phys. Rev. Lett.* 2001, *86*, 2423-2426.
[107] Sumption, M. D.; Bhatia, M.; Dou, S. X.; Rindfliesch, M.; Tomsic, M.; Arda, L.; Ozdemir, M.; Hascicek, Y. and Collings, E. W. *Supercond. Sci. Technol.* 2004, *17*, 1180-1184.
[108] Giunchi, G.; Ceresara, S.; Ripamonti, R.; Di Zenobio, A.; Rossi, S.; Chiarelli, S.; Spadoni, M.; Wesche, R. and Bruzzone, P. L. *Supercond. Sci. Technol.* 2003, 16, 285-291.
[109] Togano, K.; Nakane, Y.; Fujii, H.; Takeya, H. and Kumakura, H. *Supercond. Sci. Technol.* 2006, *19*, L17-L20.
[110] Ribeiro, R. A.; Bud'ko, S. L.; Petrovic, C. and Canfield, P. C. *Physica C* 2003, *385*, 16-23.
[111] Bardeen, J.; Copper, L. N. and Schriefffer, J. R. *Phys. Rev.* 1957, *108*, 1175-1204.
[112] Soltanian, S.; Wang, X. L.; Horvat, J.; Qin, M. J.; Liu, H. K.; Munroe, P. R. and Dou, S. X. *IEEE Trans. Appl. Supercond.* 2003, *13*, 3273-3276.
[113] Larbalestier, D. C.; Cooley, L. D.; Rikel, M. O.; Polyanskii, A. A.; Jiang, J., Patnaik, S.; Cai, X. Y.; Feldmann, D. M.; Gurevich, A.; Squitieri, A. A.; Naus, M. T.; Eom, C. B; Hellstrom, E. E.; Cava, R. J.; Regan, K. A.; Rogado, N.; Hayward, M. A.; He, T.; Slusky, J. S.; Khalifah, P.; Inumaru, K. and Haas, M. *Nature* 2001, 410, 186-189.

[114] Kitaguchi, H.: Matsumoto, A.; Kumakura, H.; Doi, T.; Yamamoto, H.; Saitoh, K.; Sosiati, H. and Hata, S. *Appl. Phys. Lett.* 2004, *85*, 2842-2844.
[115] Song, X. Y.; Babcock, S. E.; Eom, C. B.; Larbalestier, D. C.; Regan, K. A.; Cava, R. J.; Bud'ko, S. L.; Canfield, P. C. and Finnemore, D. K. *Supercond. Sci. Technol.* 2002, *15*, 511-518.
[116] Cooley, L.; Song, X. Y. and Larbalestier, D. *IEEE Trans. Appl. Supercond.* 2003, *13*, 3280-3283.
[117] Yamamoto, A.; Shimoyama, J.; Ueda, S.; Katsura, Y.; Iwayama, I.; Horii, S. and Kishio, K. *Appl. Phys. Lett.* 2005, *86*, 212502-212504.
[118] Serquis, A.; Zhu, Y. T.; Peterson, E. J.; Coulter, J. Y.; Peterson, D. E. and Mueller, F. M. *Appl. Phys. Lett.* 2001, *79*, 4399-4401.
[119] Xiao, H.; Peng, W.; Song, W. H.; Ma, R. C.; Zhang, L.; Du, J. J. and Sun, Y. P. *Physica C* 2003, *386*, 648-652.
[120] Franco, C. M.; Ferreira, B.; Dos Santos, C. A. M.; Ghivelder, L.; Filho, H. J. I. and Machado, A. J. S. *Physica C* 2004, *130*, 408–410.
[121] Ribeiro, R.; Bud'ko, S. L.; Petrovic, C. and Canfield, P. C. *Physica C* 2002, *382*, 194-202.
[122] Zhou, S. H.; Pan, A. V.; Horvat, J.; Qin, M. J. and Liu, H. K. *Supercond. Sci. Technol.* 2004, *17*, S528-S532.
[123] Lezza, P.; Abacherli, V.; Clayton, N.; Senatore, C.; Uglietti, D.; Suo H. L. and Flukiger, R *Physica C* 2004, *401*, 305-309.
[124] Jiang, C. H.; Nakane, T. and Kumakura, H. *Appl. Phys. Lett.* 2005, *87*, 252505-252507.
[125] Fujii, H.; K. Togano, K. and Kumakura, H. *Supercond. Sci. Technol.* 2002, *15*, 1571-1576.
[126] Yamada, H.; Hirakawa, M.; Kumakura, H.; Matsumoto, A. and Kitaguchi, H. *IEEE Trans. Appl. Supercond.* 2005, *15*, 3337-3340.
[127] Fang, H.; Padmanabhan, S.; Zhou, Y. X. and Salama, K. *Appl. Phys. Lett.* 2003, 82, 4113-4115.
[128] Chen, S. K.; Yates, K. A.; Blamire, M. G. and MacManus-Driscoll, J. L. *Supercond. Sci. Technol.* 2005, 18, 1473-1477.
[129] Xu, X.; Qin, M. J.; Konstantinov, K.; Dos Santos, D. I; Yeoh, W. K.; Kim, J. H. and Dou, S. X. *Supercond. Sci. Technol.* 2006, *19*, 466-469.
[130] Haßler, W.; Birajdar, B.; Gruner, W.; Herrmann, M.; Perner, O.; Rodig, C.; Schubert, M.; Holzapfel, B.; Eibl, O. and Schultz, L. *Supercond. Sci. Technol.* 2006, *19*, 512–520.
[131] Jiang, J.; Senkowicz, B. J.; Larbalestier, D. C. and Hellstrom, E. E. *Supercond. Sci. Technol.* 2006, *19*, 512–520.
[132] Horvat, J.; Soltanian, S.; Pan, A. V. and Wang, X. L. *J. of Appl. Phys.* 2004, *96*, 4342-4351.
[133] Takenobu, T.; Ito, T.; Chi, D. H.; Prassides, K. and Iwasa, Y. *Phys. Rev.* B 2001, *64*, 1345131-1345131.
[134] Paranthaman, M.; Thompson J. F. and Christen D. K. *Physica* C 2001, *355*, 1-5.
[135] Ahn, J. S. and Choi, E. J. (2001) Carbon substitution effect in MgB_2 *Preprint* cond-mat/0103169
[136] Mehl, M. J.; Papaconstantopoulos, D. A. and Singh, D. J. *Phys. Rev. B* 2001, 64, 1405091-1405094.

[137] Mickelson, W.; Cumings, J.; Han, E. Q. and Zettl, A. *Phys. Rev. B* 2002, *65*, 0525051-0525053.

[138] Maurin, I.; Margadonna, S.; Prassides, K.; Takenobu, T.; Iwasa, Y. and Fitch, A. N. *Chem. Mater.* 2002, *14*, 3894-3897.

[139] Maurin, I.; Margadonna, S.; Prassides, K.; Takenobu, T.; Ito, T.; Chi, D. H.; Iwasa, Y. and Fitch, A. *Physica B* 2002, *318*, 392-397.

[140] Bharathi, A.; Balaselvi, S. J.; Kalavathi, S.; Reddy, G. L .N; Sastry, V. S.; Haritharan, Y. and Radhakrishnan, T. S. *Physica C* 2002, *370*, 211-218.

[141] Yan, Y. and Al-Jassim, M. M. *J. of Appl. Phys*. 2002, *92*, 7687-7689.

[142] Ribeiro, R. A.; Budko, S.L.; Petrovic, C. and Canfield, P. C. *Physica C* 2003, *384*, 227-236.

[143] Lee, S.; Masui, T.; Yamamoto, A.; Uchiyama, H. and Takama, S. *Physica C* 2003, *397*, 7-13.

[144] Samuely, P.; Holanova, Z.; Szabo, P.; Kacmarcık, J.; Ribeiro, R. A.; Budko, S. L. and Canfield, P. C. *Phys. Rev. B* 68, 2003, 020505-1- 020505-4.

[145] Schmidt, H.; Gray, K. E.; Hinks, D. G.; Zasadzinski, J. F.; Avdeev, M.; Jorgensen, J. D. and Burley, J. C. *Phys. Rev. B* 2003, *68*, 060508-1- 060508-4.

[146] Avdeev, M.; Jorgensen, J. D.; Ribeiro, R. A.; Bud'ko, S. L. and Canfield, P. C. *Physica* C 2003, *387*, 301-306.

[147] Singh, P. P. *Solid State Comm.* 2003, *127*, 271–274.

[148] Papagelis, K.; Arvanitidis, J.; Margiolaki, I.; Brigatti, K.; Prassides, K.; Schenck, A.; Lappas, A.; Amato, A.; Iwasa, Y. and Takenobu, T. *Physica B* 2003, *326*, 346-349.

[149] Pissas, M.; Stamnopoulos, D.; Lee, S. and Tajima, S. *Phys. Rev. B* 2004, *70*, 134503-1-134503-8.

[150] Masui, T.; Lee, S. and Tajima, S. *Phys. Rev. B* 2004, *70*, 024504-1-024504-7.

[151] Ohmichi, E.; Masui, T.; Lee, S.; Tajima, S. and Osada, T. *Phys. Rev. B* 2004, 70, 174513-1-174513-5.

[152] Wilke, R. H. T.; Bud'ko, S. L.; Canfield, P. C.; Finnemore, D. K.; Suplinskas, R. J. and Hannahs, S. T. *Phys. Rev. Lett.* 2004, *92*, 217003-1-217003-4.

[153] Lee, S.; Masui, T.; Yamamoto, A.; Uchiyama, H. and Tajima, S. *Physica C* 2004, *31*, 412–414.

[154] Ohimichi, E.; Masui, T.; Lee, S.; Tajima, S. and Osada, T. *J. of the Phys. Soc. of Jap.* 2004, *73*, 2065-2068.

[155] Balaselvi, S. J.; Gayathri, N.; Bharathi, A.; Sastry, V. S. and Hariharan, Y. *Physica C* 2004, *407*, 31-38.

[156] Holanova, Z.; Kacmarcık, J.; Szabo, P.; Samuely, P.; Sheikin, I.; Ribeiro, R. A.; Bud'ko, S. L. and Canfield, P. C. *Physica C* 2004, *404*, 195-199.

[157] Masui, T.; Lee, S.; Yamamoto, A.; Uchiyama, H. and Tajima, S. *Physica C* 2004, 412–414, 303.

[158] Ummarino, G. A.; Daghero, D.; Gonnelli, R. S. and Moudden, A. H. *Phys. Rev. B* 2005, 71, 134511-1- 134511-6.

[159] Wilke, R. H. T.; Budko, S. L.; Canfield, P .C.; Kramer, M. J.; Wu, Y. Q.; Finnemore, D. K.; Suplinskas, R. J.; Marzik, J. V. and Hannahs, S. T. *Physica C* 2005, 418, 160-167.

[160] Kazakov, S. M.; Puzniak, R.; Rogacki, K.; Mironov, A. V.; Zhigadlo, N. D.; Jun, J.; Soltmann, Ch.; Batlogg, B. and Karpinski, J. *Phys. Rev. B* 2005, *71*, 024533-1- 024533-10.

[161] Kasinathan, D.; Lee, K. W. and Pickett, W. E. *Physica C* 2005, *424*, 116-167.

[162] Kortus, J.; Dolgov, O. V.; Kremer, R. K. and Golubov, A. A. *Phys. Rev. Lett.* 2005, 94, 027002-1-027002-4.

[163] Wilke, R. H. T, Budko, S. L.; Canfield, P. C.; Finnemore, D. K.; Raymond, D.; Suplinskas, J and Hannahs, S. T. *Physica C* 2005, *424*, 1-16.

[164] Gonnelli, R. S.; Daghero, D.; Calzolari, A.; Ummarino, G. A.; Dellarocca, V.; Stepanov, V. A.; Kazakov, S. M.; Zhigadlo, N. and Karpinski, *J. Phys. Rev. B* 2005, *71*, 060503-1-060503-4.

[165] Tenne, D. A.; Xi, X. X.; Pogrebnyakov, A. V. and Redwing, J. M. *Phys. Rev. B* 2005, *71*, 132512-1-132512-4.

[166] Huang, X.; Mickelson, W.; Regan, B. C. and Zettl, A. *Solid State Communications* 2005, *136*, 278-282.

[167] Sakuntala, T.; Bharathi, A.; Deb, S. K.; Gayathri, N.; Sundar, C. S. and Hariharan, Y. *J. Phys.: Condens. Matter* 2005, *17*, 3285-3292.

[168] Samuely, P.; Szabó, P.; Hol'Anov,á Z.; Budko, S. and Canfield, P. *Physica C* 2006, *435*, 71-73.

[169] Ueda, S.; Shimoyama, J. I.; Yamamoto, A.; Katsura, Y.; Iwayama, I.; Horii, S. and Kishio, K. *Physica C* 2005, *1225*, 426-431.

[170] Yamamoto, A.; Shimoyama, J. I.; Ueda, S.; Katsura, Y.; Horii, S. and Kishio, K. *IEEE Trans. Appl. Supercond.* 2005, *15*, 3292-3295.

[171] Yamamoto, A.; Shimoyama, J. I.; Ueda, S.; Iwayama, I.; Horii, S. and Kishio, K., *Supercond. Sci. Technol.* 2005, 18, 1323-1328.

[172] Lezza, P.; Senatore, C. and Flukiger, R. *Supercond. Sci. Technol.* 2006, *19*, 1030–1033.

[173] Yamada, H.; Hirakawa, M.; Kumakura, H. and Kitaguchi, H. *Supercond. Sci. Technol.* 2006, *19*, 175–177.

[174] Zhou, S. H.; Pan, A. V.; Dou, S. X. and Wexler, D. 'Sugar coating of boron powder for efficient carbon doping of MgB2 with enhanced current-carrying performance' *Advanced Materials* in press.

[175] Kim, J. H.; Zhou, S.; Hossain, M. S. A.; Pan, A. V. and Dou, S. X. *Appl. Phy. Lett* 2006, *89*, 142505-142507.

[176] Iijima, S. *Nature* 1991, *354*, 56-58.

[177] Li, Y.; Kinloch, I. A. and Windle, A. H. *Science* 2004, *304*, 276-278.

[178] Choi, E. S.; Brook,s J. S.; Eaton, D. L.; Al-Haik, M. S.; Hussaini, M. Y.; Garmestani, H.; Li D. and Dahmen, K. *J. of Appl. Phys.* 2003, *94*, 6034-6039.

[179] Fossheim, K.; Tuset, E. D.; Ebbesen, T. W.; Treacy, M. M. J. and Schwartz, J. *Physica C* 1995, *248*, 195-202.

[180] Huang, S.; Koblischka, M. R.; Fossheim, K.; Ebbesen, T. W. and Johansen, T. H. *Physica C* 1999, *311*, 172-186.

[181] Yang, P. and Lieber, C. M. *Science* 1996, *273*, 1836-1840.

[182] Yang, P. and Lieber, C. M. *Appl. Phy. Lett.* 1997, *70*, 3158-3160.

[183] Yang, P. and Lieber, C. M. *J. of Mat. Res.* 1997, *12*, 2981-2996.

[184] Wei, J.; Li, Y.; Xu, C.; Wei, B. and Wu, D. *Materials Chemistry and Physics* 2003, *78*, 785-790.

[185] Wei, B.Q.; Vajtai, R. and Ajayan, P. M. *Appl. Phys. Lett.* 2001, 79, 1172-1175.
[186] Kim, P.; Shi, L.; Majumdar, A. and McEuen, P. L. *Phys. Rev. Lett.* 2001, *87*, 215502-1-215502-4.
[187] Treacy, M. M. J.; Ebbesen, T. W. and Gibson, J. M. *Nature* 1996, *381*, 678-680.
[188] Dou, S. X.; Yeoh, W. K.; Horvat, J. and Ionescu, M. *Appl. Phys. Lett.* 2003, *83*, 4996-4998.
[189] Yeoh, W. K.; Horvat, J.; Dou, S. X. and Keast, V. *Supercond. Sci. Technol.*, 2004, *17*, S572-S577.
[190] Kim, J. H.; Yeoh, W. K.; Qin, M. J.; Xu, X. and Dou, S. X. *J. of Appl. Phys.* 2006, *100*, 013908-1 - 013908-6.
[191] Kim, J. H.; Yeoh, W. K.; Qin, M. J.; Xu, X.; Dou, S. X.; Munroe, P.; Kumakura, H.; Nakane, T. and Jiang, C. H. *Appl. Phys. Lett.* 2006, *89*, 122520-122523.
[192] Yeoh, W. K.; Horvat, J.; Dou, S. X. and Munroe, P. *IEEE Trans. Appl. Supercond.* 2005, *15*, 3284-3287.
[193] Galvan, D. H.; Li, S.; Yuhasz, W. M.; Kim, J. H.; Maple, M. B. and Adem, E. *Physica C* 2004, *403*, 145-150.
[194] Yeoh, W. K.; Kim, J. H.; Horvat, J.; Dou, S. X. and Munroe, P. *Supercond. Sci. Technol.* 2006, *19*, L5–L8.
[195] Chen, S. K.; Tan, K. S.; Glowacki, B. A.; Yeoh, W. K.; Soltanian, S.; Horvat, J. and Dou, S. X. *Appl. Phys. Lett.* 2005, *87*, 182504-182506.
[196] Dou, S. X.; Yeoh, W. K.; Shcherbakova, O.; Li, Y.; Ren, Z. M.; Wexler, D.; P. Munroe, P.; Chen, S. K.; Tan, K. S.; Glowacki, B. A. and MacManus-Driscoll, J. L. *Advanced Materials* 2006, *18*, 785-788.
[197] Yeoh, W. K.; Kim, J. H., Horvat, J.; Xu, X.; Qin, M. J.; Dou, S. X.; Jiang, C. H.; T. Nakane, T.; Kumakura, H. and Munroe, P. *Supercond. Sci. Technol.* 2006, *19*, 596-599.
[198] Ma, Y.; Zhang, X.; Nishijima, G; Watanabe, K.; Awaji, S. and Bai, X. *Appl. Phys. Lett* 2006, *88*, 072502-072504.
[199] Yeoh, W. K.; Kim, J. H., Horvat, J.; Xu, X.; Dou, S. X.; Rindfleisch, M.; Tomsic, M.; Nakane, T. and Kumakura, H. ASC 2006 poster presentation "Effect of Carbon Substitution on the Superconducting Properties of multi-walled carbon nanotube and nano carbon doped MgB_2"
[200] Soltanian, S.; Wang, X. L.; Horvat, J.; Dou, S. X.; Sumption, M. D.; Bhatia, M.; Collings, E. W.; Munroe, P. and Tomsic, M. *Supercond. Sci. Technol.* 2005, *18*, 658-666.
[201] Serquis, A.; Civale, L.; Hammon, D. L.; Serrano, G. and Nesterenko, V. F. *IEEE Trans. Appl. Supercond.* 2005, *15*, 3188-3191.
[202] Xu, H. L.; Feng, Y.; Xu, Z.; Yan, G.; Cao, L .Z. and Li, X. G. *Chin. Phys. Lett.* 2004, *21*, 2511-2513.
[203] Ban, E.; Sakaguchi, R.; Matsuoka, Y.; Goto, T.; Watanabe, K. and Nishijima, G. *Physica C* 2005, *1249*, 426–431.
[204] Soltanian, S.; Horvat, J.; Wang, X. L.; Munroe, P. and Dou, S. X. *Physica C* 2003, *390*, 185-190.
[205] Cheng, Z. H.; Shen, B. G.; Zhang J.; Zhang, S. Y.; Zhao, T. Y. and Zhao, H. W. *J. Appl. Phys.* 2002, 91, 7125-7127.

[206] Pogrebnyakov, A. V.; Xi, X. X.; Redwing, J. M.; Vaithyanathan, V.; Schlom, D. G.; Soukiassian, A.; Mi, S. B.; Jia, C. L.; Giencke, J. E.; Eom, C. B.; Chen, J.; Hu, Y. F.; Cui, Y. and Li, Q. *Appl. Phys. Lett.* 2004, *85*, 2017-2019.

[207] Kumakura, H.; Kitaguchi, H.; Matsumoto, A. and Hatakeyama, H. *IEEE Trans. Appl. Supercond.* 2005, *15*, 3184-3187.

[208] Zhao, Y.; Cheng, C. H.; Rui, X. F.; Zhang, H.; Munroe, P.; Zeng, H. M.; Koshizuka, N. and Murakami, M. *Appl. Phys. Lett.* 2003, *83*, 2916-2919.

[209] Ueda, S.; Shimoyama, J.; Yamamoto, A.; Horii, S. and Kishio, K. *Supercond. Sci. Technol.* 2004, *17*, 926-930.

[210] Xiang, J. Y.; Zheng, D. N; Li, J. Q.; Li, S. L.; Wen, H. H. and Zhao, Z. X. *Physica C* 2003, *386*, 611-615.

[211] Cimberle, M. R.; Novak, M.; Manfrinetti, P. and Palenzona, A. *Supercond. Sci. Technol.* 2002, *15*, 43-47.

[212] Toulemonde, P.; Musolino, N. and Flukiger, R. *Supercond. Sci. Technol.* 2003, *16*, 231-236.

[213] Berenov, A.; Serquis, A.; Liao, X. Z.; Zhu, Y. T.; Peterson, D. E.; Bugoslavsky, Y.; Yates, K. A.; Blamire, M. G.; Cohen, L. F. and MacManus-Driscoll, J. L. *Supercond. Sci. Technol.* 2004, *17*, 1093-1096.

[214] Xu, G. J.; Grivel, J. C.; Abrahamsen, A. B.; Chen, X. P. and Andersen, N. H. *Physica C* 2004, *403*, 113-118.

[215] Martínez, E.; Angurel, L. A.; Navarro, R.; Millán, A.; Rillo, C. and Artigas, M. *IEEE Trans. Appl. Supercond.* 2003, *13*, 3210-3213.

[216] Choi, E. M.; Lee, H. S.; Kim, H. J.; Lee, S. I.; Kim, H. J. and Kang, W. N. *Appl. Phys. Lett.* 2004, *84*, 82-84.

[217] Wang, S. F.; Dai, S. F.; Zhou, Y. L.; Zhu, Y. B.; Chen, Z. H.; Lu, H. B. and Yang, G. Z. *J. Supercond.* 2004, *17*, 397-400.

[218] Gato, D.; Machi, T.; Zhao, Y.; Koshizuka, N.; Murakami, M. and Arai, S. *Physica C* 2003, *272*, 392–396.

[219] Zhao, Y.; Feng, Y.; Huang, D. X.; Machi, T.; Cheng, C. H.; Nakao, K.; Chikumoto, N.; Fudamoto, Y.; Koshizuka, N. and Murakami, M. *Physica C* 2002, *378*, 122-126.

[220] Fu, B. Q.; Feng, Y.; Yan, G.; Zhao, Y.; Pradhan, A. K.; Cheng, C. H.; Ji, P.; Liu, X. H.; Liu, C. F.; Zhou, L. and Yau, K. F. *J. Appl Phys*. 2002, *92*, 7341-7344.

[221] Zhao, Y.; Feng, Y.; Machi, T.; Cheng, C. H.; Huang, D. X.; Fudamoto, Y.; Koshizuka, N. and Murakami, M. *Europhys. Lett.* 2002, *57*, 437-443.

[222] Fu, B. Q.; Feng, Y.; Zhao, Y.; Pradhan, A. K.; Cheng, C. H.; Ji, P.; Liu, H.; Liu, C. F.; G. Yan, and Zhou, L. *Physica C* 2003, *386*, 659-662.

[223] Zhou, S. H.; Liu, H. K.; Horvat, J. and Dou, S. X. *J. Low Temp. Phys*. 2003, *131*, 687-692.

[224] Anderson, Jr. N. E.; Straszheim, W. E.; Budko, S. L.; Canfield, P. C.; Finnemore, D. K. and Suplinskas, R. J. *Physica C* 2003, *390*, 11-15.

[225] Finnemore, D. K.; Straszheim, W. E.; Budko, S. L.; Canfield, P. C.; Anderson, Jr. N. E. and Suplinskas, R. J. *Physica C* 2003, *385*, 278-285.

[226] Prikhna, T. A.; Gawalek, W.; Savchuk, Y. M.; Moshchil, V. E.; N. V. Sergienko, N. V.; Habisreuther, T.; Wendt, M.; Hergt, R.; Schmidt, Ch.; Dellith, J.; Melnikov, V. S.; Assmann, A.; Litzkendorf, D. and Nagorny, P. A. *Physica* C 2004, *402*, 223-233.

[227] Haigh, S.; Kovac, P.; Prikhna, T. A.; Savchuk, Y. M.; Kilburn, M. R.; Salter, C.; Hutchison, J. and Grovenor, C. *Supercond. Sci. Technol.* 2005, *18*, 1190-1196.

[228] Zhao, Y.; Huang, D. X.; Feng, Y.; Cheng, C. H.; Machi, T.; Koshizuka, N. and Murakami, M. *Appl. Phys. Lett.* 2002, *80*, 1640-1642.

[229] Dou, S. X.; Soltanian, S.; Zhao, Y.; Getin, E.; Chen, Z.; Shcherbakova, O.; and Horvat, J. *Supercond. Sci. Technol.* 2005, *18*, 710-715.

[230] Shekhar, C.; Giri, R.; Tiwari, R. S.; Rana, D. S.; Malik, S. K. and Srivastava, O. N. *Supercond. Sci. Technol.* 2005, *18*, 1210-1214.

[231] Kimishima, Y.; Uehara, M.; Kuramoto, T.; Takano, S. and Takami, S. *Physica C* 2004, *402*, 412–414.

[232] Kovac, P.; Husek, I.; Melisek, T.; Grovenor, C. R. M.; Haigh, S. and Jones, H. *Supercond. Sci. Technol.* 2004, *17*, 1225-1230.

[233] Feng, Y.; Zhao, Y.; Sun, Y. P.; Liu, F. C.; Fu, B. Q.; Zhou, L.; Cheng, C. H.; Koshizuka, N. and Murakami, M. *Appl. Phys. Lett.* 2001, *79*, 3983-3985.

[234] Feng, Y.; Zhao, Y.; Pradhan, A. K.; Cheng, C. H.; Yau, J. K. F.; Zhou, L.; Koshizuka, N. and Murakami, M. *J. Appl. Phys.* 2002, *92*, 2614-2619.

[235] Prikhna, T. A.; Gawalek, W.; Surzhenko, A. B.; Moshchil, V. E.; Sergienko, N. V.; Savchuk, Y. M.; Melnikov, V. S.; Nagorny, P. A.; Habisreuther, T.; Dub, S. N.; Wendt, M.; Litzkendorf, D.; Dellith, J.; Schmidt, C.; Krabbes, G. and Vlasenko, A. V. *Physica* C 2002, *372*, 1543-1545.

[236] Tachikawa, K.; Yamada, Y.; Enomoto, M.; Aodai, M. and Kumakura, H. *IEEE Trans. Appl. Supercond.* 2003, *13*, 3269-3272.

[237] Tachikawa, K., Yamada, Y.; Enomoto, M.; Aodai, M. and Kumakura, H. *Physica C* 2003, *1030*, 392–396.

[238] Wang, X. L.; Soltanian, S.; James, M.; Qin, M. J.; Horvat, J.; Yao, Q. W.; Liu, H. K. and Dou, S. X. *Physica C* 2004, *63*, 408–410.

[239] Ko, J. W.; Yoo, J.; Kim, Y. K.; Kim, H. D.; Oh, K. H.; Choe, S. J.; Chung, H.; Chung, S. J.; Kumakura, H.; Matsumoto, A. and Togano, K. *IEEE Trans. Appl. Supercond.* 2003, *13*, 3214-3216.

[240] Delfany, M.; Wang, X. L.; Soltanian, S.; Horvat, J.; Liu, H. K. and Dou, S. X. *Ceramics International* 2004, *30*, 1581-1606.

[241] Rui, X. F.; Chen, J.; Chen, X.; Guo, W. and Zhang, H. *Physica C* 2004, *412*, 312-315.

[242] Shen, T. M.; Li G.; Zhu, X. T.; Cheng, C. H. and Zhao, Y. *Supercond. Sci. Technol.* 2005, *18*, L49-L52.

[243] Prozorov, T.; Prozorov, R.; Snezhko, A. and Suslick, K. S. *Appl. Phys. Lett.* 2003, *83*, 2019-2021.

[244] Kovac, P.; Husek, I.; Melisek, T.; Grivel, J. C.; Pachla, W.; Strbıkl, V.; Diduszko, R.; Homeyer, J. and Andersen, N. H., *Supercond. Sci. Technol.* 2004, *17*, L41-L46.

[245] Jiang, C. H.; Hatakeyama, H. and Kumakura, H. *Physica C* 2005, *423*, 45-50.

[246] Perner, O.; Hassler, W.; Eckert, J.; Fischer, C.; Mickel, C.; Fuchs, G.; Holzapfel, B. and Schultz, L. *Physica C* 2005, *432*, 15-24.

[247] Ma Y.; Kumakura H.; Matsumoto A.; Hatakeyama H. and Togano K. *Supercond. Sci. Technol.* 2003, *16*, 852-856.

[248] Matsumoto A.; Kumakura H.; Kitaguchi H. and H. Hatakeyama H. *Supercond. Sci. Technol.* 2004, *17*, S319-S323.

[249] Ma, Y.; Kumakura, H.; Matsumoto, A. and Togano, K. *Appl. Phys. Lett.* 2003, *83*, 1181-1183.
[250] Jiang, C. H.; Nakane, T. and Kumakura, H. *Supercond. Sci. Technol.* 2005, *18*, 902–906.
[251] Jiang, C. H.; Nakane, T. and Kumakura, H. *Physica C* 2006, *436*, 118–122.
[252] Rui, X. F.; Zhao, Y.; Xu, Y. Y.; Zhang, L.; Sun, X. F.; Wang, Y. Z. and Zhang, H. *Supercond. Sci. Technol.* 2004, *17*, 689-691.
[253] Yamamoto, A.; Takao, C.; Masui, T.; Izumi, M. and Tajima, S. *Physica C* 2002, *383*, 197-206.
[254] Nakane, T.; Takeya, H.; Fujii, H. and Kumakura, H. *Supercond. Sci. Technol.* 2005, *18*, 521-525.
[255] Kikuchi, A.; Iijima, Y.; Banno, N.; Takeuchi, T. and Inoue, K. *IEEE Trans. Appl. Supercond.* 2005, *15*, 3207-3210.
[256] Haruta, M.; Fujiyoshi, T.; Sueyoshi, T.; Miyahara, K.; Doi, T. Kitaguchi, H.; Awaji, S. and Watanabe, K. *Supercond. Sci. Technol.* 2005, *18*, 1460-1463.
[257] Xu, G. J.; Grivel, J. C.; Abrahamsen, A. B. and Andersen, N. H.; *Physica C* 2004, *406*, 95-99.
[258] Bhatia, M.; Sumption, M. D.; Collings, E. W. and Dregia, S. *Appl. Phys. Lett.* 2005, *87*, 042505-042507.
[259] Chen, S. K.; Glowacki, B. A.; MacManus-Driscoll, J. L.; Vickers, M. E. and Majoros, M. *Supercond. Sci. Technol.* 2004, *17*, 243-248.
[260] Xu, H. L.; Feng, Y.; Xu, Z.; Li, C. S.; Yan, G.; Wu, Y. F.; Chen, Z. L.; Mossang, E. and Sulpice, A. *Physica C* 2005, 1244, 426–431.
[261] Fujii, H.; Togano, K. and Kumakura, H. *Supercond. Sci. Technol.* 2003, *16*, 432-436.
[262] Chen, S. K.; Wei, M. and MacManus-Driscoll, J. L. *Appl. Phys. Lett.* 2006, *88*, 192512-1-192512-3.
[263] Awana, V. P. S.; Isobe, M.; Singh, K. P.; Takayama-Muromachi, E. and Kishan, H. *Supercond. Sci. Technol* 2006, *19*, 551–555.
[264] Sumption, M. D.; Bhatia, M.; Buta, F.; Bohnenstiehl, S.; Tomsic, M.; Rindfleisch, M.; Yue, J.; Phillips, J.; Kawabata, S. and Collings, E. W. *Supercond. Sci. Technol.* 2005, *18*, 961-965.
[265] Sumption, M. D.; Bhatia, M.; Rindfleisch, M.; Tomsic, M. and Collings, E. W. *Appl. Phys. Lett.* 2005, *86*, 102501-102503.
[266] Collings, E. W.; Lee, E.; Sumption, M. D.; Tomsic, M.; Wang, X. L.; Soltanian, S. and Dou, S. X. *Physica C* 2003, *386*, 555-559.
[267] Grasso, G.; Malagoli, A.; Modica, M.; Tumino, A.; Ferdeghini, C.; Siri, A. S.; Vignola, C.; Martini, L.; Previtali, V. and Volpini, G. *Supercond. Sci.* Technol. 2003, *16*, 271-275.
[268] Sumption, M. D.; Tomsic, M.; Bhatia, M.; Hascieck, Y.; Dou, S. X. and Collings, E. W. Presented at the ICMC Topical Conferences Feb 10-13 2004, Wollongong, Australia.
[269] Serquis, A.; Civale, L.; Coulter, J. Y.; Hammon, D. L.; Liao, X. Z.; Zhu, Y. T.; Peterson, D. E.; Mueller, F. M.; Nesterenko, V. F. and Indrakanti, S. S. *Supercond. Sci. Technol.* 2004, *17*, L35-L37.
[270] Kumakura, H. Short Course presentation in the Applied Superconductivity Conference 2004 Jacksonville, Florida, USA.
[271] Service, R. F. *Science* 2002, 295, 786-788.

[272] Seeber, B.; In *Handbook of Applied Superconductivity*; Seeber B. Ed.; 0750303778; Institute of Physics Publishing: Bristol 1998; Vol. 2, pp. 307.

[273] Tschopp, W. H. and Laukien, D. D. In *Handbook of Applied Superconductivity*; Seeber B. Ed.; 0750303778; Institute of Physics Publishing: Bristol 1998; Vol. 2 pp 1191

[274] Suo, H. L.; Lezza, P.; Uglietti, D.; Beneduce, C.; Abächerli, V. and Flükiger, R. *IEEE Trans. Appl. Supercond* 2003, *13*, 3265-3268.

[275] Flükiger, R.; Lezza, P.; Beneduce, C.; Musolino, N. and Suo, H. L. *Supercond. Sci. Technol.* 2003, *16*, 264-270.

[276] Kitaguchi, H.; Matsumoto, A.; Hatakeyama, H. and Kumakura, H. *Physica C* 2004, *401*, 246-250.

[277] Kováč, P.; Hušek, I. and Melišek, T. *Physica C* 2004, *401*, 282-285.

[278] Goldacker, W.; Schlachter, S. I.; Liu, B.; Obst, B. and Klimenko, E. *Physica C* 2004, 401, 80-86.

In: Progress in Advanced Materials Research
Editor: Nicolas H. Voler, pp. 79-117
ISBN: 978-1-60021-893-4

Chapter 3

NOVEL SELF-ASSEMBLED POLYELECTROLYTE CAPSULES: PREPARATION, PROPERTIES AND PROSPECTS

Wen-fei Dong[1] and An-Wu Xu[2]
[1] Key Laboratory of Automobile Materials, Ministry of Education, Jilin University, Changchun, 130025, China.
[2] Division of Nanomaterials and Chemistry, Hefei National Laboratory for Physical Sciences at Microscale, the Structure Research Laboratory of CAS, the School of Chemistry and Materials, University of Science and Technology of China, Hefei 230026, China

ABSTRACT

Polyelectrolyte capsules, composed of oppositely charged polyelectrolytes using the alternating layer-by-layer (LBL) self-assembly onto colloidal templates followed by core removal, have received widespread interests owing to their ease of fabrication, versatility with regards to selection of the materials, well-defined shell structure and tunable functionality. Although these capsules are structurally controlled at the nanometer scale, their robustness and high tolerance to imperfections offer significant promises for many applications in optics, electronics, sensors and biotechnology. In this chapter, we have given an overview of the recent progress and highlighted the potential developments in this field. Firstly, the core removal mechanisms and the structure-property relationships of capsules are outlined and elucidated. A variety of sacrificed templates had been utilized for capsule preparation, including polymeric colloids (i.e. melamine formaldehyde and polystyrene colloids), inorganic particles (i.e. silica and metal carbonate particles), and living materials (i.e. cells) etc. It is well documented that the wall thickness and the core decomposition conditions are two critical factors to govern the stability, integrity, and mesh size of polyelectrolyte capsules templated on pH-degradable melamine formaldehyde microparticles. After optimization of the core removal conditions, the diffusion and permeability of polyelectrolyte capsules are shown to be strongly dependent on the thickness, which is controlled by the variation of solvent quality (i.e. salt concentration) or the number of bilayers. The mesh size of intact capsules is also shown to decrease from meso- (2 ~ 50 nm) to microporous (less than 2 nm) as thickness increases. On the contrary, polyelectrolyte capsules made by manganese carbonate particles exhibit different permeability behaviors. Secondly, since many smart materials in bioengineering, nanotechnology and medicine require a depot of encapsulated drugs that are released on demand at a specific location by an external stimulus, polyelectrolyte capsules are very promising systems in the development of microencapsulation technologies with permeation control by changing environments. For example, electrochemical stimuli are very promising, since e.g. the redox potential near a desired organ where a pharmaceutical drug should be released may differ from other locations in the body or also, a corrosion pit may self-anneal due to the release of an inhibitor by local potential change. However, conventional, organic polyelectrolyte based microcapsules have their limitations due to their slow response to trace amounts of trigger or restricted freedom of choice of stimuli. We present a novel redox-sensitive capsules which show the dramatically changes of their permeability and diffusion behaviors upon triggered by redox agents. Consequently, many efforts had been devoted to fabricate various novel intelligent micro- and nanocapsules with unique controlled, responsive and switchable permeability. Although these responsive capsules, such as pH-, light-, or temperature-sensitive etc., had been developed, many strict barriers for their real applications still exist and have to be overcome in the future. Concerning to the practical issues, the comments and suggestions will be summarized and discussed in the end.

1. Introduction

Hollow polymeric microcapsules (or hollow microspheres) [1-3] present a unique architecture having a continuous wall with a hollow interior, and exhibit unique encapsulation and delivery properties for sensitive contents or active agents and have potential application in the diverse fields such as drug delivery, food industry, biology and biomedicine. Such ideal microcapsules should be mechanically tough and elastic, also be stable in a wide range of conditions, and have high encapsulation efficiency, controlled or switched permeability, and can be prepared by an easy and time-effective synthetic route.

Besides lipid vesicles or inorganic hollow spheres, several approaches have been used to achieve hollow microcapsules, including the self assembly of block copolymers [4,5], interfacial emulsion polymerization [6], self-organization of polymer micelles by non-covalent or covalent bonds [7], spontaneous colloidosome assembly of polymers and nanoparticles, etc. [8].

In 1998, a novel route to prepare microcapsules had been achieved by layer-by-layer (LBL) [9] adsorption of oppositely charged polyelectrolytes on decomposable particles and followed by core removal [10]. The electrostatic self-assembly layer by layer technique (LBL), first proposed by Iler in1966, was developed by Decher and coworkers [9]. They used anionic and cationic bipolar amphiphiles to build up multilayers by alternating dipping of a substrate into solutions of oppositely charged molecules. In the same year, they extended the method to polyelectrolytes, as shown in Scheme 1.

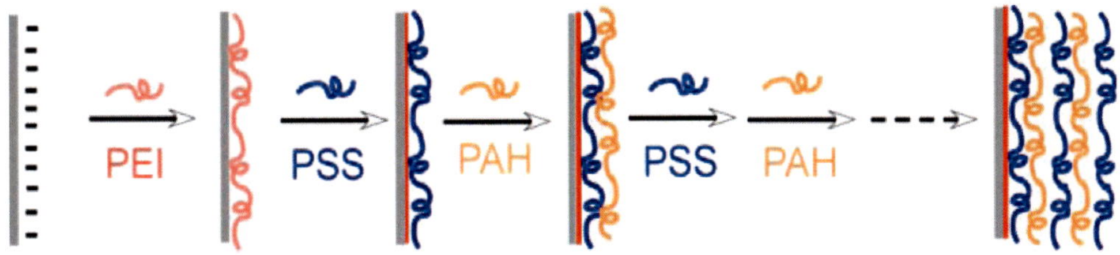

Scheme 1. Scheme of the LBL assembly of PEMs on flat substrate. Here, PSS: Poly(sodium 4-styrenesulfonate), and PAH: Poly(allylamine hydrochloride).

The crucial factor for successful deposition is the surface charge reversal upon deposition of the layers. The advantage of this LBL technique is that: 1) The thickness of the adsorbed layers can be tuned precisely in the range of nanometers; 2) There are no restrictions on the type of polyelectrolytes; 3) The film shows a unique and selective permeability; 4) Functional films could be designed to monitor the change of external environment. Because of this, research on the LBL technique is progressing rapidly.

Once colloidal particles are coated by LBL multilayers, the surface-modified colloidal particles show more unique features than those flat multilayers, for example, they have potential application in medical or drug delivery systems and bio-nanosensors; in addition, the high amount of material on the surface of colloidal particle enables direct investigation by modern methods, such as electrophoresis, calorimetric measurements and light scattering etc., and the controlled arrangement of colloidal particles exhibits superior electronic and optical properties [11,12].

The most commonly used colloidal templates include silica, polymer latex, inorganic or organic crystal, and biological blood cells, etc. [2]. All these templates can be degraded or dissolved. The important question of "What happens to the shell material after core removal?" was first addressed in the first "capsules" paper in 1998 [10]. The authors claimed that after removing the core by dissolution or calcination, a hollow capsule would be left in the solution with a continuous polymer shell (Scheme 2). The capsules are achieved with high yield. The particular interests stem from their tunable functionality, well-defined structure, accurate thickness control, and versatile and aqueous-based preparation procedures. Compared with other approaches, this method is rather simple and facile.

Further research demonstrated that these capsules have unique properties, which are different from those of flat multilayers [9,13,14]: the unique selective permeability, which cannot be observed in other kinds of capsules or vesicles. The planar polyelectrolyte multilayers (PEMs) of poly (styrene sulfonate) (PSS) and poly(allylamine hydrocloride) (PAH) $(PSS/PAH)_n$ (n is the number of bilayers), exhibited a very dense and robust film [14]: they enable limiting of ion diffusion across the layers, and show selective permeability for different ions but have the ability to resist proteins or other macromolecules. It had been reported that the penetration of dyes into flat PEMs occurs with very low diffusion coefficient of the order of 10^{-19} m^2/s [15]. However, unlike the planar PEMs, melamine formaldehyde (MF) capsules of $(PSS/PAH)_n$ exhibit a semi-permeable behavior since they can exclude PSS oligomers with molecular weight above 5 kDa (4200g/mol and upward) [16]. The diffusion coefficient of dye through MF capsules has been observed to be on the order of 10^{-14} m^2/s [17], which is faster than those of planar PEMs. In principle, the semi-permeability is a very important feature for these capsules. It allows the efficient loading of the polymer or biomacromolecule inside the capsules and then the polymer inside can be used to encapsulate drugs or to synthesize particles by variation of the environmental conditions in the capsule interior. On the other hand, one can easily switch off their permeability by reducing the mesh size when additional layers adsorb on the capsule surfaces. For instance, the diffusion coefficient of dye would decrease to 10^{-16} m^2/s, when four bilayers of PEM are fabricated on these semi-permeable capsules [17]. Meanwhile, the favorable encapsulation properties that can be achieved by either chemical [18,19] or physical pathways [20]. These modified capsules can be used as stable and robust substrate for model bio-membranes, and the capsule shells can serve as effective barrier to yield different chemical environments between the two sides of the capsule wall as well as different interfacial properties. These novel properties prompt further study on the physicochemical behavior of these capsules.

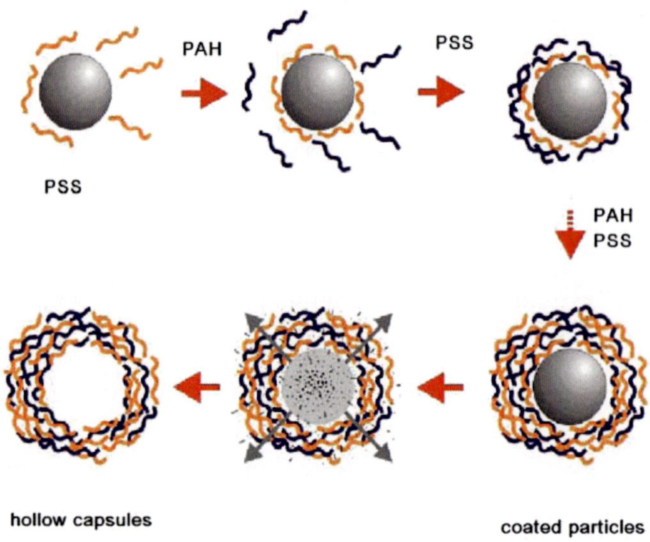

Scheme 2. The preparation procedures of polyelectrolyte microcapsules.

In this chapter, we will mainly focus on recent progress about the preparation and smart application of polyelectrolyte capsules templated on various organic or inorganic colloids.

2. Structure and Property

Among these capsules, weakly cross-linked melamine formaldehyde (MF) particles have been employed as sacrificial cores, because MF particles can be obtained with a narrow size distribution and have a good acid-solubility. However, to drive these capsules towards smart applications, an essential prerequisite is to understand their intrinsic properties, such as the structure-permeability relationship. In this section, we will systemically investigate the physicochemical behavior of capsules templated on MF. The pair of PSS and PAH is selected to fabricate the capsule wall since it produces a stable and robust multilayer. After core decomposition in low pH solution, stable capsules (MF capsules) are obtained.

As above mentioned, MF capsules of $(PSS/PAH)_n$ exhibit a semi-permeable behavior since they can exclude PSS oligomers with molecular weight above 5 kDa (4200g/mol and upward) [16]. However, up to date, little is known about the fundamental principles of the semi-permeability of MF capsules. The reason is three-fold: firstly, the semi-permeability of MF capsules is sensitive to preparation conditions and core quality; secondly, the permeability is strongly influenced by shell materials and shell thickness, as well as the core material; thirdly, there are no standard measurements to evaluate the permeability. Therefore, researchers lead to some different conclusions. For instance, the charged PAH with high molecular weight hardly permeate through MF capsules, whose surface is terminated with PAH. This is due to electrostatic repulsion of capsule shell. However, it is not in agreement with the self-spontaneous phenomena [20], which is demonstrated recently that PAH could diffuse though the MF capsules (with PAH terminated surface). Although the reason for permeation of PAH is due to the adsorption of the negative charged complex inside capsules, the capsule shell structure is still unknown. In general, one acceptable reason of the semi-permeability is the formation of holes or pores on the shell after core removal. However, the mechanism is not made clear because the formation of these holes is very sensitive to preparation and environmental conditions.

Since the semi-permeability [16] is a complicated issue to understand, here we use the standard pairs of PSS and PAH as the shell and the acid-soluble MF particles as the cores to prepare capsules. The core decomposition can be investigated as a function of pH value with the novel measurement of confocal Raman microscopy (CRM) [21]. More details on the structure and the permeability can be investigated by surface force microscopy and confocal laser scanning microscopy (CLSM).

We find that two critical factors control the permeability properties of MF capsules [22]: one is the core decomposition condition (the pH value or the amount of acid solution) and the other is the PEM wall thickness (referring to the capsule thickness before core decomposition) dominated by the variation of solvent quality (i.e. salt concentration) or the number of bilayers.

2.1. Mechanism and Optimization of Core Removal

One of the most common substrates used to synthesize hollow polyelectrolyte multilayer capsules are melamine resin microparticles, which are a network of MF cross-linked by ether linkages (III) or methylene bridges (IV), as shown in Scheme 3. After adsorption of the polymer onto the (MF), the core is dissolved by tuning the external solution. Although the ether linkages are decomposed by acid hydrolysis, the methylene bridges remain intact. Thus, the core is decomposed into oligo- as well as polymeric units.

Here CRM is used to study the chemical structure of the MF core *in-situ*. Figure 1a shows a typical Raman spectrum of a suspension of MF particles (V):

1. The sharp and strong Raman active mode at 981 cm^{-1} is attributed to the ring breathing in the melamine group,
2. The peak at 671 cm^{-1} is that of the melamine monomer (I) which broadens upon condensation,

3. The shoulder at 906 cm^{-1} is from the C-O-C symmetric stretch in the ether bridge which disappears after acid hydrolysis,
4. The bands at 1394 cm^{-1}, 1453 cm^{-1} and 1555 cm^{-1} are assumed to be the different ring breathing modes in the methylolmelamine.

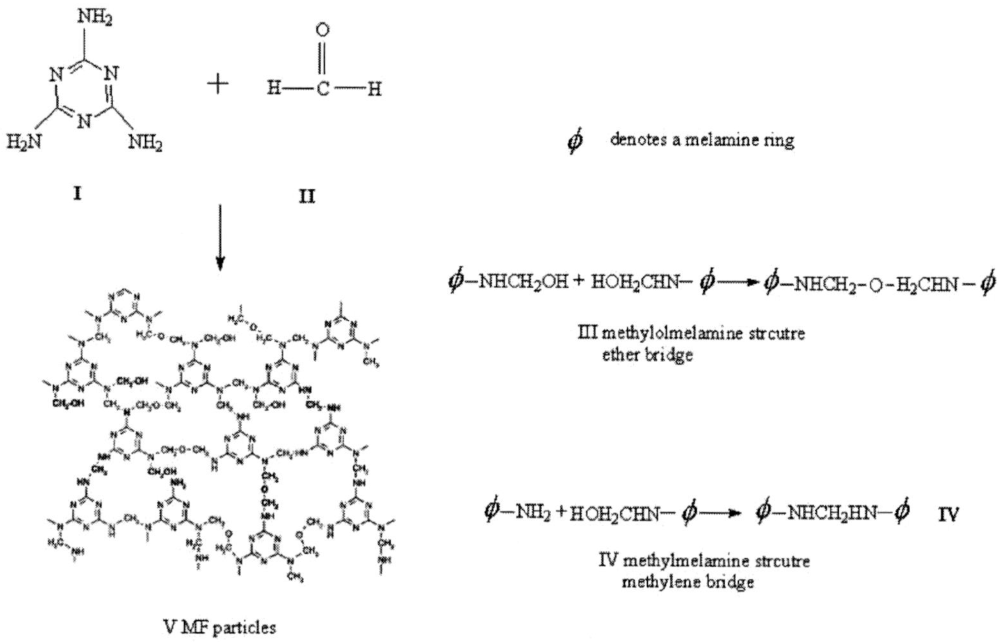

Scheme 3. Chemical structure of (I) melamine, (II) formaldehyde, (III) the methylolmelamine structure by the ether bridge, (IV) the methylmelamine structure by the methyl bridge and (V) MF particles. Reproduced from ref. 22. Copyright © 2005, American Chemical Society.

First consider the decomposition of naked MF microparticles by acid hydrolysis in aqueous solution. Suspensions of MF particles were exposed to hydrochloric acid solutions with varying pH and the Raman spectrum of the suspension was recorded at equilibrium. The spectrum of MF particles in (pH = 2.0) solution is shown in figure 1b. It can be readily observed that there is no appreciable difference between this and the reference spectrum in figure 1a, indicating the stability of the particles under these conditions. When the MF particles are dispersed in solution (pH = 1.0), the Raman signal is attenuated greatly due to rapid particle decomposition (The minimum pH for MF decomposition was found to be 1.7). To obtain a measurable spectrum, a drop of solution was evaporated to form a solid film, the spectrum of which is shown in figure 1c. The typical change in this spectrum is the band at 906 cm^{-1}, which disappears as expected due to the acid hydrolysis of the ether band. Also the band at 1453 cm^{-1} (the NCH_2O entity) in the methylolmelamine (III) has also decreased upon hydrolysis. Meanwhile, some new Raman active modes in the methyl-bridged melamine structure become more intense, such as at 1031 cm^{-1} (tentatively attributed to the motion of the triazine ring), 1435 cm^{-1} (the CH_2 vibration in NCH_2N), and at 1513 cm^{-1} (the CH_2 vibration of the methyl bond). Additional bonds can be observed in the range between 1339 cm^{-1} and 1537 cm^{-1}.

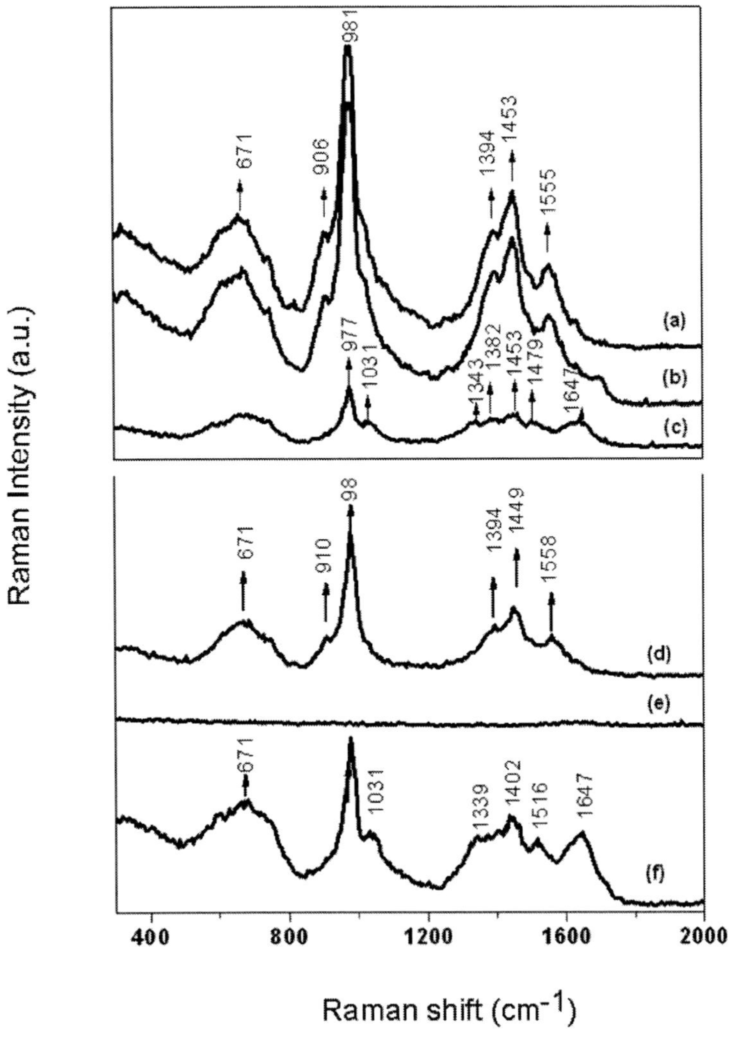

Figure 1. Raman spectrum of the MF core as a function of pH value at (a) pH = 7, (b) pH = 2, (c) pH = 1.0 of HCl acid solution (MF particles are dissolved in pH = 1.0 acid solution, and dried solutions are measured). The Raman spectrum of the capsule interior after the core decomposition at (d) pH = 2, (e) pH = 1.0, (f) pH = 0.5 of PEM coated MF particles. Reproduced from ref. 22. Copyright © 2005, American Chemical Society.

The diameter of the MF degradation product (i.e. MF oligomers) is estimated to about 2.5 nm by an analytical ultracentrifuge measurement, which is in agreement with previous studies [23]. This suggests that the degradation product (MF oligomers) are on average composed of melamine formaldehyde polymers with a degree of polymerization (n = 60).

The acid hydrolysis decomposition of MF resin becomes a more complex function of pH in the presence of encapsulating polyelectrolyte multilayers. For simplicity, this discussion is restricted to results obtained when four bilayers, $(PSS/PAH)_{n=4}$, were fabricated by the layer by layer technique onto MF particles at a salt concentration of C_{NaCl} = 0.25 M. The Raman spectrum of the capsule interior was obtained using confocal imaging at a high magnification (i.e. 100x) so that the diameter of the laser ($D_l \approx 1$ μm) was smaller than the capsule diameter ($D_c \approx 5$ μm).

As for the naked MF, the encapsulated MF cannot be dissolved for a pH value greater than 2.0, which can be seen by comparison of the spectrum shown in figure 1d with those in figure 1b (naked particles at pH = 2.0) and figure 1a (naked particles at pH = 7.0)

When the acidity was increased to pH 1.0, the turbidity of the suspension decreased rapidly suggesting decomposition of the MF core. The absence of Raman active modes in the capsule interior, as shown in figure 1e, signifies complete core removal. Additional spectra obtained by scanning across the capsule wall confirmed the absence of MF oligomer in the polymer matrix.

If core removal proceeds at pH 0.5 ~ 0.8, strong Raman bands can be observed within the capsule, as shown in figure 1f. By comparison with figure 1c, it can be concluded that the fraction of Raman activity by MF oligomers has increased, thus prompting one question: Why there are so many oligomers inside the capsules? As discussed previously, the MF oligomers are formed due to the hydrolysis of the ether bridges of MF particles. The hydrolysis product remains active at low pH and can easily form supramolecular structures due to hydrogen bonding. The same effect can be observed in the decomposition of naked particles (After exposure to a low pH, the particles are degraded into oligomers with a size of 3 nm. After incubation of the decomposition product for one week at room temperature, the oligomers self-aggregate into complexes with an average size of 160 nm, as measured by analytic ultracentrifuge). In addition, it also has been reported that melamine can be hydrolyzed to ammeline. Therefore, if the pH is too low (pH < 0.5 ~ 0.8), a second reaction (acid catalyzed re-polymerization) occurs in parallel with hydrolysis and results in the formation of C-N-C linkages. Since, the second reaction becomes increasingly during the hydrolysis, the reaction equilibrium favors the formation of an MF gel although some of the MF oligomers escape from the shell. Both of these mechanisms underscore the role of confinement of the hydrolysis product by the polyelectrolyte shell on the core release process and suggest an optimum pH window (1.0 < pH < 2.0) for core removal.

2.2. Capsule Stability and Yield

The formation of hollow capsules after core removal is a function of the number of polyelectrolyte (PSS/PAH) bilayers and salt concentration, and hence the structure and thickness of the shell. This results from a competition between the osmotic stress from in the capsule interior and the mechanical resistance conferred by the polymer shell. For capsule stability, the elastic modulus must be the same order of magnitude as the stress in the capsule wall. Typically, removal of the MF core results in a low yield of stable capsules when the number of bilayers, n, is less than four. When ($n \geq 4$), intact capsules can be observed. This dependence of stability on the number of bilayers suggests that there is a transition in the structure (and mechanical properties) of the shell as a function of its thickness.

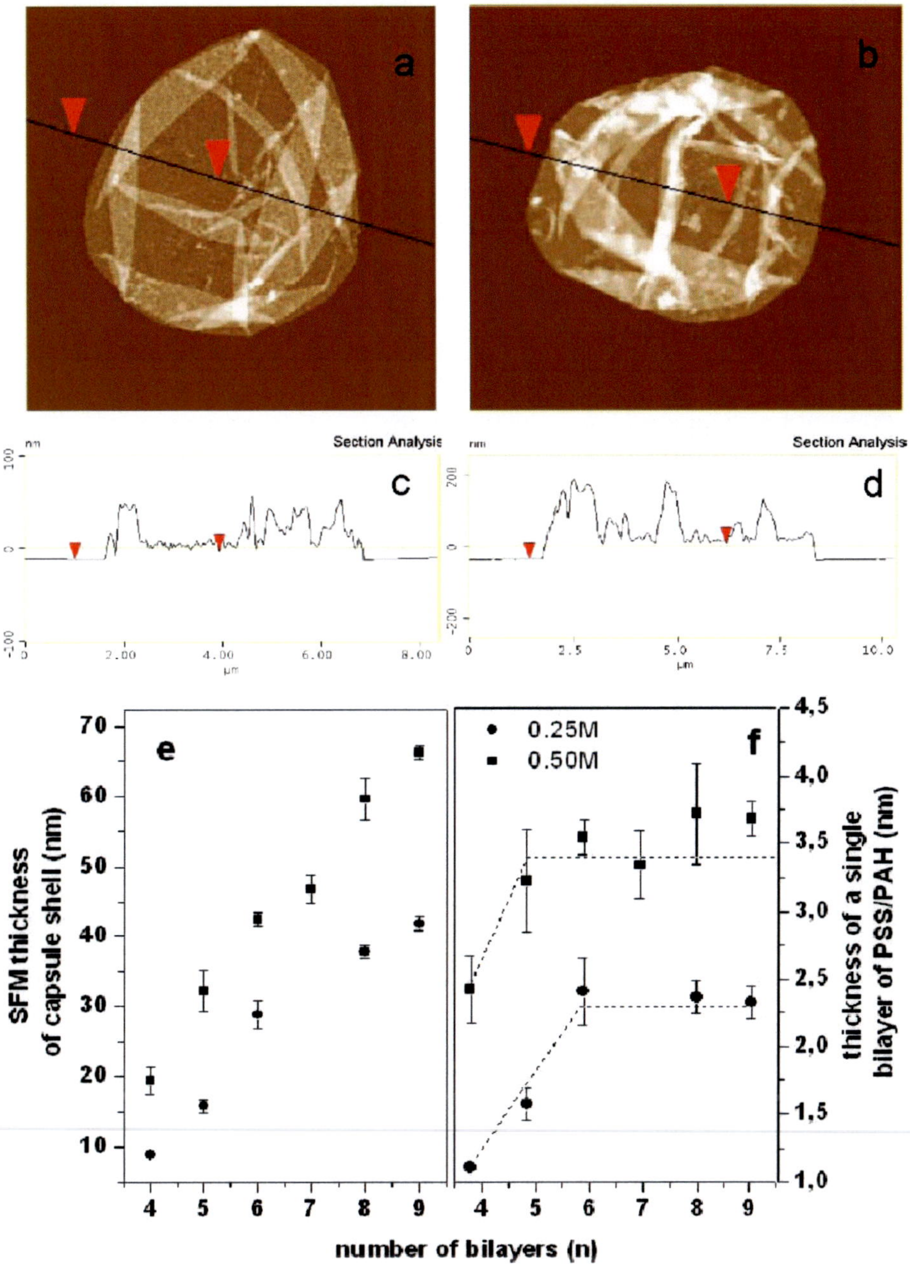

Figure 2. SFM images of capsules (a) four bilayers at C_{NaCl} = 0.25 M and (b) nine bilayers at C_{NaCl} = 0.25 M after removing the core by HCl at pH = 1.0. The SFM thickness of the capsules is measured by the difference in the heights corresponding to the arrow in (a) or (b), where the heights along the line are shown in (c) and (d); (e) The SFM thickness and (f) the average single bilayer thickness of the capsules as a function of the number of bilayers at different salt concentration of C_{NaCl} = 0.25 M (h and j), and C_{NaCl} = 0.5 M (g and i) (see the text). Reprinted with permission from ref. 22. Copyright © 2005, American Chemical Society.

In this work, the shell thickness of capsules was measured by Surface Force Microscopy (SFM) using a standard method introduced elsewhere. The capsules were first dried on mica and then minimum height in a fold free region was measured. Figure 2a and 2b show a typical top view SFM images of capsules with (n = 4) and (n = 9) at C_{NaCl} = 0.25 M, respectively. In an area free of folds, the minimum thickness $h_{SFM\ thickness,\ min}$ is regarded as twice the wall thickness. Figure 2e shows the increase of $h_{SFM\ thickness,\ min}$ as a function of the number of bilayers (n) at C_{NaCl} = 0.25 M and 0.5 M, respectively. The average thickness $h_{single\ bilayer}$ is then found by normalizing $h_{SFM\ thickness,\ min}$ by the twice the number of bilayers. Figure 2f shows the dependence of $h_{single\ bilayer}$ on the bilayer number and salt concentration. If the film growth were linear, a constant value in the thickness per bilayer would be expected; i.e. the thickness per bilayer would be independent of the total thickness. However, the plateau in the thickness as a function of layer number indicates the presence of a phase transition between expanded and condensed bilayers at both salt concentrations. The onset of this transition occurs earlier for a higher salt concentration, which also implies that this transition depends on the total film thickness. Similar transitions were also found for films grown on planar substrates and free interfaces and will be reported in a future work.

This suggests that $h_{single\ bilayer}$ can be used to evaluate the capsule stability. In general, when the capsule wall is too thin ($h < 10$ nm), it cannot sustain such a high osmotic pressure, and then the capsule shell is destroyed. As the thickness increases between ($12 < h < 16$ nm), a transition in the shell structure, as evidenced in thickness of a single bilayer, occurs. Thereafter, the capsule stability and yield is increased.

2.3. Capsule Integrity

It is important to differentiate the capsule stability and yield from capsule integrity. The presence of stable capsules does not imply that all capsules are bounded by a continuous surface. A method for the identification of intact and broken capsules by osmotic deformation has been developed previously [23]. With being subjected to a solution of high molecular weight polyelectrolyte having an osmotic pressure much greater than that of the capsule interior, intact capsules deform and broken capsules do not. There are two requisites for the probe polymer; it should not adsorb onto the capsule exterior and should not permeate through the capsule wall. Therefore, capsule integrity, i.e. fraction of unbroken capsules, was studied using polycation-terminated capsules and dye-labeled polyallyamine hydrochloride, M_W = 70 kDa, as a probe.

Figure 3 shows capsule integrity as a function of the number of bilayers (n) and salt concentration. For n = 5 and at C_{NaCl} = 0.25 M, the unbroken capsule ratio is about 96% and decreases monotonically as a function of layer number. Similar results are found at C_{NaCl} = 0.5 M. Because the increase of polyelectrolyte multilayer thickness as a function of salt concentration is well established [10], these data suggest that the number of broken capsules increases with h.

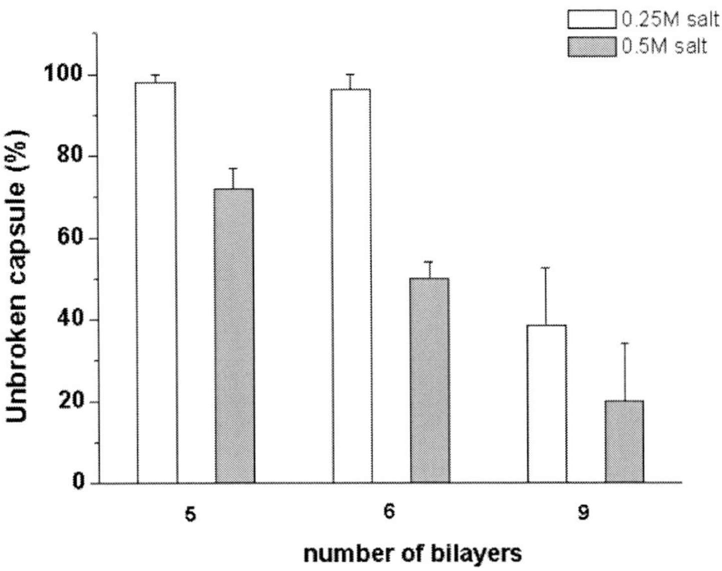

Figure 3. The fraction of unbroken capsules as a function of the number of layers at different salt concentrations (C_{NaCl} = 0.25 M and C_{NaCl} = 0.5 M). Reproduced from ref. 22. Copyright © 2005, American Chemical Society.

In an earlier study, a higher capsule integrity for (approximately 97% for $n = 5$, $C_{NaCl} =$ 0.25 M) was reported. These differences arise because of differences in preparation conditions. In the previous study, the synthesis was performed at room temperature using the filtration method. The capsules in this study were prepared using the centrifugation method at a relatively low temperature (283K) to avoid the re-polymerization of the MF oligomers.

Clearly, there is a trade-off between capsule stability and capsule integrity. Below the critical thickness required for mechanical stability, core release results in destruction of the shell. As the shell thickness ($12 < h < 16$ nm) increases either by bilayer number (n) or salt concentration, stable capsules, denoted here as (Type II), can be observed. Further increase in thickness ($h > 16$ nm) results in capsules of high stability, but includes both intact (Type III) and broken (Type IV) capsules. The capsule types are shown in Scheme 4 as a function of shell thickness. The porosity of each capsule type is discussed below.

2.4. Capsule Mesh Size

Capsule permeability is governed by a combination of hydrodynamic sieving, electrostatic interactions, and Donnan effects in the polymeric shell. By measuring the equilibrium distribution of fluorescently labeled probe materials of varying size across intact capsules, bounding values of the average capsule mesh size, ξ, can be determined. The transport rates of different probe materials are the subject of a future publication.

The hydrodynamic radius, R_H, of each probe can be calculated from its bulk diffusion coefficient according to the Stokes-Einstein relation:

$$D = \frac{kT}{6\pi\eta R_H} \tag{1}$$

The determination of the diffusion coefficient, D, by PFG NMR with stimulated echo is described in detail elsewhere. Values of D and R_H for each probe in this study are listed in Table 1.

Table 1. Diffusion coefficient and hydrodynamic radius of materials used in this work

Material	Diffusion coefficient m^2/s	Hydrodynamic Radius nm
Rhodamin	2.84e-10	0.76
4.4 kDa TRITC-Dextran	1.05e-10	2.05
66 kDa TRITC-Dextran	2.40e-11	8.98
464 kDa FITC-Dextran	1.26e-11	17.1
70 kDa PSS	1.8e-11	12.0
70 kDa PAH	2.8e-11	7.7

We first consider the low molecular weight dye TRITC (tetramethylrhodamine isothiocyanate). Figures 4a-c show that permeation of this probe occurs in both capsule Type II and III, independent of shell thickness. This suggests that the mesh size of these capsules must be at least the hydrodynamic radius of TRITC to negate the electrostatic repulsion of this probe from the capsule shell. The *in-situ* measurement of the diffusion of organic penetrants into multilayers of (PSS/PAH) synthesized on planar substrates have yielded mesh sizes much smaller, i.e. $\xi = 0.5$ nm [30]. Thus, it can be inferred that MF core release results in pore formation in the capsule wall. The use of a neutral analyte eliminates electrostatic effects and allows exclusive focus on mesh size-dependent permeability as a function of shell thickness. For this purpose, two different polysaccharides (Dextran) with molecular weights of 4.4 kDa and 66 kDa were used. Capsules with differing thickness were prepared by the variation of the number of bilayers at a constant salt concentration $C_{NaCl} = 0.25$ M. Figure 4d shows that only Type II capsules, ($3 < n < 6$), are permeable to 4 kDa Dextran. As thickness increases, the mesh size is reduced, and the probe is excluded. The presence of both capsule Type III and IV can be confirmed in figures 4e and 4f by increasing the concentration of probe material to induce osmotic deformation. Similar results for the molecular weight 66 kDa Dextran are shown in figure 4g-4i.

Some conclusions regarding the porosity as a function of film thickness can be reached. The osmotic deformation of Type II capsules by polyallylamine hydrochloride places an upper bound on the mesh size of at least several multiples of the hydrodynamic radius (R_H = 7.7 nm) of the polymer. However, passage of 66 kDa Dextran (R_H = 9.0 nm) occurs freely. This suggests that Type II capsules are "mesoporous", i.e. possess an average pore size between two and fifty nanometers. As previously mentioned, these pores are the result of the burst release of MF oligomers having an average size of 2-3 nm. It has been shown that the pore size of the Type II can be reduced (or healed) by additional layer adsorption and other methods [17]. Therefore, the permeability of Type II capsules can be controlled easily and will be studied in further work.

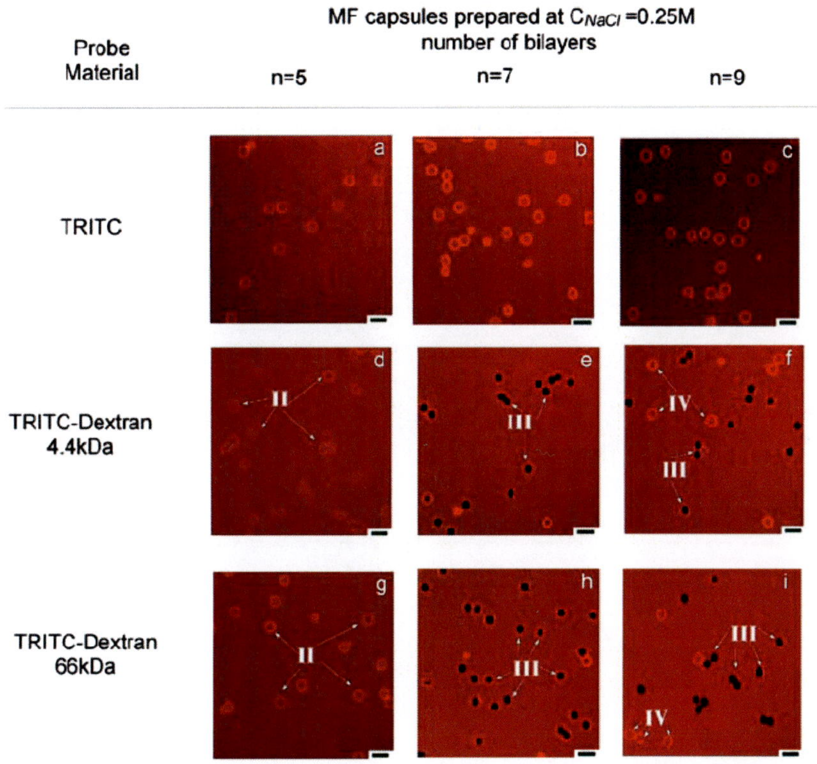

Figure 4. CLSM images of MF capsules of PSS/PAH by the different probe material, (a-c) FITC, (d-f) TRITC labeled dextran 4 kDa, (g-i) TRITC labeled dextran 66 kDa as a function of the number of bilayers. (The scale bar is 10 μm, and measured in neutral water). Indicated are also the types of capsules belonging to the different observations. Reproduced with permission from ref. 22. Copyright © 2005, American Chemical Society.

Type III capsules are impermeable to 4 kDa Dextran (R_H = 2.1 nm), but permeable for TRITC (R_H = 0.76 nm). This indicates that the nomenclature "microporous", i.e. average pore size less than two nanometers, is applicable. This smaller mesh size is the result of entrainment of released (MF) oligomers by the multilayers during the core release. Prior work concludes that as much as 20 % (w/w) MF oligomers of released oligomens remain on the shell. Most existing studies of capsules prepared on sacrificial MF particles are Type III. Since the capsule shell thickness is around 16 nm, which is in the transition region between Type II and III, the macroscopic properties, i.e. permeability and elasticity, depend strongly on the environment (temperature or solvent quality, etc.) and decomposition conditions.

Figure 5 shows a confocal microscopy sequence of the removal of MF inside the capsules. The time interval is about 2 seconds from a to d. One clearly observes the removal of the material first from the surface of MF particles and then to the central region. The process takes on the order of several seconds. Meanwhile an increase of the capsule diameter will appear at the beginning and a decrease later.

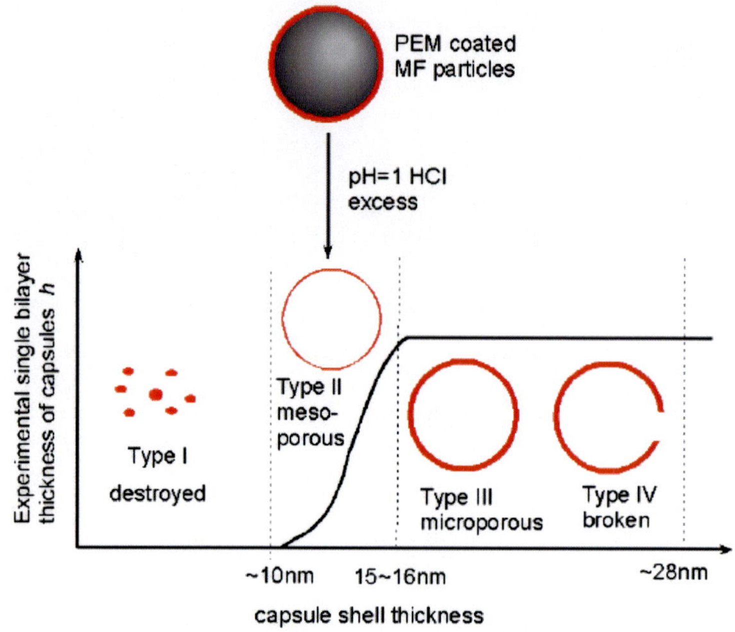

Scheme 4. Capsule type is evaluated by the capsule porosity and integrity. The curve describes the experimentally derived single bilayer thickness versus the AFM shell thickness of the PEM. There are at least four different capsule types of different structure and permeability along the curves, including Type II (intact, mesoporous capsule), Type III (intact, microporous capsule) and Type IV (broken). Reproduced from ref. 22. Copyright © 2005, American Chemical Society.

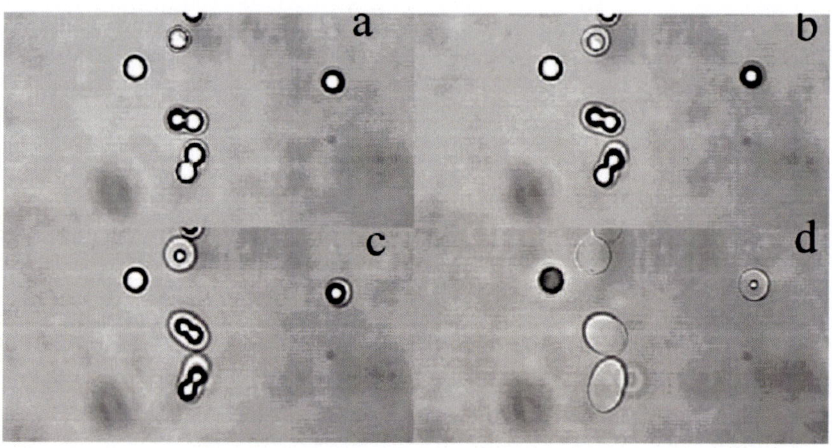

Figure 5. A confocal microscopy sequence of the removal of MF inside the capsules. The time interval is about 2 seconds from a to d. (scan bar is 10 μm) [24].

This indicates a built-up of osmotic pressure that is later relaxed due to material permeation into the external medium. If the osmotic pressure of these oligomers is high enough, it may lead to shell rupture and capsule destruction. The maximum osmotic pressure might be built up if core material would not be released. Assume the molecular weight of MF oligomers is about 200 g/mol, this would lead to 18 MPa according to the van't Hoffs law. This very high value may be reduced due to volume increase. At the beginning of core release, the capsule wall increases quickly with the core diameter decreasing fast. Let us assume that the increasing volume is equal to the reduction of core volume: the osmotic pressure will reach 10 MPa at the initial stage. This osmotic pressure of 10 MPa leads to a tension of 20 N/m in the wall of a particle of 2 μm in radius. The tension is rather large given the very small thickness of the layer. It exceeds the known limit of stability of a lipid bilayer by a factor of 200. Therefore, if the capsule wall is too thin (< 10 nm), it cannot sustain such a high osmotic pressure, and then the capsule shell is destroyed. If the thickness increases to 12 nm ~16 nm, the capsule shell can keep the stability against the core release, but some of the shell material will be removed from the wall during the release of core. Here, if low salt concentration is used for PEM preparation, the capsule shell thickness is lower than that at high salt concentration. One will lose more material than at high salt concentration. Once the capsule thickness exceeds 16 nm, the capsule shell may have enough strength to resist the pressure developed on core release.

As shown in scheme 4, many broken capsules can be observed when increasing the thickness. There are at least one or two big holes on the broken capsules, confirmed by SFM (figure 6). These holes are too big and cannot be healed by another method.

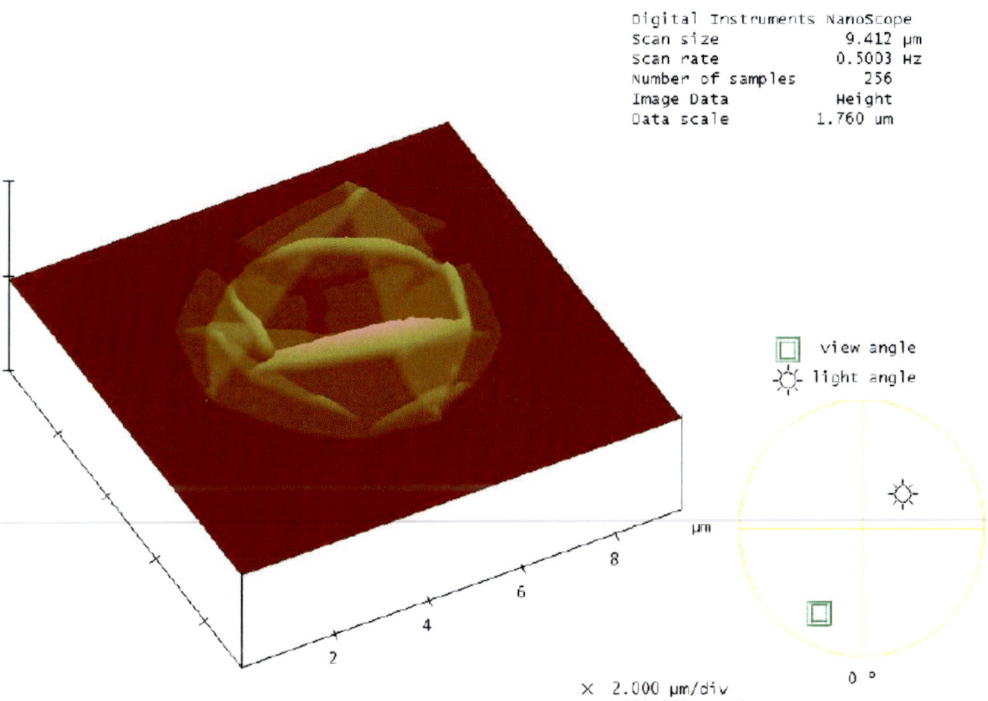

Figure 6. The broken capsules (n = 9, CNaCl = 0.25 M) [24].

As known, the driving force for core release is the concentration gradient between the interior of the capsules and the bulk solution. Consider a capsule of volume $\frac{4\pi}{3}r^3$, surface

$4\pi r^2$, permeability P and solute concentration c_i. From $\frac{-dc_i}{dt}\cdot\frac{4\pi}{3}r^3 = P\cdot 4\pi\cdot r^2\cdot c_i$, we obtain

$$c_1 = c_i^o \exp\left(\frac{-3P}{r}\right)t \tag{2}$$

Increasing the wall thickness, the permeability of MF oligomers will decrease. It leads to a C_1 raise according to eq (2), which results in a high pressure inside the capsules.

To release the high pressure, the capsule wall swells to decrease C_1. Increasing the capsule size results in a big mesh size of the shell, which can allow a fast release of the core residue. For the thin capsule, the burst release of the core residue will break through the wall and leave a big hole on the shell. But for the thicker capsules, the walls have enough strength to resist the release of oligomers. Consequently, more and more oligomers stay inside and wait for the release from the shell.

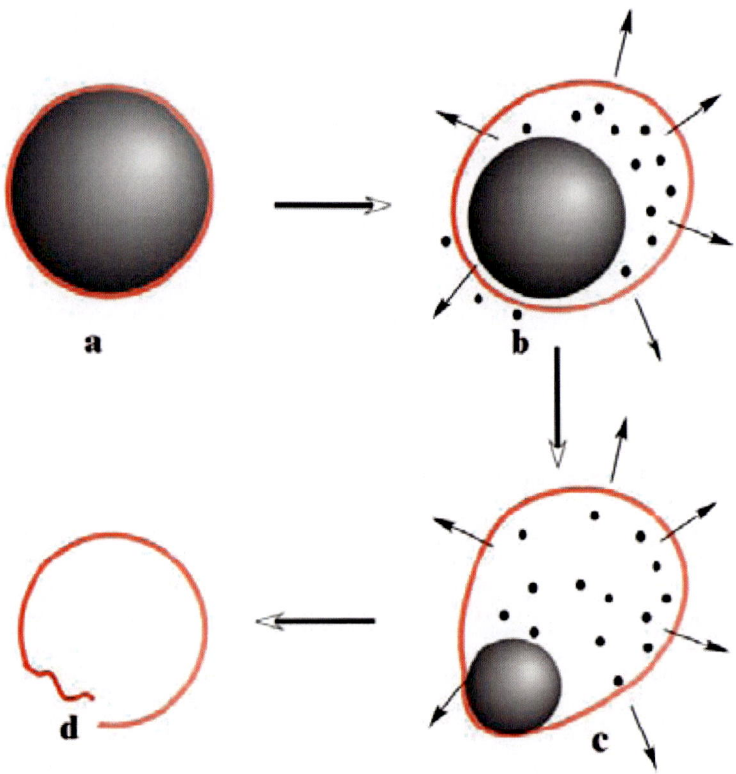

Schemes 5. The mechanism of broken capsules breaking.

If there are defects on the capsule shells, the pressure then may succeed to tear the shell. In this case, not only MF oligomers but the non-dissolved core residue will pass through the defect and form a big hole on the shell, as shown in Scheme 5. However, if the MF oligomer concentration difference between bulk solution and capsule interior can be at equilibrium, those capsules can allow the slow core release from the shell; it will help to reduce the broken capsule ratio. One can realize this by reducing the amount of acid for core decomposition. In most cases, the capsule with five bilayers can be obtained with a small broken capsule ratio. However, the core residue is also reported to stick on the shell, which can influence the permeability of the capsule shell.

Briefly, for those thicker capsules, they have a good capsule stability, but they can be either intact capsules or broken capsules. With an increasing the wall thickness, more broken capsules can be observed in the system. However for the thin capsules, they have a low stability. But they can keep the shell intact although some material of shell is lost during the core release.

2.5. Paradigms for Control of Capsule Integrity

Although Type II capsules have high integrity, their stability limits the yield and ultimately their application. Type III capsules have high stability and, but their integrity is low. Thus, the improvement of the integrity, i.e. decrease of the fraction of capsules ruptured during core release, is required to drive these materials to application.

To suppress the fraction of Type IV capsules, controlled core decomposition is proposed. Normally, a particle concentration, 0.5% (w/w), is used during core release in acid solution (pH = 1.0). Low particle concentration (or the excess of acid) ensures a rapid and complete dissolution of MF, however, it increases the maximum osmotic pressure that a capsule is exposed to during release. By increasing the particle concentration, or decreasing the excess acid, the kinetic rate of decomposition is slowed, thus decreasing the number of oligomers in the capsule at any time. Under these conditions, the driving force for diffusion of oligomers out of the capsule would remain approximately the same, because the capsule interior would remain essentially saturated with oligomers.

To demonstrate efficiency of this method, core release of MF through PSS/PAH multilayers (n = 4-9, C_{NaCl} = 0.25 M) at elevated particle concentration, 2% (w/w), was performed. Under these conditions, high capsule integrity (> 96%) was measured among all values of n, indicating significant improvement. Capsules of (n = 4) are freely permeable to 4 kDa Dextran (figure 7a), but capsules with (n = 6, 9) are impermeable (figure 7b and 7c). In addition, FITC-PAH is inhibited through the capsules in figure 7d. Since all capsules are intact, this indicates that the capsules (n > 4) are Type III according to the Scheme 4. Moreover, figure 7b and 7c show that the average diameter of the capsules (n = 6, 9) are larger than that of capsules prepared by uncontrolled release, because of the hindered release of MF oligomers.

The MF oligomers inside capsules were investigated by the confocal Raman spectroscopy. Figure 8 shows the Raman spectra of MF capsules (here the capsules were measured in the air-dried state). Figure 8a shows the spectra of MF capsules (n = 9), which were prepared by the normal method. Figure 8b shows the results of MF capsules with the same number of bilayers, but prepared by the controlled core decomposition method. Comparing these two spectra, a Raman band at 981 cm^{-1} can be easily distinguished in figure 8b. This peak is assigned to the MF oligomers. Even when the number of bilayers (n) is reduced to 6, the same peak is still found in the capsules, as shown in figure 6c but as expected the intensity is lower as compared with figure 8b.

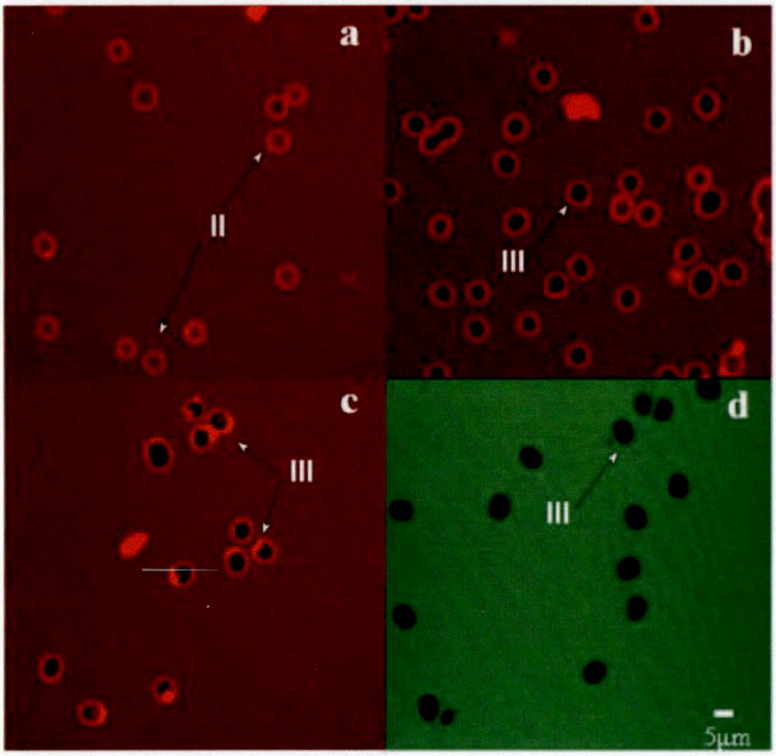

Figure 7. CLSM image of (a) MF capsules with n = 4, (b) n = 6,(c) n = 9, at C_{NaCl} = 0.25 M, where the probe material is TRITC-dextran with 4 kDa molecular weight. (d) Image of MF capsule (n = 9, at C_{NaCl} = 0.25 M) with probe material of FITC-PAH. All capsules are prepared by the controlled core decomposition method. (The bar is 5 μm measured in neutral water). Indicated are also the types of capsules belonging to the different observations. Reproduced from ref. 22. Copyright © 2005, American Chemical Society.

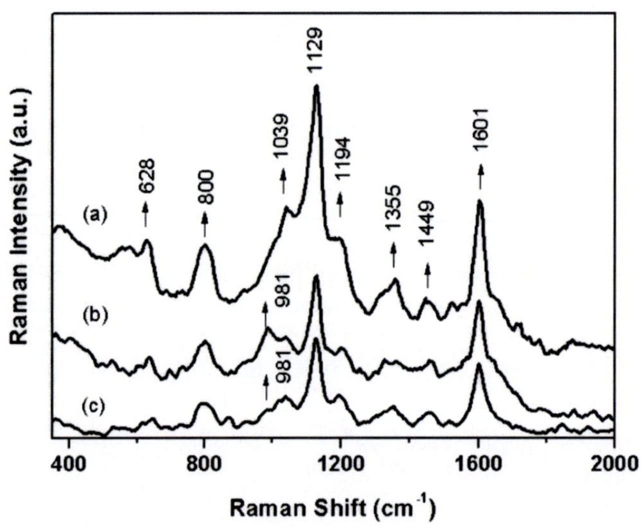

Figure 8. Continued on next page.

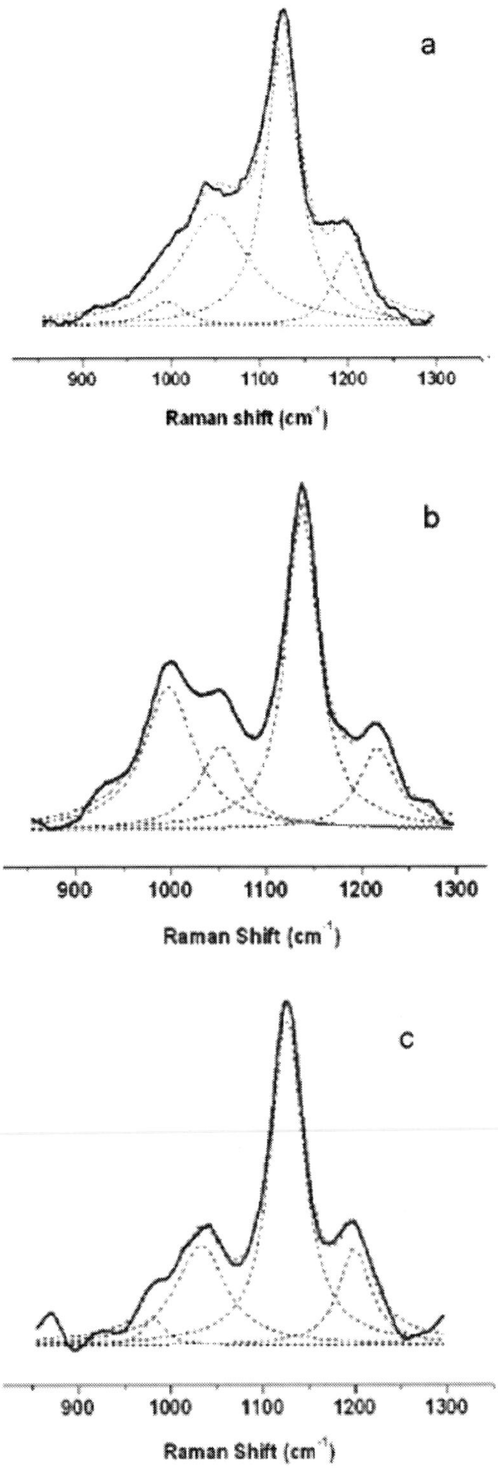

Figure 8. Raman spectrum of (a) MF capsules with n = 9 (normal method, C_{NaCl} = 0.25 M), (b) MF capsule with n=9 (the controlled core decomposition method, C_{NaCl} = 0.25 M) and (c) MF capsule with n = 6 (the controlled core decomposition method, C_{NaCl} = 0.25 M). Reprinted with permission from ref. 22. Copyright © 2005, American Chemical Society.

This “controlled core decomposition” results in more MF oligomers inside the capsules. These oligomers can improve the capsule integrity, but they also influence the permeability of the capsules via a strong decrease in the mesh size of the capsules. The presence of MF oligomer is demonstrated by analysis of the enlarged Raman spectrum (from 800 to 1350 cm^{-1}), as shown in figure 8a-c. Shown are four different peaks: 981 cm^{-1} attributed to MF oligomers and 1030 cm^{-1}, 1129 cm^{-1}, and 1194 cm^{-1} due to PSS [21]. Obviously, the peak of MF oligomer in figure 6b is higher than that in figure 6a. After normalizing the peak area of oligomer by the area of one of the PSS peaks, i.e. $\frac{A_{981}}{A_{1129}}$, the relative percentages of MF can be calculated: 9.5% in figure 8a, 68.8% in figure 8b, and 5.3% in figure 8c, respectively. These values demonstrate that at the same number of bilayers, the MF that remains in capsules made by “controlled core decomposition” is about seven times higher than those of capsules made by normal way. Apparently, the “controlled core decomposition” method is helpful to surmount the high broken ratio of capsules when the wall thickness is increased. The capsules prepared by this method are all Type III, however, this method does not change the critical multilayer thickness for capsule preparation.

3. Smart Applications

External stimuli has been utilized to control the mass transport and morphology within polyelectrolyte multilayers (PEMs) or polyelectrolyte capsules, for instance ionic strength [25,26], pH [27-30], temperature [31-33], post chemical reaction [34-39], metal ion coordination [40], and electric field [41,42]. In this section, organic solvent treatment has been investigated in order to demonstrate its effects on the structure, mechanic properties, and permeability of the thin films [21,43-46].

As above discussed, stable and robust capsules had been produced by the polyelectrolyte pair of poly (styrene sulfonate) (PSS) and poly(allylamine hydrocloride) (PAH) on weakly cross-linked melamine formaldehyde (MF) particles as sacrificial cores. By controlling the PEMs film thickness and core removal conditions, two different intact capsules are obtained in the term of mesh size, which are so called microporous and mesoporous capsules. Mesoporous capsules have a big mesh size to allow large macromolecule to pass easily, but microporous capsules does not.

Here, we focus on controlled permeability of mesoporous capsules by a simple and easy method of “PEM coating and solvent treatment”, as shown in figure 9. Different from other reports [44,46], we use the porous capsules as a template for growth of additional PEM layers. These PEMs films play an important role on the permeability of mesoporous capsules (as well as microporous capsules [17]). Upon solvent treatment, microporous capsules become the mesoporous ones due to the microstructure changes within PEM films. The diffusion of large biomacromolecules (i.e. dextran) to capsules is measured by CLSM measurements. In this experiment, a typical probe material TRITC-dextran is used because it maintains a stable spherical shape in water whose hydrodynamic radius is controlled by its molecular weight, i.e. TRITC-dextran of 66 kDa has a hydrodynamic radius of 9.0 nm while the radius of 4 kDa TRITC-dextran is about 2.1 nm [22]. On the other hand, TRITC emits fluorescence and shows a red color in the CLSM images when excitated by a 552 nm laser.

Prior to the measurement, TRITC-dextran is added to the capsules solution. After 24 hours, a droplet of the solution is placed on a glass slide for the CLSM measurement.

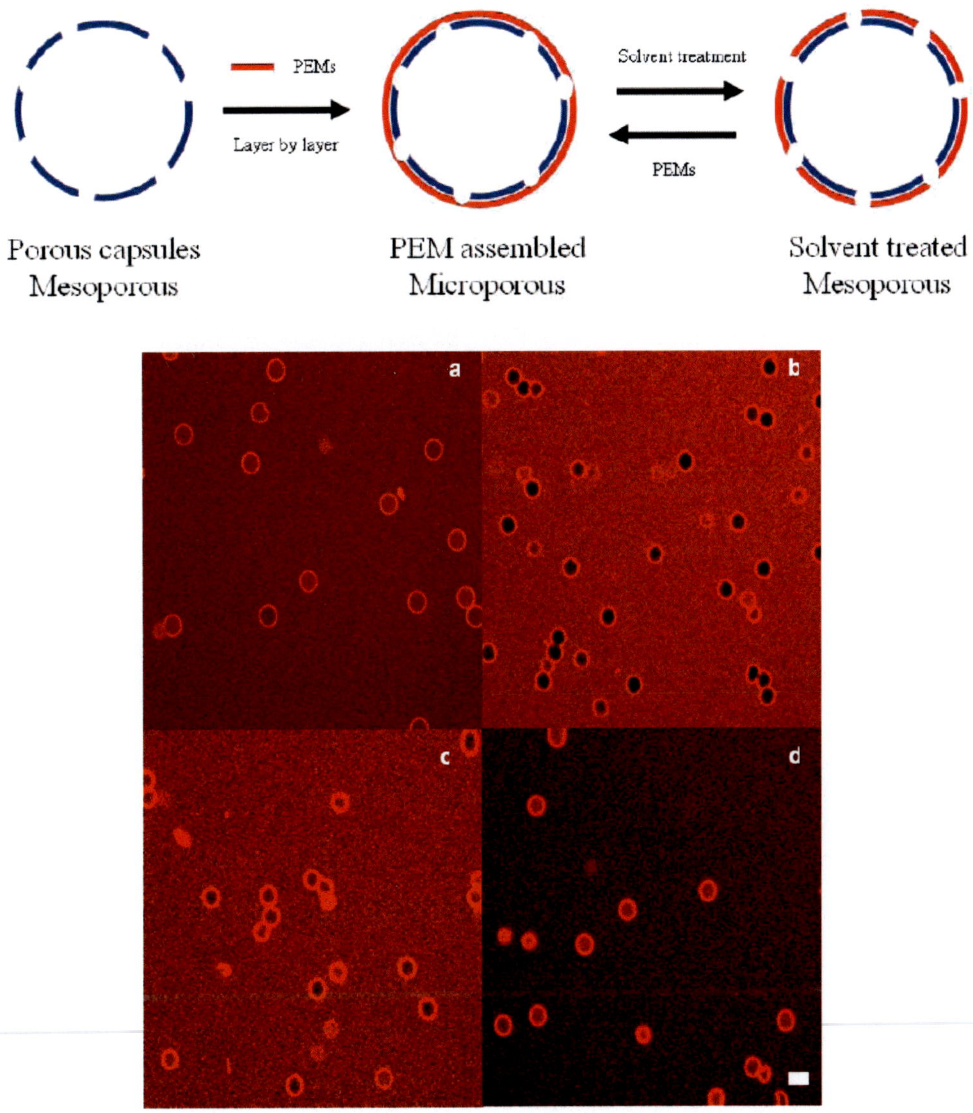

Figure 9. CLSM imaging of (a) mesoporous capsule $(PSS/PAH)_5$, (b) capsule (a) modified by additional layers of PSS/PAH, (c) capsule(b) after one hour treatment with 50% ethanol solution, (d) capsule(b) after 12 h solvent treatment. The probe material is the TRITC-Dextran with 66 kDa. Scan bar is 5 μm. The cartoon above the figure shows the way of "PEM coating and solvent treatment" to control the permeability of capsules (from microporous to mesoporous by switching the mesh sizes of capsules). The details are reported in the text. Reprinted with permission from Chem. Mater. 2005, 17, 4992 (ref. [47]). Copyright © 2005, American Chemical Society.

In figure 9a, mesoporous capsules can be easily visualized by the circular red ring due to TRITC-dextran adsorption on the capsule wall. The emission from outside of the capsule demonstrates there is TRITC-dextran in the solution. The red color still appears inside the capsule indicating that the capsule allows the diffusion of TRITC-dextran of 66 kDa, because the mesoporous $(PSS/PAH)_5$ capsules have mesh size larger than 18 nm [21,22].

Since the wall of the as-made mesoporous capsule remains charged to attract adsorption of oppositely charged polyelectrolytes, the additional polyelectrolyte multilayer are easily built up on these porous capsules. However, after coated one bilayer (or several bilayers up to 4 bilayers of PSS/PAH), capsules show the different permeability behaviors. Figure 9b show that TRITC-dextran can not penetrate into PEM coated capsules because no florescence intensity from TRITC is observed in the interior of capsules. The result shows that the mesh size of capsule wall is greatly reduced by additional deposition of polyelectrolyte layer. Smaller sized dextran (4.4 kDa, radius 2.1 nm) is also prevented from diffusing through the PEM-modified capsule wall indicating that the mesh size is roughly less than 4 nm. Therefore, after surface modification, these capsules change from the type of mesoporous to microporous capsules, in consistent with the previous reports [17].

In order to enlarge the mesh size, the PEM-modified capsules were immersed in the mixed ethanol and water solvent with 60% ethanol for a longer time (more than 12 h). Then the permeability of dextran to capsules is characterized by CLSM. Figure 9d shows a red color in the interior region of capsules after 12 hours after solvent treatment, which indicates TRITC-dextran is present in the interior of the capsules. It confirms that the solvent-treated capsules allow the passing of TRITC-dextran with 66 kDa (radius of 9 nm). Obviously, microporous capsules become mesoporous ones upon the treatment. In addition, this solvent effect is time-dependent. In a short time (i.e. one hour,), only part of capsules can allow the passing of probe material, as shown in figure 9c. However after 12 hours, probes are permeable for all capsules, in consistent with the in situ AFM results (see below).

As shown in the cartoon in figure 9, solvent effect can be utilized as a tool to control the diffusion of rigid and charge-free molecules such as dextran through the capsule wall by manipulating the mesh size of the capsules. Several cycles of "PEM coating and solvent treating" can be dealt with the mesoporous capsules. However, the capsule will be deformed eventually because of too many centrifuge processes and solvent-soften effect [44,48].

The enhanced permeability of polyelectrolyte multilayer either on flat substrate or porous template by solvent treatment is demonstrated. The alteration of film permeability reflects the strong changes of microstructure within the polymeric thin films. Now the following question is how to find the evidence to show the changes in morphology and how to understand the mechanism of solvent effects. In the following text, in-situ AFM measurement is utilized to explore the phenomena.

Figure 10a shows a typical AFM image of the $PEI/(PSS/PAH)_2$ film obtained in deionized water. The film is devoid of large aggregates and has a surface roughness of R_q = 0.57 nm for a scan area of 500×500 nm^2. The same film imaged in 96% ethanol (figure 10b) and 2-butanol (figure 10c) shows aggregates and hole formation accompanied by a considerable increase in surface roughness. The holes in figure 10b and 10c are quite uniform with an average diameter of 17 nm and depth of 2.5 nm, which is consistent with the CLSM results.

The film morphology is almost unchanged in the dry state after solvent removal as imaged in air (figure 10d). Occasional large holes due to uneven evaporation are observed in the dried samples. However, these big holes are not responsible for the solvent-controlled permeability behaviors.

The solvent-induced roughening of the PEM surface seems to be irreversible or at least exhibiting relaxation time longer than several hours because the film morphology does not revert back to the original state after reintroducing water into the AFM liquid cell.

AFM images have also been obtained in the mixed solvents of ethanol and water. Ethanol and water mixtures allow gradual changes of the solvent dielectric constant and polarity. When the ethanol volume percentage is less than 60%, no apparent morphological changes are observed on the PEM film surface by in situ AFM. The surface becomes rougher with holes developing above 60% ethanol. Figure 10e shows a typical AFM image of PEI/(PSS/PAH)$_4$ obtained in the mixed ethanol and water solvent with ethanol percentage greater than 60%. The surface roughness and the dimensions of the holes are similar to those in figure 10b and 10c. Our results suggest that the onset of solvent effect on film structure is around 60% ethanol by volume. However, the time dependent of film morphology changes is still involved in the system. For the film with four bilayer PSS/PAH, the meso-pores can be clearly observed after 3h, which is consistent with CLSM experiments.

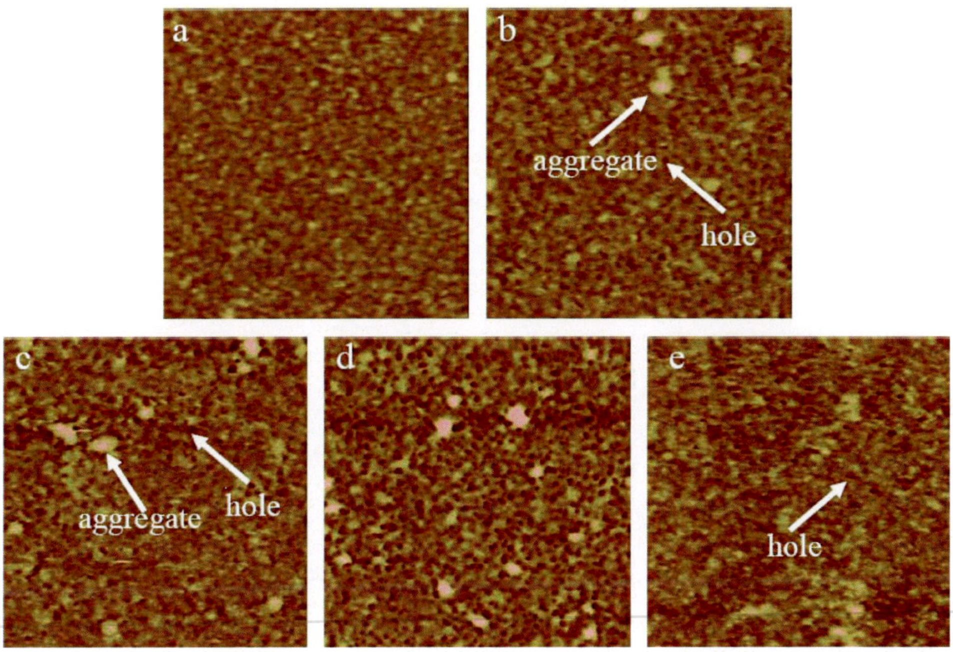

Figure 10. AFM height images with 1×1 μm^2 scan area and 10 nm z range of PEI-(PSS/PAH)$_2$ on mica captured in different solvents: (a) water (R_q = 0.57 nm); (b) 96% ethanol (R_q = 0.99 nm); (c) 2-butanol (R_q = 0.92 nm); (d) film of (c) imaged in air after 2-butanol evaporation (R_q = 0.95 nm); and (e) PEI/(PSS/PAH)$_4$ in 65% ethanol z range = 12 nm (R_q = 0.86 nm). Reproduced from Chem. Mater. 2005, 17, 4992. Copyright © 2005, American Chemical Society.

With the addition of alcohol, pore free and dense PEMs films have strong responses to the changes in local environment: The increase in surface roughness of PEM films and hole formation are observed. The fast diffusion of probe (small molecular weight ion) and the ion-entrapment are demonstrated. The switched mesh size of capsules is benefit for controlling the penetration of large macromolecular. These results can not be concluded due to polyelectrolyte dissolution and removal from the PEM film. The main and possible reason is due to the solvent quality decreases by increasing alcohol content. It is known that increasing organic solvents content decreases the dielectric constants of mixture solution. The dielectric constants of water, ethanol, and 2-butanol are 78.5, 24.3, and 15.8 respectively. If using ethanol instead of water, the electrostatic attraction between polyelectrolyte segments will increase by almost a factor of three, which is described by Coulomb force in eq. (3)

$$F = \frac{Q_1 Q_2}{4\pi\varepsilon_0 \varepsilon r^2} \quad (3)$$

Such strong attractive forces between the nearest opposite charge groups cause the polyelectrolyte complex to contract, to collapse and to coagulate, as shown in scheme 6. The rearrangement of the polyelectrolyte chains induce the formation of big aggregates and holes within the film, as revealed in the AFM images. In fact, the other interaction, for example dipole-dipole interaction, could be also involved in the microstructure changes due to the decreasing of the dielectrically constant of solvent. Therefore, the increased film inhomogeneity is responsible for higher permeability of the PEM. The effects of solvent are summarized in scheme 6. These findings show that solvent is another controllable parameter for the tuning of film structure and permeability, which have potential applications in smart encapsulation and controlled delivery.

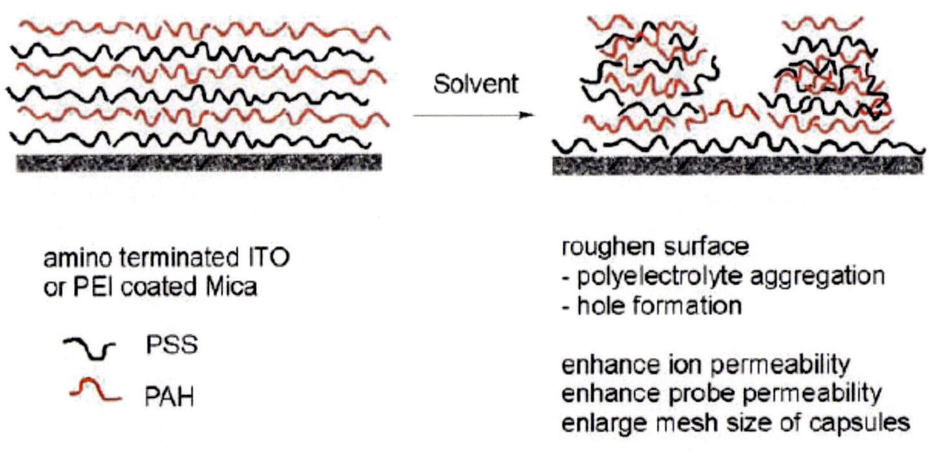

Scheme 6. Schematic illustration of solvent influencing on microstructure of polyelectrolyte multilayers. Reprinted with permission from ref. [47]. Copyright © 2005, American Chemical Society.

However, solvent effect is not a reversible process since the film morphology and permeability can not revert back to the original states for a long time. Several reasons are pointed to this issue: Firstly, it could be due to the long term relaxation of polymer complex in the confined system. Secondly, it could be due to the strong electrostatic force between opposite charged groups in the polyelectrolyte complex. For instance, during polymer chain rearrangement process, unpaired charged groups have a priority to find the other opposite ones to bind together quickly. After the solvent is changed back to a higher dielectric constant one, these new pairs can not separate due to the strong interaction. Thirdly, several other parameters could also be involved in these irreversible behaviors, such like temperature, moisture, and the hydrogen-boding (The degree of protonation of PAH will affect its H-bonding capacity with ethanol if charge density of PAH has to depend on the pH of the solutions). To realize the reversible process by solvent treatment, more efforts must be directed towards seeking the other repulsive forces or interactions (i.e. covalent cross-links or pH-responsive etc) in order to balance the strong attractive Coulomb force.

3.1. Smart Encapsulation

Various applications of capsules may be derived from the fact that the inner volume is loaded via a chemical or a physical process. It may also be attractive to prepare functional polymers inside. They can serve as nucleation sites for subsequent precipitation or as adsorption sites for various materials or a template for the synthesis of nanoparticles as well. Therefore, the principle of encapsulation in cages is quite promising. In principle, the capsule shell exhibits semi-permeability, making use of the fact that generally the wall is impermeable for polyelectrolytes.

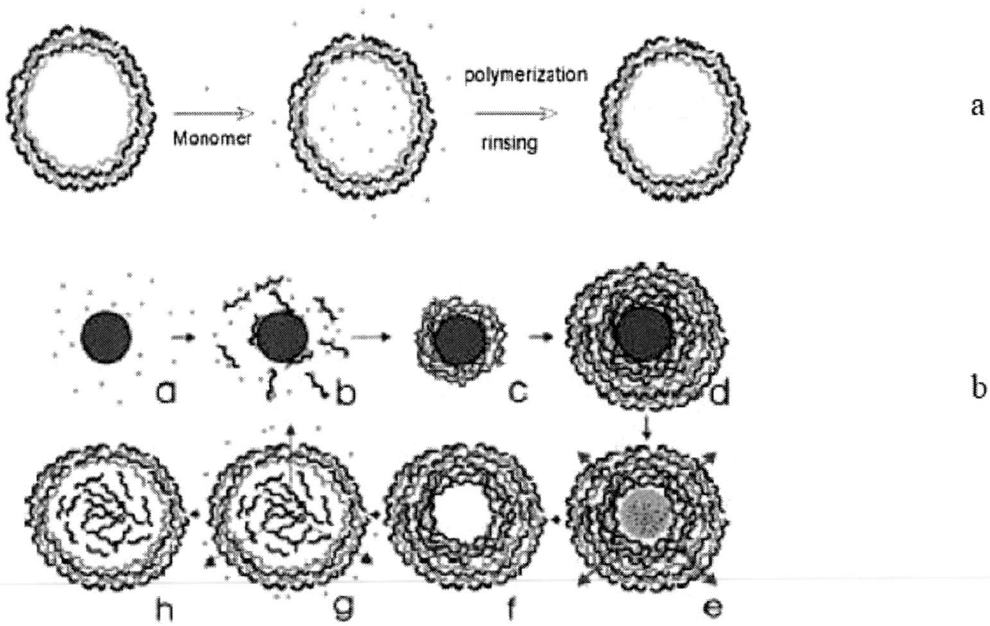

Scheme 7. Polymer filled capsules by (a) polymerization method and (b) the precipitation method. Reprinted with permission from ref. [49]. Copyright © 2000, Elsevier Science Publishers.

3.1.1. LOADING POLYELECTROLYTE INTO MF CAPSULES

The presence of positively or negatively charged polyelectrolyte molecules in the capsule interior establishes a pH gradient across the capsule shell and creates the necessary conditions to carry out pH-sensitive synthesis in the capsule microreactor. Positively charged poly (allylamine hydrochloride) increases the pH inside the polyelectrolyte capsules to 9 from that outside the capsules (pH 6).

On the contrary, loading the capsule volume with negatively charged poly (styrene sulfonate) molecules decreases the pH inside the capsules. Therefore it opens the possibility of precipitating inorganic material in acidic media. In literature, there are two major methods to realize the encapsulation of PSS in capsules, one is the polymerization approach and the other is the precipitation method, as shown in Scheme 7. Here, Scheme 7a shows the polymer filled capsule by the polymerization approach.

The polymerization method [18]: Firstly, prepare the intact capsules (II and III), with mesh size big enough to allow the monomer diffusing into and confine the high molecular weight PSS released through the shell. Polymerization is carried out in the solution of monomer (SS), capsules, initiator (potassium peroxodisulfate, 1% related to the monomer concentration), for 4 h at 80°C under nitrogen atmosphere. The reaction at high temperature 80°C will lead to the aggregation of capsules. This could be due to the link of PSS between capsules if the reaction starts on the capsule surface. Therefore, we can use ammonium peroxodisulfate and N,N,N',N'-tetramethylethylene-diamine (TMEDA) to perform the polymerization at 40°C, which can be helpful to obtain the monodispersed polymer filled capsule. But it leads to a low PSS yield at 40°C. Figure 11 shows the Raman spectrum of the PSS powder after dialysis. The typical Raman spectrum of PSS shows the active mode at 1125 and 1600 cm^{-1}, which is assigned to the vibration of sulfates $\nu(SO_2)$ and aromatic ring quadrant stretching. Figure 11b shows the spectrum from inside of the PSS filled capsules, which is prepared by the polymerization method. Obviously, both Raman spectra are similar. This demonstrates that the material inside the capsules is PSS exclusively. Here the PEM shell does not contribute to the spectrum. There are two reasons: firstly, with help of the confocal technique, the effective sample region is reduced to sub-micrometer. One can easily focus on the central of capsules without collecting any signal from the PEM shell. Secondly, the Raman signal of the MF capsule shell is too low. Hence it can be ignored. Until now, the polymers of poly NIPAAM and PSS can be easily synthesized inside capsules in our group.

The precipitation method [49] is one of the physical methods, which is shown in Scheme 7b; Firstly assembling the complex of the polymer with metal ion on the shell (step Scheme 7b a-c) at the beginning stage and then coating polyelectrolyte multilayer (Scheme 7b step d). After removal of the core, the capsules remain in the solution. Then EDTA or high salt are used to destroy the complex of polymer with the metal ion (Scheme 7b step g) until the ion is released from the shell and consequently the free polymer will disperse inside the capsules (Scheme 7b step h). In contrast to the polymerization method, the precipitation approach is a very simple and universal method, not only for the positively charged polymer but also for the negatively charged material (which is hard to be synthesized by the polymerization method). In addition, the amount of precipitated polymer can be controlled during the PEM assembly.

By the precipitation method, one can get the capsule filled with PSS or PAH. For preparation of PAH filled capsules, the suspension of MF-particles was mixed with 4-pyrene sulfate (4-PS^{4-} ions) (Scheme 7b-*a*), and then the polycation PAH was added, which leads to the formation of precipitates 4-PS^{4-}/PAH (Scheme 7b-*b*). After 10-15 minutes the particles were centrifuged (Scheme 7b-*c*), and the excess of unbound PAH molecules was removed. It should be noted that such complexes of multi-charged ions and PAH (4-PS^{4-}/PAH and Citrate/PAH) are not absolutely stable. 4-PS^{4-}/PAH complexes could be destroyed by high ionic strength. For dissolution of the PAH/Citrate complex, a pH higher than 10 or lower than 2 could be used. Continuous PEMs of PAH/PSS will be coated on these particles by the LBL method. The MF-particles were dissolved in 0.1M HCl leaving hollow capsules(Scheme 7b-*e*). The precipitated polymer might be dissolved in the capsule interior after treatment in 2M NaCl where the outer shell is stable (Scheme 7b-*g*). Multivalent ions are small enough to penetrate through polyelectrolyte multilayers comprising the outer shell while polymers used for inner shell build-up can't be expelled due to their high molecular weight. Thus, these polymers are captured inside the capsule as freely floating molecules (Scheme 7 b-*h*).For preparation of the PSS filled capsule, we can use the complex of PSS with Y^{3+} to form the first layer on MF particles and the other step is the same with PAH filled capsules.

However, PSS filled capsules exhibit a different Raman spectrum in figure 11d compared with figure 11b, which is prepared by the polymerization method. In figure 11d, there are several strong Raman peaks. As compared with the Raman spectrum of MF particles and MF oligomers, the materials inside the capsules probably are from the core residues, (named as MF gel). This MF gel exhibits a slight difference with MF oligomers and MF particles in the region of 1300 ~ 1600 cm^{-1}. If we compare figure 11c (PAH filled capsules) with 11a (PSS filled capsules), the spectra are similar with each other. It means that in both cases, the MF core is hard to be dissolved completely so that MF gel is formed inside the capsules. But since the MF gel shows a strong Raman spectrum, it is difficult to draw a conclusion that PSS or PAH deposit inside the capsules by the Raman studies. Therefore, the polymerization method is suited for polymer filled MF capsules.

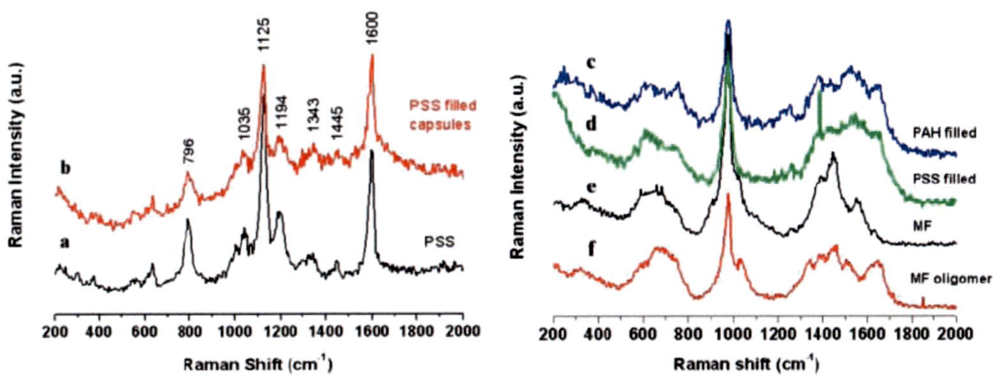

Figure 11. Raman spectrum of (a) PSS powder, (b) inside of the PSS filled capsules prepared by the polymerization method. Raman spectrum of (c) inside of PAH filled capsule and (d) inside of PSS filled capsule which are prepared by the precipitation method. Raman spectrum of (e) MF particles and (f) MF oligomers [24].

3.2.2. LOADING POLYELECTROLYTE IN $MNCO_3$ CAPSULES

By the precipitation method, one cannot remove the MF gel inside capsules after core removal. To solve this problem, one can use an inorganic core ($MnCO_3$ particles). These particles had been developed recently in our group instead of those MF particles [50]. The advantage of these carbonate particles is relative to that they can be dissolved completely by acid or EDTA and the manganese ion are easily diffusing through the shell. Therefore there are no core residues inside the capsules.

We study the polymer encapsulation in the capsules by the controlled precipitation method [19,50]. The manganese carbonate template is prepared by a simple mixing method described here: Acidic manganese sulfate solution ($9x10^{-3}$ M, pH = 4.2 adjusted by sulfuric acid) is added at one-to-one volume ratio to $2.25x10^{-3}$ M NH_4HCO_3. Then the stirred mixture is aged at 50 ^{0}C for 16 hours. The resulting $MnCO_3$ particles have a round shape with a diameter of 3.5 ± 0.3 μm. PSS filled $MnCO_3$ capsules $(PAH/PSS)_4$ are prepared by two stages. In the first stage a mixture of template $MnCO_3$ suspension ($5{\cdot}10^8$ cm^{-3}) with 0.01 M solution of $Y(NO_3)_3$ was made and then the PSS solution (1 mg/ml) was dropped (10 μl each droplet) stirring continuously until the final PSS concentration reached 0.7 mg/ml. Then, the assembly of PAH/PSS multilayers on the top of the obtained $MnCO_3$/PSS particles was arranged by the layer-by-layer technique. After formation of PAH/PSS shells the $MnCO_3$ core was dissolved in 0.1 M HCl. To destroy the inner PSS/Y^{3+} complex the capsules were exposed to 2 M NaCl solution for 7 days. As a result, freely floating PSS molecules were uniformly distributed inside the PAH/PSS polyelectrolyte capsule.

Figure 12a presents the Raman spectra of $MnCO_3$ particles. A typical peak of $MnCO_3$ is at 1084 cm^{-1} assigned to the carbonate group. After dissolution by acid, this peak disappeared because of the decomposition of carbonate and hollow capsules are left in the solution. After loading PSS inside the capsule, its Raman spectrum from the middle of the capsules can be observed in figure 12b. Clearly, there are no $MnCO_3$ peaks from inside of the capsules. It indicates that PSS can be encapsulated inside capsules by the precipitation method, which is totally different from MF capsules.

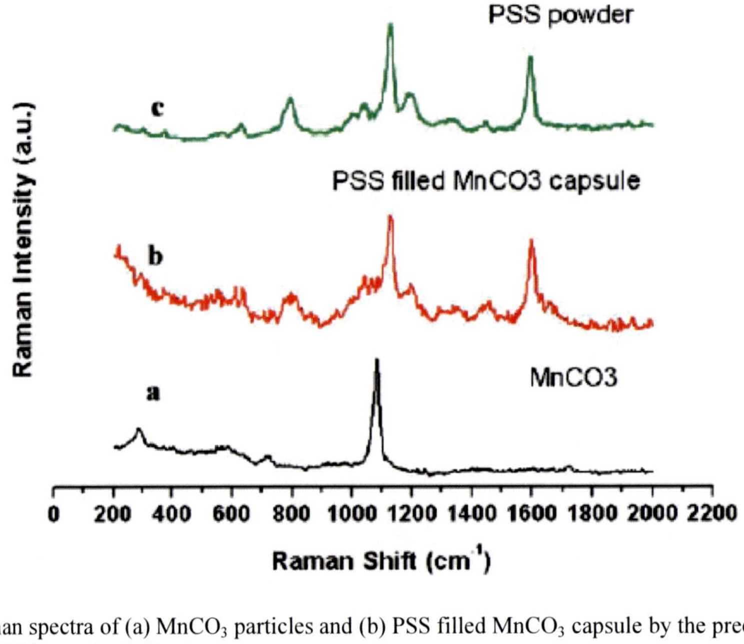

Figure 12. Raman spectra of (a) $MnCO_3$ particles and (b) PSS filled $MnCO_3$ capsule by the precipitation method and (c) PSS powder [24].

3.2.3. LOADING POLYELECTROLYTE IN POROUS $CACO_3$ CAPSULES

The potential use of the capsules may be increased if one substitutes the classical MF particles or PS latex with cheap and biocompatible inorganic particles. Thus, carbonate particles, like $MnCO_3$ (3.7 μm), $CaCO_3$ (2.1 μm), $CdCO_3$ (5.6 μm), etc. had been developed in our group. Among them, $CaCO_3$ particles are mainly considered in this section. These $CaCO_3$ particles were prepared by the precipitation crystallization method. Briefly, 0.33 M Na_2CO_3 solution was rapidly poured into an equal volume of 0.33 M solution of $CaCl_2$ at room temperature with vigorous stirring. After drying the solution in air, spherical $CaCO_3$ particles were obtained. It has been reported that these particles have a pore size of about 30 nm, which allows the macromolecules to penetrate into the particles [52]. To encapsulate BSA (Serum albumin) in $CaCO_3$ particles, we first incubated the BSA solution with $CaCO_3$ microparticles for 24 h to let BSA diffuse into the cavities and pores. For coating the BSA loaded particles with several bilayers of PSS/PAH pairs, the same approach as for the capsule preparation is followed. Core dissolution was achieved with 0.2M EDTA solution at pH = 7.0 for 24 h with repeating twice or three times. After rinsing these capsules with water, they were dispersed in neutral solution.

Confocal Raman spectroscopy was used to evaluate the encapsulation of BSA inside $CaCO_3$ capsules. Figure 13 shows the Raman spectrum of BSA filled capsules and BSA solution. The spectra are from the shell (figure 13 c), interior (figure 13 b) and outside (figure 13 a) of BSA filled capsules. The weak peak around 1600 cm^{-1} in figure 13a comes from the glass slides, which can be neglected in the other Raman spectra. There are no Raman active bands outside the capsules. Figure 13 c shows distinct Raman active modes at 1125 cm^{-1} and 1600 cm^{-1} from the capsule shell, which is from the PSS group in PEMs. Figure 13 b shows the Raman spectrum from inside the capsules. It is obvious that there is no carbonated peak inside the capsules, indicating that the $CaCO_3$ core is completely removed. However, some new Raman peaks from BSA can be observed from inside the capsules if one compares with the BSA spectrum in figure 13 d. This demonstrates that BSA is indeed encapsulated inside the capsules.

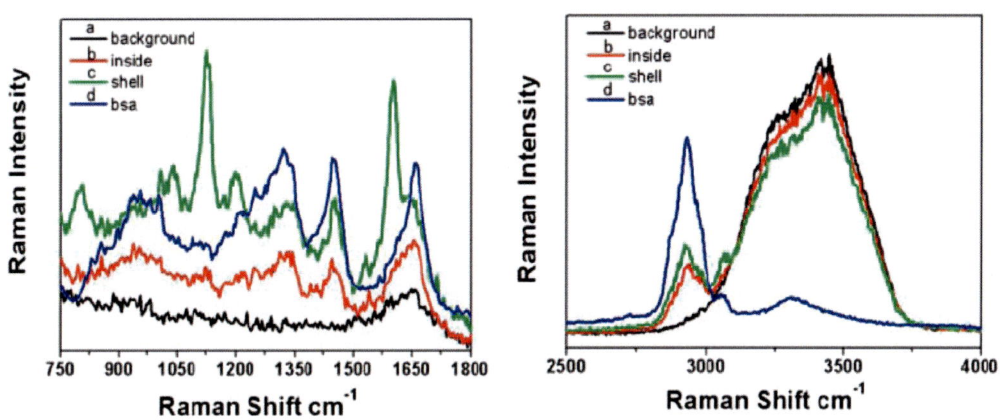

Figure 13. Raman spectra of (a) outside, (b) inside and (c) shell of BSA loaded $CaCO_3$ Capsules and (d) BSA solution [24].

Raman images can be used to detect the distribution of loaded BSA inside the capsules. We choose the Raman bands at 1126 cm^{-1} and at 2933 cm^{-1} from figure 13 used for acquiring Raman images. The results are shown in figure 14. The band at 1126 cm^{-1} is from the PSS group. Thus the images in figure 14b indicate the PSS distribution in the capsules. One clearly observes that PSS exists on the capsules shell because of the high intensity of the PSS spectrum. To evaluate the PSS distribution, first a plot of the intensity along a line is shown in figure 14c. The capsule size can be defined as the distance between two sharp peaks. The capsule size thus derived is roughly about 9 μm. The highest shell peak intensity is about 32 counts. The intensity from inside of the capsules is quite low (about 5 ~ 6 counts), which means very little PSS is inside the capsules. It is calculated that about 14% (intensity ratio of

$R_I = \frac{I_{inside}}{I_{inside} + I_{shell}}$) or 20% (area ratio of $\boldsymbol{R}_A = \frac{A_{inside}}{A_{inside} + A_{shell}}$) PSS are found inside the capsule, which results in a very low Raman intensity inside the capsules.

If BSA has a sharp peak, which is not appearing in the Raman spectrum of the capsule shell, we can use this peak to determine the position of BSA in the capsules. We find most of the BSA peaks are combined with PEM peaks. In figure 13c, the strongest peaks of BSA are at 2933 cm^{-1}, but it can also be attributed to PEMs. Thus, the Raman image of BSA cannot be measured directly.

However, we can describe the BSA distribution inside the capsules by the Raman Images indirectly. For example, figure 14b shows the Raman images at 2933 cm^{-1}. We can easily observe that bright regions stay on the shell. One can observe the shell of the capsules with highest intensity of 32 counts. However, the intensity from inside capsules is obviously higher than from outside. The intensity ratio $\boldsymbol{R}_I$ is about 35% and the area ratio $\boldsymbol{R}_A$ is about 78% at 2933 cm^{-1}, which is higher than 14% and 20% at 1126cm^{-1}. As a result of figure 14 at 1126 cm^{-1}, PEM content of the shell is about four times higher than that inside the capsules (depending on $\boldsymbol{R}_A$). But at 2933 cm^{-1}, the results show that 22% of the CH group is on the shell and 78% is inside the capsules. Since the CH group is not only from PEM but also from BSA, one can estimate that there are at least 72% CH inside the capsules are contributed to BSA. The same conclusion can be made upon using R_I for the calculation.

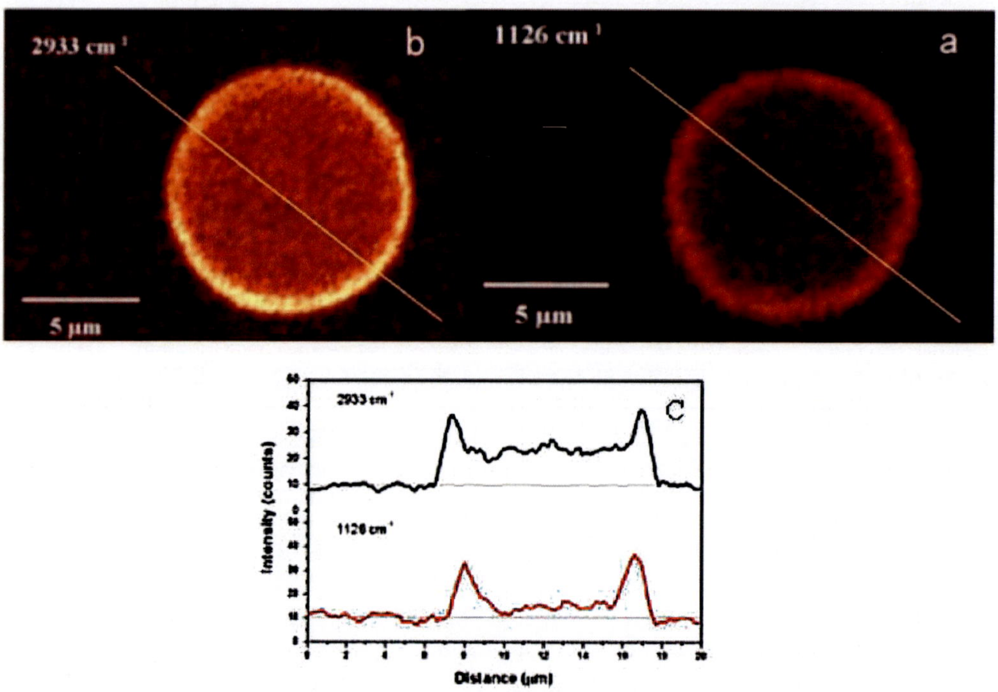

Figure 14. Raman contrast images of the BSA loaded $CaCO_3$ Capsule [24].

In summary, the encapsulation behavior of $CaCO_3$ particles is dominated by the rough surface and the pores inside the particles. Therefore, water-soluble macromolecules, like proteins or polyelectrolytes can be encapsulated inside the particles. After coating with several bilayers of PEMs, they will perverse the charged polymer inside the capsule within a dense polymer shell. Thus, these functionalized capsules can be used for drug delivery or biomaterial encapsulation.

3.3. Intelligent Release [53]

"Smart" capsules can recognize a stimulus as a signal and then significantly alter for instance their chain conformation in response to small changes in the environmental conditions [11]. Molecular structures that are composed of such materials, with the ability to be triggered to contract or expand in a controlled fashion, have a great application potential in numerous fields, including nanotechnology, biochemistry, organic and physical chemistry, and materials science. Conventionally, organic polyelectrolyte-based microcapsules show the controlled pattern by ultrasound [54], radiation [55], pH trigger [56], etc. However, all these display limitations in some significant applications due to their slow response to trace amounts of trigger or restricted freedom of choice of stimuli [57-60].

Electrochemical stimuli are very promising, since the redox potential near a desired organ where a pharmaceutical drug should be released may differ from other locations in the body, or that a corrosion pit may self-anneal due to the release of an inhibitor by local potential change. This, however, requires the construction of capsules with redox-active compounds. Recently, we had prepared poly(ferrocenylsilanes) (PFS) [58-60], composed of organometallic units in the main chain, belong to a class of stimulus-responsive materials [53]. Due to the ferrocene-bearing repeating units of the polymer, PFS can be reversibly oxidized and reduced by chemical as well as electrochemical means.

The water-soluble poly(ferrocenylsilane) polyanion 1 and polycation 2 are shown in Scheme 8. PFS capsules were fabricated by the electrostatic layer-by-layer assembly of these polyelectrolytes onto colloidal templates. These templates were then removed after the completion of the assembly. Complementary to the MF cores, more stable microcapsules, solely composed of PFS polyanions and polycations with a better defined integrity, were fabricated in high yield using manganese carbonate ($MnCO_3$) templates. Metal carbonate crystals that can be removed by EDTA solutions have been intensively employed in recent years due to their suitability in the fabrication of clean capsules, dissolution in mild conditions and relative ease of synthesis.

Figure 15c displays an example of a dried capsule with four bilayers of PFS polyanion/polycation. An average bilayer (PFS^-/PFS^+) thickness of 6.0 ± 0.3 nm for adsorption from 0.5 M NaCl was calculated. This value was larger than the typical bilayer thickness (~ 4.5 nm) obtained on planar (silicon or gold) substrates under the same adsorption conditions. A possible explanation for this variation in film thickness may be related to the dependence of multilayer film growth on the characteristics of the templating materials (e.g. surface charge density). The surface of the collapsed capsules (as shown in AFM images) exhibited a roughness (rms, in an area of 3 x 3 μm^2) of ~ 5 nm, which was significantly larger than the respective rms values obtained for multilayer films on planar substrates (in the sub nm range).

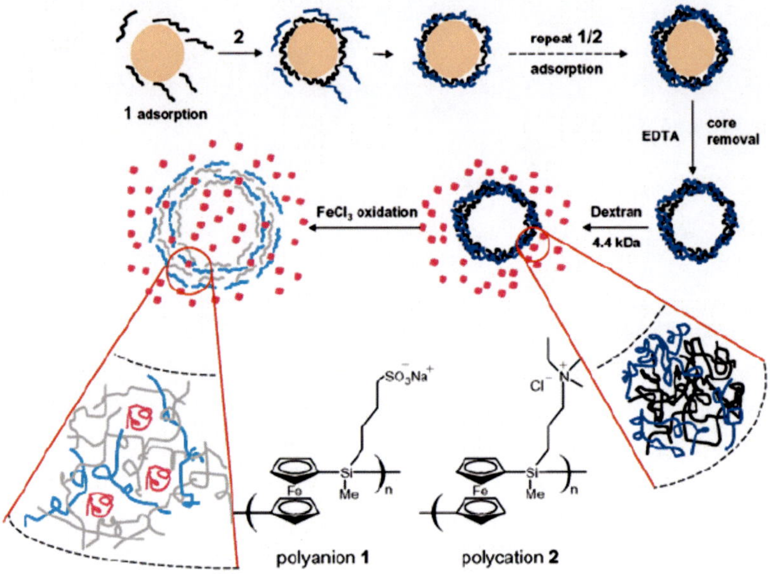

Scheme 8. Schematics of organometallic multilayer capsule formation and permeability control. Polyanion 1 and polycation 2 were used in the electrostatic self-assembly onto curved substrates followed by core removal. The permeability of the capsules obtained can be tuned via chemical oxidation e.g. by $FeCl_3$. The different colors of polymer chains in the bottom left represent their different oxidational/conformational states. Reproduced with permission from Nat. Mater., 2006, 5, 724. Copyright © 2006, Nature Publishing Group.

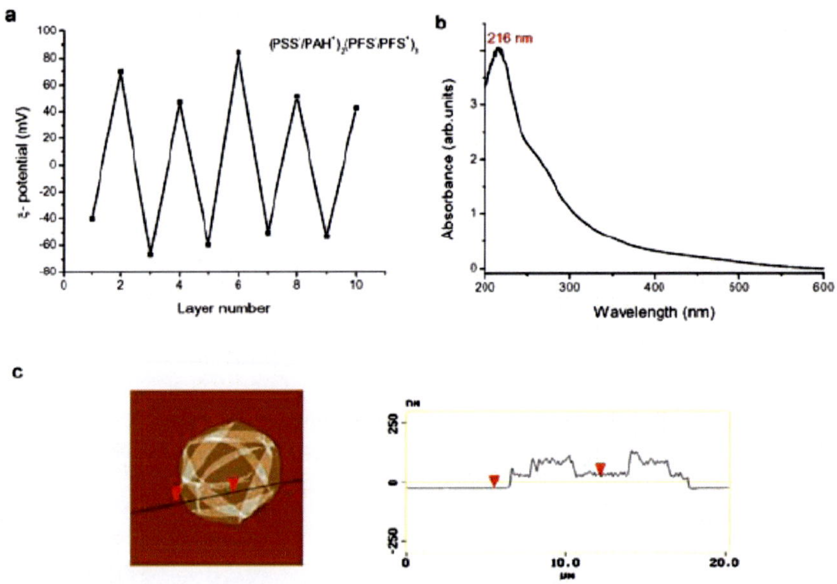

Figure 15 Characterization of poly(ferrocenylsilane) multilayer microcapsules. a, Zeta-potential measurements following the layer-by-layer deposition process of PFS polyions on MF cores. b, UV/Vis spectrum of the aqueous dispersion of $(PSS^-/PAH^+)_2(PFS^-/PFS^+)_3$ capsules. The strong absorption peak at λ = 216 nm is characteristic for PFS. c, One collapsed PFS capsule (four bilayers of (PFS^-/PFS^+)) imaged by TM AFM in the dry state; the AFM thickness of the capsules is measured by the difference in the heights corresponding to the arrows shown. Reproduced from ref. [53]. Copyright © 2006, Nature Publishing Group.

The unique chemical constitution of these capsules ensured their peculiar redox-responsive permeability (Scheme 8). In order to study the response of our PFS capsules to redox-stimuli, ferric chloride ($FeCl_3$) was chosen as an oxidant due to its effectiveness in oxidizing PFS and its water solubility (under slightly acidic conditions). By using tetramethylrhodamine isothiocyanate (TRITC) labeled dextran (4.4 kDa) as a fluorescence probe, PFS capsules that were originally impermeable to these molecules were mixed on-site with an aqueous $FeCl_3$ solution (1 mmol, pH = 4, tuned by HCl) during in-situ confocal laser scanning microscopy (CLSM) imaging.

Upon exposure to Fe^{3+} ions originating from $FeCl_3$, the PFS capsules were observed to display a continuous expansion accompanied by an increasing permeability of the capsules. figure16a shows the CLSM images of PFS capsules (with 6 bilayers) before and after oxidation. Figure16b reveals that the oxidized capsules were completely permeable to 4.4 kDa dextran molecules, as it was no longer possible to differentiate the fluorescence intensity from outside and inside the capsules. Following the exposure to $FeCl_3$, a substantial change in the capsule size accompanied the onset of permeability, as observed for the oxidized capsules in figure16. Due to the specific chemical constitution of these capsules, they exhibited a unique redox-stimulus responsive permeability, schematically captured and explained in Scheme 8.

Capsule expansion will proceed in a continuous manner until the final disappearance of the capsule. The process of expansion and dissolution was monitored as a function of time to describe the kinetics of the capsule swelling under these conditions. The oxidation-induced expansion occurs following an induction period of ca. 15 min. The long induction period is a result of a slow diffusion of the Fe^{3+} ions to the location of the capsules and to the capsule interior, and thus not a result of a slow oxidation reaction. After the induction period, data were collected on the mean diameter values of a capsule from the onset of the addition of $FeCl_3$ until the final disappearance of the capsule. The continuous capsule expansion could actually be stopped by adding effective PFS reducing agents, e.g. dithiothreitol (DTT) during the oxidation process. The degree of capsule expansion was calculated (averaged over a population of ca. 20 capsules) as:

$$\textit{capsule expansion (\%)} = \frac{\overline{d_t - d_0}}{d_0} \times 100\%$$

where d is the diameter of the capsules observed by CLSM, with subscripts 0 and t denoting values prior to and at "t", respectively.

Figure16b summarizes the expansion data as a function of time. The circular points in figure 16b represent the experimental data that were fitted by a simple exponential growth curve as depicted by the red line. The data indicated that a capsule expanded in an accelerated manner towards the final disintegration, following kinetics similar to a first-order reaction, where the rate law can be described as:

$$\frac{A}{A_0} - 1 = e^{kt} \tag{4}$$

with k as the rate constant. A fit of the data resulted in a typical k-value of 0.010 s^{-1} in the present case, and this value was found to strongly depend on the concentration of the oxidant. Our study on the local oxidation of $(PFS^{-}/PFS^{+})_5$ capsules showed that when the concentration of the $FeCl_3$ solution decreased from 3 to 1 mmol, the duration for complete capsule disintegration increased from 1.5 to 15 minutes. These times are more than one order of magnitude shorter than those measured for main chain scissoring, and it was therefore concluded that they were determined by the build-up of charges inside the films.

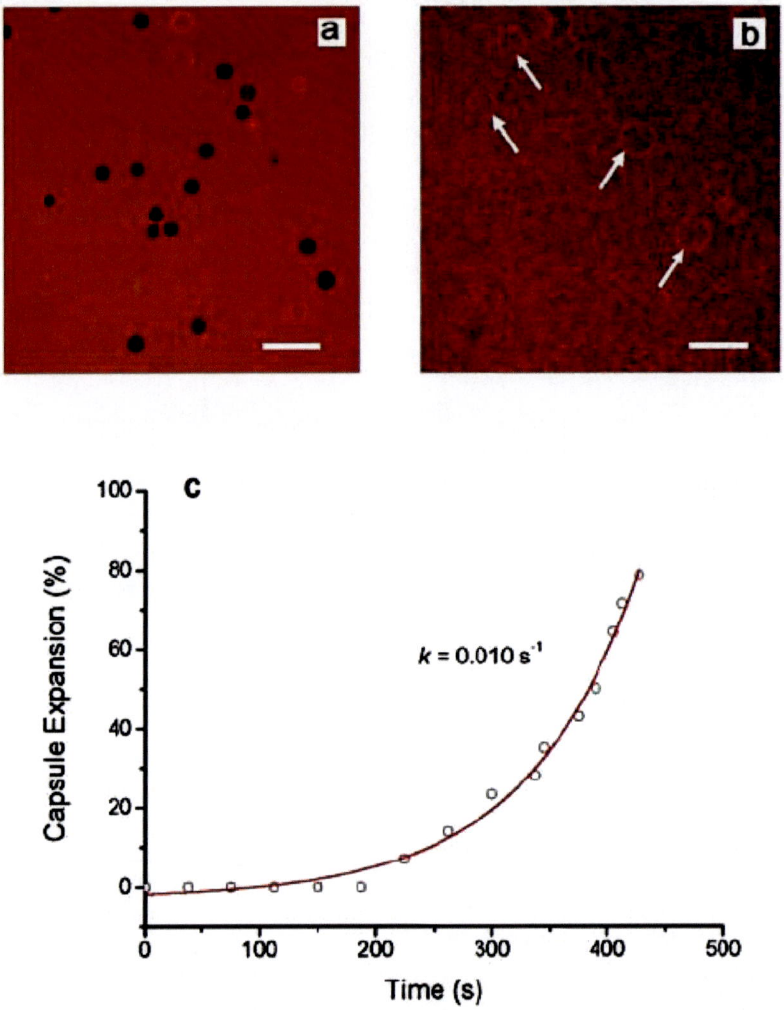

Figure 16. Redox-responsive permeability accompanied by expansion of PFS capsules. a, Confocal fluorescence micrographs of a capsule solution containing 4.4 kDa TRITC labelled dextran before (1) and after (2) $FeCl_3$ oxidation. b, PFS capsule expansion with time in the final oxidized states. The circles are experimental data and the red line is an exponential fit, which gives a first order expansion rate constant of $k = 0.017\ s^{-1}$ for the example shown. Reproduced from Nat. Mater., 2006, 5, 724 [53]. Copyright © 2006, Nature Publishing Group.

This oxidation-induced capsule permeability change and disintegration were extremely peculiar, especially since a very small trigger ($FeCl_3$ with a sub-mmol concentration) was sufficient to cause the change. This expansion was interpreted as a result of several factors. Firstly, oxidation of the ferrocenyl groups on the polymer caused the amount of positive charges along the main chain to increase. Thus, in a multi-layered structure containing the same polymer backbone, electrostatic repulsion due to the excess positive charges on the chains would increase the distance between like-charged segments, i.e. an expansion both along the chain and in the multilayer growth direction would be favored. As aforementioned, we have previously shown that a stiffening (i.e. an increase of the Kuhn segment length and segment elasticity) accompanied the oxidation. This effect is likely to increase the average pore size and enhance the permeability accordingly. Finally, with the continuous increase of the inner- and inter-molecular electrostatic repulsion with an increasing degree of oxidation, the charges on the differently charged PFS^- and PFS^+ polyelectrolytes species could not hold the capsule wall together. This led to the disintegration of the original charge-compensated layer-by-layer structure.

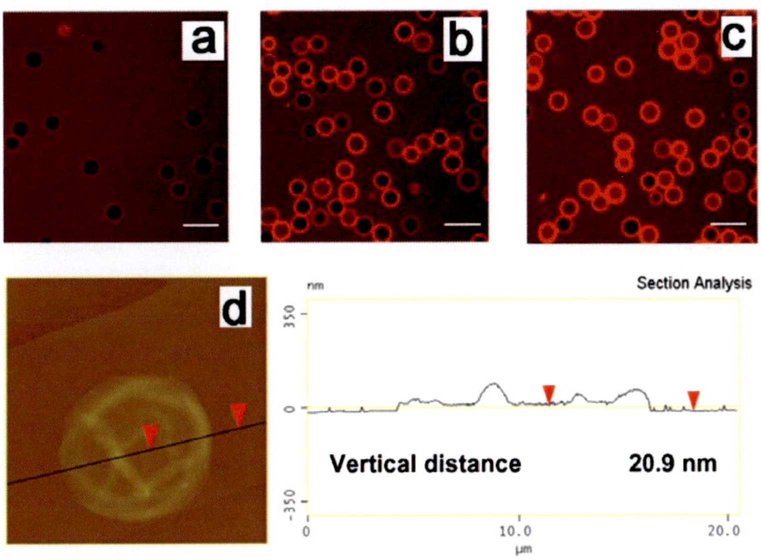

Figure 17. Redox-responsive permeability of multilayer capsules with organometallic-organic composite-wall structures. a, Local oxidation of $(PFS^-/PFS^+)_5(PSS^-/PAH^+)_1$ microcapsules by $FeCl_3$ (1 mmol, pH = 4) monitored by CLSM. Capsules that are originally impermeable (1) to 4.4 kDa dextran molecules show partial permeability (2) in the early oxidized stage (10 min) and almost complete permeability (3) after oxidation for more than 1h. Scale bar = 20 μm. b, TM AFM height image of a $(PFS^-/PFS^+)_5(PSS^-/PAH^+)_1$ capsule after oxidation for over 2h by $FeCl_3$ (1mmol, pH = 4). The integrity of the capsules has been preserved. Reprinted with permission from ref. [53]. Copyright © 2006, Nature Publishing Group.

In order to maintain the integrity of the capsules, and at the same time effectively change the capsule permeability, composite-wall capsules with PFS^-/PFS^+ polyion pairs as inner layers and redox-insensitive polyelectrolyte pairs (PSS^-/PAH^+) in the outer layers were fabricated. Examples of CLSM micrographs of capsules composed of $(PFS^-/PFS^+)_5(PSS^-/PAH^+)_1$ are demonstrated in figure 17. These capsules were robust and essentially impermeable to 4.4 kDa dextran molecules in the neutral state. After adding $FeCl_3$ (1 mmol) solution, the capsules started to become permeable, indicated by the strong increase of the fluorescence intensity in the capsule interior (figure 17b). During the course of the oxidation process of PFS, an almost complete permeability of all capsules was observed (figure 17c). CLSM images also exhibited a gradual accumulation of dye molecules in the capsule wall (see also scheme 8). Similar dye accumulation was reported earlier [4]. The capsule integrity was preserved, as shown in figure 17d. However, the double wall thickness of the capsules decreased from 69 ± 3 nm before oxidation to 22 ± 3 nm after oxidation for over 2 h, indicating some loss of material. Neither CLSM nor AFM measurements showed any measurable capsule size change prior to and following oxidation.

In addition to the concentration of oxidants, the rate of permeability change could also be tuned by the number of PSS^-/PAH^+ bilayers on the outer-wall of the capsules. Increasing the number of outer-layers slowed down the permeability change of the capsules. Obviously the capping non-redox active bilayers acted as “blocking” layers on the capsule surface. When varying the number of capping PSS^-/PAH^+ bilayers from n = 1-3 on the same $(PFS^-/PFS^+)_5$ inner-layer structure and using the same oxidant concentration, the recorded time scale for more than 80% of impermeable capsules to become permeable increased from 1 h to 6 h.

4. PROSPECTS

Although the structure–property relationship of polyelectrolyte microcapsules had been systematically demonstrated, their smart applications have not yet been fully developed. Several emerging fields will appear in the near future:

1. One challenging project concerns the applications of these polyelectrolyte multilayers or polyelectrolyte capsules as the special nanoreactors or nanocontainers since many free and immobile counter ions or polymer segments are entrapped inside memberane. The special environment gives a chance for studies of complicated structure of multilayers under a nonequilibrium state. Therefore, by adjusting the conditions of the bulk solution, like pH or solvent quality, one can find changes of PEM structure (hydrophobic or hydrophilic, and pK value).
2. Hollow capsules are good candidates of stable and soft supporting templates, either for the self- assembly of lipids or other proteins in order to build up a new generation of biosensors or bio-membranes, or for shape-forming agents to control the non-classical crystallization of inorganic meso-particles together with the functional polymer.
3. This is an essential prerequisite to drive this promising system towards smart applications. New tools will be developed. As a good tool, NMR has the ability to distinguish these changes by the diffusion measurement or the 2D NMR spectrum. For example, when the capsules are treated with solvent, one could find a tiny peak from capsules in the NMR spectrum, which could not be detected before. In addition, by time resolved Raman spectroscopy, we can also investigate the inner structure change in the PEMs. It will be helpful to understand the intrinsic properties of these thin films or capsules.

ACKNOWLEDGEMENTS

The authors wish to acknowledge the financial support from the National Natural Science Foundation of China (20671096).

REFERENCES

[1] Caruso, F. *Advanced Materials.* 2001, *13*, 11.
[2] Peyratout, C. S.; Dahne, L. *Angewandte Chemie-International Edition.* 2004, *43*, 3762.
[3] Sukhorukov, G. B.; Fery, A.; Brumen, M.; Mohwald, H. *Physical Chemistry Chemical Physics.* 2004, *6*, 4078.
[4] Discher, B. M.; Won, Y. Y.; Ege, D. S.; Lee, J. C. M.; Bates, F. S.; Discher, D. E.; Hammer, D. A. *Science.* 1999, *284*, 1143.
[5] Antonietti, M.; Forster, S. *Advanced Materials.* 2003, *15*, 1323.
[6] Jang, J.; Ha, H. *Langmuir.* 2002, *18*, 5613.
[7] Huang, H.; Remsen, E. E.; Kowalewski, T.; Wooley, K. L. *J. Am. Chem. Soc.* 1999, *121*, 3805.
[8] Wong, M. S.; Cha, J. N.; Choi, K. S.; Deming, T. J.; Stucky, G. D. *Nano Letters.* 2002, *2*, 583.
[9] Decher, G.; Hong, J. D. *Makromolekulare Chemie-Macromolecular Symposia.* 1991, *46*, 321.
[10] Donath, E.; Sukhorukov, G. B.; Caruso, F.; Davis, S. A.; Mohwald, H. *Angewandte Chemie-International Edition.* 1998, *37*, 2202.
[11] Caruso, F.; Caruso, R. A.; Mohwald, H. *Science.* 1998, *282*, 1111.
[12] Caruso, F.; Donath, E.; Mohwald, H. *Journal of Physical Chemistry. B* 1998, *102*, 2011.
[13] Decher, G.; Hong, J. D.; Schmitt, J. *Thin Solid Films.* 1992, *210-211*, 831.
[14] Liu, X. Y.; Bruening, M. L. *Chemistry of Materials.* 2004, *16*, 351.
[15] Klitzing, R. v.; Mohwald, H. *Macromolecules.* 1996, *29*, 6901.
[16] Sukhorukov, G. B.; Brumen, M.; Donath, E.; Mohwald, H. *Journal of Physical Chemistry. B* 1999, *103*, 6434.
[17] Ibarz, G.; Dahne, L.; Donath, E.; Mohwald, H. *Chemistry of Materials.* 2002, *14*, 4059.
[18] Dahne, L.; Leporatti, S.; Donath, E.; Mohwald, H. *Journal of the American Chemical Society.* 2001, *123*, 5431.
[19] Shchukin, D. G.; Dong, W. F.; Sukhorukov, G. B. *Macromolecular Rapid Communications.* 2003, *24*, 462.
[20] Gao, C. Y.; Donath, E.; Mohwald, H.; Shen, J. C. *Angewandte Chemie-International Edition.* 2002, *41*, 3789.
[21] Dong, W. F.; Sukhorukov, G. B.; Mohwald, H. *Physical Chemistry Chemical Physics.* 2003, *5*, 3003.
[22] Dong, W. F.; Ferri, J. K.; Adalsteinsson, T.; Schonhoff, M.; Sukhorukov, G. B.; Mohwald, H. *Chemistry of Materials.* 2005, *17*, 2603.
[23] Gao, C. Y.; Moya, S.; Donath, E.; Mohwald, H. *Macromolecular Chemistry and Physics.* 2002, *203*, 953.
[24] Dong, W. F. *unpublised results.* 2006.

[25] McAloney, R. A.; Sinyor, M.; Dudnik, V.; Goh, M. C. *Langmuir.* 2001, *17*, 6655.

[26] Hammond, P. T. *Advanced Materials.* 2004, *16*, 1271.

[27] Hiller, J.; Mendelsohn, J. D.; Rubner, M. F. *Nature Materials.* 2002, *1*, 59.

[28] Sui, Z. J.; Schlenoff, J. B. *Langmuir.* 2004, *20*, 6026.

[29] Lulevich, V. V.; Vinogradova, O. I. *Langmuir.* 2004, *20*, 2874.

[30] Sukhorukov, G. B.; Antipov, A. A.; Voigt, A.; Donath, E.; Mohwald, H. *Macromolecular Rapid Communications.* 2001, *22*, 44.

[31] Tan, H. L.; McMurdo, M. J.; Pan, G. Q.; Van Patten, P. G. *Langmuir.* 2003, *19*, 9311.

[32] Gao, C. Y.; Leporatti, S.; Moya, S.; Donath, E.; Mohwald, H. *Chemistry-a European Journal.* 2003, *9*, 915.

[33] Dautzenberg, H.; Gao, Y. B.; Hahn, M. *Langmuir.* 2000, *16*, 9070.

[34] Vuillaume, P. Y.; Jonas, A. M.; Laschewsky, A. *Macromolecules.* 2002, *35*, 5004.

[35] Welsh, E. R.; Schauer, C. L.; Santos, J. P.; Price, R. R. *Langmuir.* 2004, *20*, 1807.

[36] Richert, L.; Engler, A. J.; Discher, D. E.; Picart, C. *Biomacromolecules.* 2004, *5*, 1908.

[37] Dai, J. H.; Jensen, A. W.; Mohanty, D. K.; Erndt, J.; Bruening, M. L. *Langmuir.* 2001, *17*, 931.

[38] Harris, J. J.; Bruening, M. L. *Langmuir.* 2000, *16*, 2006.

[39] Harris, J. J.; Stair, J. L.; Bruening, M. L. *Chemistry of Materials.* 2000, *12*, 1941.

[40] Krass, H.; Papastavrou, G.; Kurth, D. G. *Chemistry of Materials.* 2003, *15*, 196.

[41] Kurth, D. G.; Lopez, J. P.; Dong, W. F. *Chemical Communications.* 2005, 2119.

[42] Sun, J.; Jin, P.; Wang, Z. G. *Nanotechnology.* 2004, *15*, 1763.

[43] Poptoshev, E.; Schoeler, B.; Caruso, F. *Langmuir.* 2004, *20*, 829.

[44] Kim, B. S.; Lebedeva, O. V.; Koynov, K.; Gong, H. F.; Glasser, G.; Lieberwith, I.; Vinogradova, O. I. *Macromolecules.* 2005, *38*, 5214.

[45] Dubas, S. T.; Schlenoff, J. B. *Macromolecules.* 1999, *32*, 8153.

[46] Lvov, Y.; Antipov, A. A.; Mamedov, A.; Mohwald, H.; Sukhorukov, G. B. *Nano Letters.* 2001, *1*, 125.

[47] Dong, W. F.; Liu, S. Q.; Wan, L.; Mao, G. Z.; Kurth, D. G.; Mohwald, H. *Chemistry of Materials.* 2005, *17*, 4992.

[48] Vinogradova, O. I.; Andrienko, D.; Lulevich, V. V.; Nordschild, S.; Sukhorukov, G. B. *Macromolecules.* 2004, *37*, 1113.

[49] Radtchenko, I. L.; Sukhorukov, G. B.; Leporatti, S.; Khomutov, G. B.; Donath, E.; Mohwald, H. *Journal of Colloid and Interface Science.* 2000, *230*, 272.

[50] Sukhorukov, G. B.; Shchukin, D. G.; Dong, W. F.; Mohwald, H.; Lulevich, V. V.; Vinogradova, O. I. *Macromolecular Chemistry and Physics.* 2004, *205*, 530.

[51] Antipov, A. A.; Shchukin, D.; Fedutik, Y.; Petrov, A. I.; Sukhorukov, G. B.; Mohwald, H. *Colloids and Surfaces a-Physicochemical and Engineering Aspects.* 2003, *224*, 175.

[52] Sukhorukov, G. B.; Volodkin, D. V.; Gunther, A. M.; Petrov, A. I.; Shenoy, D. B.; Mohwald, H. *Journal of Materials Chemistry.* 2004, *14*, 2073.

[53] Ma, Y.; Dong, W. F.; Hempeniusa, M. A.; Mohwald, H.; Vancso, G. J. *Nature Materials.* 2006, asap.

[54] Shchukin, D. G.; Gorin, D. A.; Mohwald, H. *Langmuir.* 2006, *22*, 7400.

[55] Skirtach, A. G.; Javier, A. M.; Kreft, O.; Kohler, K.; Alberola, A. P.; Mohwald, H.; Parak, W. J.; Sukhorukov, G. B. *Angewandte Chemie-International Edition.* 2006, *45*, 4612.

[56] De Geest, B. G.; Dejugnat, C.; Sukhorukov, G. B.; Braeckmans, K.; De Smedt, S. C.; Demeester, J. *Advanced Materials.* 2005, *17*, 2357.

[57] Sukhorukov, G.; Fery, A.; Mohwald, H. *Progress in Polymer Science.* 2005, *30*, 885.

[58] Zou, S.; Korczagin, I.; Hempenius, M. A.; Schonherr, H.; Vancso, G. J. *Polymer.* 2006, *47*, 2483.

[59] Giannotti, M. I.; Lv, H.; Ma, Y.; Steenvoorden, M. P.; Overweg, A. R.; Roerdink, M.; Hempenius, M. A.; Vancso, G. J. *Journal of Inorganic and Organometallic Polymers and Materials.* 2005, *15*, 527.

[60] Korczagin, I.; Lammertink, R. G. H.; Hempenius, M. A.; Golze, S.; Vancso, G. J. Surface nano- and microstructuring with organometallic polymers. In *Ordered Polymeric Nanostructures at Surfaces*; SPRINGER-VERLAG BERLIN: Berlin, 2006; Vol. 200; pp 91.

In: Progress in Advanced Materials Research
Editor: Nicolas H. Voler, pp. 119-137

ISBN: 978-1-60021-893-4

Chapter 4

PREPARATION AND CHARACTERIZATION OF PEROVSKITE TYPE SEMICONDUCTING THIN FILMS

R. Parashkov[1*], A. Brehier[1], A. Georgiev[1], S. Bouchoule[2], X. Lafosse[2], J.-S. Lauret[1], C. T. Nguyen[1], M. Leroux[3] and E. Deleporte[1*]

[1] Laboratoire de Photonique Quantique et Moléculaire (UMR 8537), Institut d´Alembert (IFR 121), Ecole Normale Supérieure de Cachan, 61 avenue du Président Wilson, 94235 Cachan Cedex, France.

[2] CNRS, Laboratoire de Photonique et de Nanostructures, 91460 Marcoussis, France.

[3] CNRS, Centre de Recherche sur l'Hetero-Epitaxie et ses Applications, 06560 Valbonne, France.

ABSTRACT

Here we present the preparation of nine 2D and 3D lead halide perovskite structures whose exciton energy is between 2.05 and 3.64 eV. We investigate in details the optical properties of these excitons, by means of transmission, photoluminescence, reflectivity, and ellipsometric spectroscopies, and by atomic force microscopy. We characterize planar microcavities in which strong exciton-photon coupling occurs between the lead halide based perovskites and the standing wave. The small linewidth and large oscillator strength characteristic of perovskites make them very interesting structures to allow the observation of strong coupling for hybrid semiconductors. In this chapter we investigate the potential for application of these materials in hybrid organic-inorganic devices.

* Authors for correspondence: radoslav.parashkov@lpqm.ens-cachan.fr, emmanuelle.deleporte@lpqm-ens-cachan.fr

I. Introduction

Since the beginning of organic electronics, tremendous efforts have been made for the development of new active materials for organic transistors, organic light emitting diodes (OLEDs), and organic lasers [1,2]. Perovskite molecules are good candidates for application as active materials in hybrid organic-inorganic devices. When perovskites are deposited by spin-coating on a glass substrate, a self-organized structure is obtained: inorganic sheets alternate with organic layers. These self-organized layered perovskites are analogous to semiconductor multiple quantum wells (MQW) structures: the wells are inorganic and of very thin width (several Angstroms) whereas the barriers are organic [3]. The optical properties of exciton materials such as MQW have been intensively studied [4]. In perovskite structures, significant enhancement of the exciton binding energy is observed because of the dielectric confinement effect, the exciton binding energy reaching values as high as 300 meV. This enables strong photoluminescence even at room temperature sufficient to offer potential applications as emitter materials in electroluminescent devices [3,5]. Perovskite materials have also found application in the field of hybrid organic-inorganic transistors [6]. Until now the optical properties of the organic-inorganic perovskites have not been extensively investigated.

The basic layered 3D perovskite structures AMX_4 are schematically depicted in figure 1. Each of the inorganic layers consists of sheets of corner-sharing metal halide octahedra. In order to satisfy charge-balancing requirements, A is a organic amine cation, the M cation is generally a divalent metal and X a halide. The larger A cations fill the coordinated holes among the octahedra. A typical example for such a structure is provided by $(CH_3NH_3)SnI_3$, which has the cubic lattice constant a = 6.2397 Å [7]. Analogous lead (II) based perovskites have also been investigated [8]. For a perfectly ordered perovskite structure the geometrically imposed conditions for the A, M, and X ions is that they should be in close contact, with other words $(R_A + R_X) = t\sqrt{2}\,(R_M + R_X)$, where R_A, R_M, and R_X are the ionic radii for the corresponding ions and the tolerance factor must satisfy $t \approx 1$. Empirically it has been found that t should be between 0.8 and 0.9 for most cubic perovskites, although there is a slightly expanded range for distorted structures [9]. The limit on R_A is found to be approximately 2.6 Å, by using $t = 1$ and essentially the largest values for R_M and R_X [10]. As a consequence only small organic cations consisting three or less carbon-carbon or carbon-nitrogen bonds are expected to fit in the structure.

Usually 3D lead halide perovskite structures are stabilized by mixing 1 : 1 ratios of the metal halide with the organic cation salt, where the organic part must be small enough to fit in the holes within the inorganic network of corner-sharing metal halide octahedra. In contrast, the single-layered 2D perovskites can be formed from mixtures of metal halide and generally longer or more complex organic salts. It is very interesting when mixtures of short and long organic cations are employed. In general, several possibilities might be expected. Segregation between 2D and 3D phases can occurs. The other possibility is that the organic cations might locally segregate onto separate layers of a more complex hybrid structure, thereby creating intergrowths of the 3D perovskite and the single layered 2D perovskite structures (figure 2) [9].

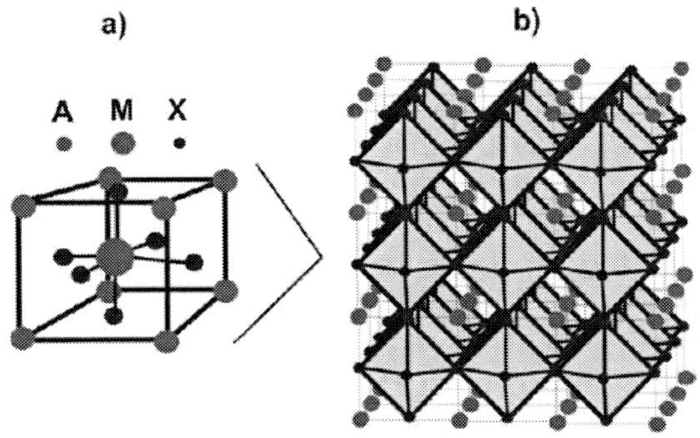

Figure 1. a) Ball and stick model of the basic AMX_3 perovskite unit cell; b) polyhedral representation of how the structure extends in three dimensions.

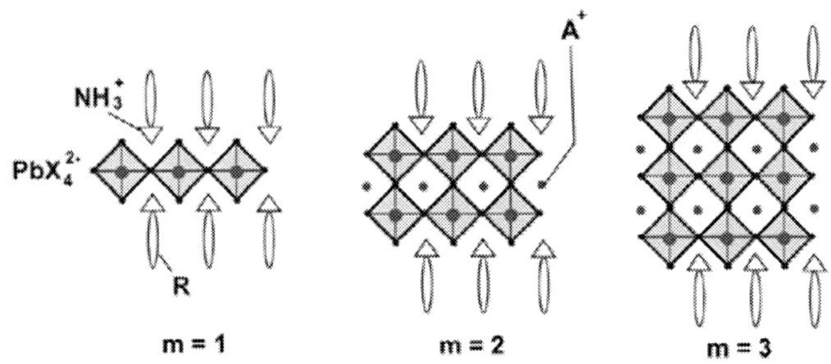

Figure 2. Family of perovskite compounds formed from parent AMX_3 (3D) and $(RNH_3)_2MX_4$ (2D) perovskites: m = 1 (monolayer); m = 2 (bilayer); m = 3 (trilayer).

In the case of lead based perovskite structures, the second of the above mentioned possibilities is observed. In this chapter we present the preparation and characterization of monolayered (m = 1), bilayered (m = 2), and trilayered (m =3) lead halide based perovskite thin-films (figure 2.).

It is also very tempting to characterize the emitters of light such as excitons inside photonic band gap structures such as microcavities. The results of such investigations could be very attractive for device applications and allow the study of new optical phenomena. The effect of the microcavity on the emission of light can be divided into two regimes. In the weak coupling regime the spontaneous emission of the materials can be dramatically enhanced or reduced, this regime is employed in applications such as vertical cavity surface emitting lasers [11] or resonant cavity light emitting diodes [12]. In the strong coupling regime, the excitonic and confined photon modes can no longer be considered independent eigenmodes of the system, the normal modes are a linear and coherent superposition of the exciton and photon states, called cavity polaritons [13]. The anti-crossing between the exciton and photon modes is related to the Rabi splitting. Since the first experimental evidence of strong coupling in a one-dimensional semiconductor microcavity over a decade ago [13], this field has been intensively studied due to the interest in coherent and stimulated effects in such systems, which can lead to optical devices [14]. In particular, the properties of polaritons to form a Bose-Einstein condensate may lead to the realization of low threshold polariton laser [15].

The field has been recently expanded to one-dimensional microcavities containing thin layers of organic materials such as J-aggregates of cyanine dyes or zinc-porphyrin [16], in which the strong coupling regime has been observed. Because of the relatively large oscillator strengths of the organic exciton, anti-crossings as large as 180 meV at room temperature have been reported [16]. Because of the increased interaction strength between organic excitons and photons, strong coupling can occur in organic microcavities having Q factors as low as 10 [17]. The large Rabi-splittings obtained in organic microcavities are of particular significance for current efforts to achieve polariton lasing in microcavities at room temperature because of the possibility of maintaining the strong coupling at 300K.

In this chapter we present the preparation and the characterization of various perovskite thin films. The materials were characterized by means of transmission, photoluminescence, reflectivity, and ellipsometric spectroscopies, and by atomic force microscopy. The strong coupling between lead halide based perovskites and the confined photon mode was studied in a planar one-dimensional microcavity.

II. Preparation of the Perovskite Thin Films

The chemical structures of the layered perovskites used in this study are shown in figure 3.

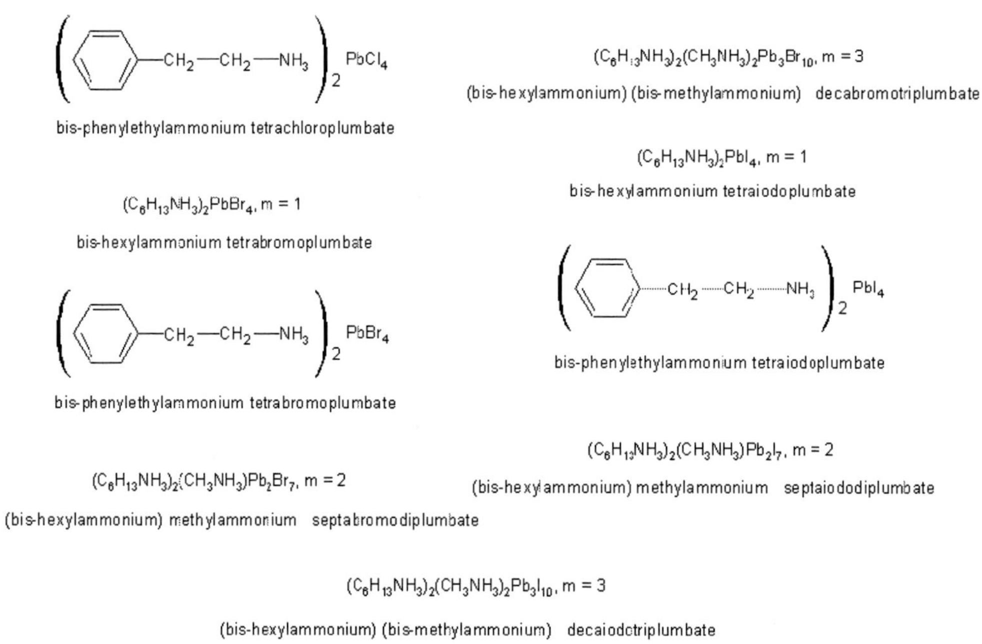

Figure 3. Chemical structures of layered perovskite compounds used in this study.

A. Amine Hydrohalides

N-alkylamines ($C_nH_{2n+1}NH_2$) or phenylethylamine ($C_6H_5C_2H_4NH_2$), and HX (X=Cl, Br, I) (aqueous solutions) are commercially available.

Stoichiometric amounts of HX and amine were mixed in a flask cooled with water in order to remove the reaction heat. The reaction was carried out at 0 °C. After the reaction, water was evaporated at elevated temperature [18].

B. Halideplumbate Perovskite Compounds

Thin films of the hybrid perovskite compound were fabricated on glass slides, silicon wafers, and dielectric mirrors by the spincoating method. To obtain 'monolayer' (i.e. $(C_6H_{13}NH_3)_2PbX_4$ or $(C_6H_5C_2H_4NH_3)_2PbX_4$ with m = 1, X = Cl, Br, I), amine hydrohalides were reacted with a stoichiometric amount of the corresponding PbX_2 in dymethylformamide (DMF). Thin films of monolayered perovskites were made by spincoating the obtained solutions on a substrate. To obtain 'bilayer' structures (m = 2), such as $(C_6H_{13}NH_3)_2(CH_3NH_3)Pb_2X_7$, stoichiometric amounts of $C_6H_{13}NH_3X$, CH_3NH_3X, and PbX_2 (X = Br, I) were reacted in DMF. On the other hand, pure 'trilayer' structures (m = 3), such as $(C_6H_{13}NH_3)_2(CH_3NH_3)_2Pb_3X_{10}$ perovskites, were not obtained by reacting stoichiometric amounts of $C_6H_{13}NH_3X$, CH_3NH_3X, and PbX_2 (X = Br, I). It has been found that, however, trilayer structures are most efficiently prepared when these materials are reacted in the following non-stoichiometric condition: $C_6H_{13}NH_3X : CH_3NH_3X : PbX_2 = 2 : 3 : 3$ in molar ratio [19].

C. Thin-Film Formation

In this experiment, acetone and DMF were used as solvents. Transparent and homogeneous solutions of $(C_6H_{13}NH_3)_2PbI_4$ were obtained from acetone. Transparent and homogeneous solutions of all other perovskites structures (figure 3) without formation of any precipitates were also formed in DMF.

$(RNH_3)_2PbX_4$ films deposited on SiO_2 glass substrates from 1.5 wt% DMF or acetone solution were homogeneous with high transparency in the visible region. These films were used for the characterization of the perovskites by absorption and PL spectroscopy, and by AFM measurements. The thickness of these films determined by profilometry and AFM measurements was between 30 and 50 nm.

$(RNH_3)_2PbX_4$ and $(RNH_3)_2(CH_3NH_3)_{m-1}Pb_mX_{m+3}$ films deposited on silicon substrates from 20 wt% DMF or acetone solution were homogeneous. These films were used for the characterization of the perovskites by ellipsometry. The thicknesses of the studied films were from 140 to 200 nm. The thickness of the films was determined by profilometry and ellipsometric measurements. $(RNH_3)_2PbX_4$, (X = Cl, Br) films were colorless. $(RNH_3)_2(CH_3NH_3)_{m-1}Pb_mX_{m+3}$, (for m = 1; X = I), and (for m = 2, X = Br) films were yellow, $(RNH_3)_2(CH_3NH_3)Pb_2I_7$ red, and $(RNH_3)_2(CH_3NH_3)_2Pb_3I_{10}$ dark red at room temperature.

III. Characterization of the Perovskite Thin Films

A. Optical Absorption and Photoluminescence

The absorption properties of the perovskite, and especially its absorption wavelength, can be tuned by changing the metal M, the ions X and A, or the organic part R (figure 2). Perovskite systems are thus very interesting: by tuning the absorption wavelength, the perovskite exciton energies can be shifted from the red to the blue range.

Absorption spectra were recorded on a Perkin-Elmer Lambda UV-VIS spectrophotometer. For the optical absorption measurements, perovskites were deposited on glass substrates. The photoluminescence spectra for the perovskites absorbing in the visible range were excited by 404 nm light from a 20 mW laser diode, and for the perovskites absorbing in UV range with a He-Cd laser at 325 nm. This light was strongly absorbed by each sample, ensuring that the observed luminescence came from the front side of the samples. The luminescence was detected by a CCD camera at room temperature. Each individual spectrum was accumulated over several seconds. For the optical absorption measurements, perovskites were deposited on glass substrates.

Figure 4 shows the optical absorption spectra of the perovskites thin films studied in this work measured at room temperature. For better presentation the spectra of the different perovskites are moved in respect to each other. The absorption coefficient ratios between different thin films are preserved.

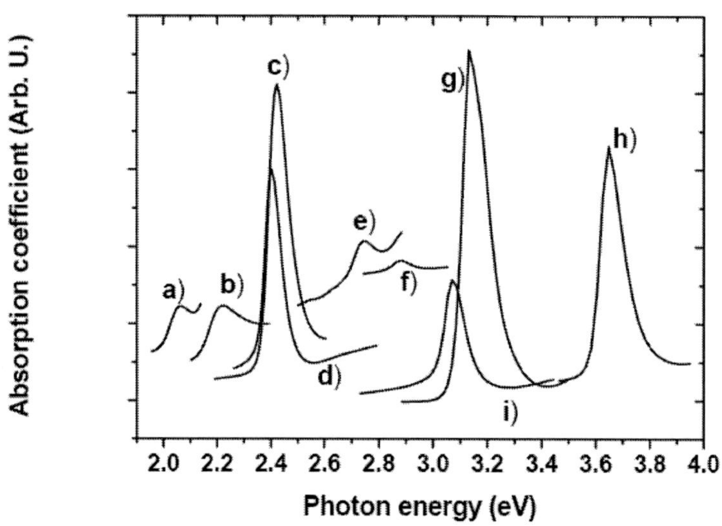

Figure 4. Optical absorption spectra of the perovskites thin films studied in this work measured at room temperature: a) $(C_6H_{13}NH_3)_2(CH_3NH_3)_2Pb_3I_{10}$; b) $(C_6H_{13}NH_3)_2(CH_3NH_3)Pb_2I_7$; c) $(C_6H_{13}NH_3)_2PbI_4$; d) $(C_6H_5C_2H_4NH_3)_2PbI_4$; e) $(C_6H_{13}NH_3)_2(CH_3NH_3)_2Pb_3Br_{10}$; f) $(C_6H_{13}NH_3)_2(CH_3NH_3)Pb_2Br_7$; g) $(C_6H_{13}NH_3)_2PbBr_4$; h) $(C_6H_5C_2H_4NH_3)_2PbCl_4$; i) $(C_6H_5C_2H_4NH_3)_2PbBr_4$.

The appearance of an intense exciton absorption bands denotes that a layered perovskite structure was formed in the spincoated films. It is seen from this figure that strong absorption bands were observed at 3.65 eV for the $(C_6H_5C_2H_4NH_3)_2PbCl_4$ films, between 2.74 and 3.12

eV for the $(RNH_3)_2(CH_3NH_3)_{m-1}Pb_mBr_{m+3}$ films, and between 2.06 and 2.41 eV for the $(RNH_3)_2(CH_3NH_3)_{m-1}Pb_mI_{m+3}$ films.

Regarding the peak positions, the exciton absorption band of lead based perovskites has been attributed to the excitons formed in the inorganic layers by the transition from 6s to 6p Pb^{2+} orbital [4, 20, 21]. In PbI_2, the top of the valence band is composed of the Pb (6s) orbital which is hybridized with the I (5p) orbital, and the bottom of the conduction band has mainly Pb (6s) character [22]. With respect to the optical transition in $[PbBr_6]^{4-}$ octahedra based compounds such as $(RNH_3)_2(CH_3NH_3)_{m-1}Pb_mBr_{m+3}$, the top of the valence band is composed of the Pb^{2+}(6s) orbital hybridized with the Br (4p) orbital, and the bottom of the conduction band has mainly Pb^{2+} (6p) character. The exciton is confined two dimensionally by the organic layers [4, 24]. In the case of chlorine lead based perovskites, similarly to the bromine based, the top of the valence band mainly has Pb (6s) character because the energy level of Cl (3p) orbital is relatively lower compared with that of the Pb (6s) orbital [23]. In the case of $(C_6H_{13}NH_3)_2PbI_4$ it should be noted that the deformation of the PbI^{2-}_4 layer also affected the degree of hybridization between Pb (6s) and I (5p) orbitals [23]. A blue shift in the exciton absorption band due to a possible phase transition has been reported in the case of $(C_{10}H_{21}NH_3)_2PbI_4$, this structural phase transition involves also the deformation of the PbI^{2-}_4 layer [3]. Shifts in the exciton absorption band (figure 4) of the perovskites due to the structural changes in the organic radical part R can in general be explained by the deformation of the PbX^{2-}_4 layer, as previously discussed. The shifts of the exciton absorption bands of the multilayered perovskites structures (figure 4) are attributed to the changes in the thickness of the quantum wells of these structures.

A theoretical model to deal with excitons in the perovskite compounds, based on the dielectric confinement effect, was proposed by Muljarov et al., [25] and the model has been further verified in details by means of two-photon and electroabsorption spectroscopies [26]. The electronic and excitonic structures of a wide range of lead halide perovskites have been extensively investigated [27-29]. By these efforts, it seems to be founded that the excitons in two-dimensional perovskite molecular structures are elucidated quite well as two-dimensional Wannier excitons, although their in-plane Bohr radii are no more than several times larger than the unit cell dimensions.

Absorption and photoluminescence properties of the studied perovskites are presented in table 1. Absorption, photoluminescence (PL), and the PL half-life stability of the films were measured at room temperature. Excitation energies of the incident light beam for the PL measurements are indicated above. The presented values correspond to the band maxima for the exciton absorption and PL spectra of the perovskites. The corresponding full width at half-maxima (FWHM) of the absorption and PL peaks and the Stokes shifts values are also presented. Stokes shift is the difference between positions of the band maxima of the absorption and luminescence spectra of the same electronic transition. PL stability (half-life) is measured as the time necessary for PL intensity to decrease to the half of its value compared with the value at the start of the measurement. Changes in the absorption and PL exciton peak position from the red to the blue range are observable when changing X from I to Br, and Cl. The decrease of the exciton absorption and PL energies in respect to the increase of the thickness of the perovskite sheets is due to the quantum confinement or dimensionality effects [30]. The FWHM of the absorption excitonic bands for the two dimensional perovskite structures $(RNH_3)_2PbX_4$ increase with the change of halogen from iodine to chlorine. The increase of the FWHM of the perovskite absorption peaks in respect to the thickness of the perovskite sheets is also observed. For $(RNH_3)_2(CH_3NH_3)_{m-1}Pb_mI_{m+3}$ thin films, the FWHM changes from 0.03 eV for m = 3 to 0.08 eV for m = 1, and for $(RNH_3)_2(CH_3NH_3)_{m-1}Pb_mBr_{m+3}$ thin films from 0.03 eV for m = 3 to 0.11 eV for m = 1. The effect of the organic layer character on the absorption and PL properties of the 2D perovskites structures is interesting. In the case of iodine based perovskite thin films, the aromatic $C_6H_5C_2H_4NH_3$ leads to larger absorption and PL FWHM, and higher absorption and PL exciton energies compared to the aliphatic $C_6H_{13}NH_3$ layer. On the contrary in the case of bromine based perovskite thin films, the aromatic $C_6H_5C_2H_4NH_3$ leads to smaller absorption FWHM and lower absorption exciton energies compared to the aliphatic $C_6H_{13}NH_3$ layer. This effect in respect to the structural changes in the organic radical part R can be attributed to the differences in the deformation of the $PbBr^{2-}_4$ and PbI^{2-}_4 layer, as discussed above.

Table 1. Absorption and photoluminescence properties of the studied perovskites. The given values are of the maximal absorption (λ_{max_Abs}) and photoluminescence (λ_{max_PL}) of the materials, and the corresponding full width half-maxima (FWHM). PL stability (half-life) is calculated as the time necessary for PL intensity to decrease to the half of its value compared with the value at the start of the measurement

Perovskite	λ_{max_Abs} (eV)	FWHM Absorption (eV)	λ_{max_PL} (eV)	FWHM PL (eV)	Stokes shift (eV)	PL stability half-life (min)
$(C_6H_{13}NH_3)_2(CH_3NH_3)_2Pb_3I_{10}$	2.05	0.03	2.03	0.08	0.02	14.21
$(C_6H_{13}NH_3)_2(CH_3NH_3)Pb_2I_7$	2.23	0.07	2.15	0.12	0.08	8.33
$(C_6H_{13}NH_3)_2PbI_4$	2.40	0.08	2.35	0.07	0.05	4.31
$(C_6H_5C_2H_4NH_3)_2PbI_4$	2.41	0.09	2.38	0.09	0.03	2.47
$(C_6H_{13}NH_3)_2(CH_3NH_3)_2Pb_3Br_{10}$	2.74	0.03	2.69	0.10	0.05	9.84
$(C_6H_{13}NH_3)_2(CH_3NH_3)Pb_2Br_7$	2.88	0.03	2.82	0.06	0.06	4.79
$(C_6H_5C_2H_4NH_3)_2PbBr_4$	3.06	0.09	n.a.	n.a.	n.a.	n.a.
$(C_6H_{13}NH_3)_2PbBr_4$	3.14	0.11	n.a.	n.a.	n.a.	n.a.
$(C_6H_5C_2H_4NH_3)_2PbCl_4$	3.65	0.11	3.62	0.03	0.02	n.a.

Characteristic properties of the PL of all of the investigated perovskites is the small Stokes shift between 20 and 80 meV, and the overlapping of the absorption and PL peaks (table 1). The absorption and corresponding PL spectra of $(C_6H_5C_2H_4NH_3)_2PbCl_4$ and $(C_6H_5C_2H_4NH_3)_2PbI_4$ are shown in figure 5. In general there are two types of PL from excitons in the ionic crystals such as PbX_4 [31-33]. The first is the sharp emission with no Stokes shift and the second is the broad PL with large Stokes shift. The former is called free-exciton emission and the latter corresponds to the self trapped excitons. Therefore we can conclude that in the case of $(RNH_3)_2PbX_4$ and $(RNH_3)_2(CH_3NH_3)_{m-1}Pb_mI_{m+3}$ (X = Cl, Br, I) perovskites thin films, the PL originates from the free exciton emission.

The PL photostability of $(RNH_3)_2PbX_4$ and $(RNH_3)_2(CH_3NH_3)_{m-1}Pb_mI_{m+3}$ (X = Br, I) was studied. From table 1 it can be seen that in general the iodine based perovskites are more stable to irradiation than the bromine based. We observed the decrease of the photostability with the decrease of the thickness of the perovskite sheets, and a higher stability of the perovskites with aliphatic organic part than those with aromatic organic part. The mechanism of the degradation and possible photooxidation of the above listed materials is a subject of ongoing studies in our laboratory. In general, the decrease of the PL intensity of the materials after irradiation can be attributed to the degradation of the layers in perovskite structures.

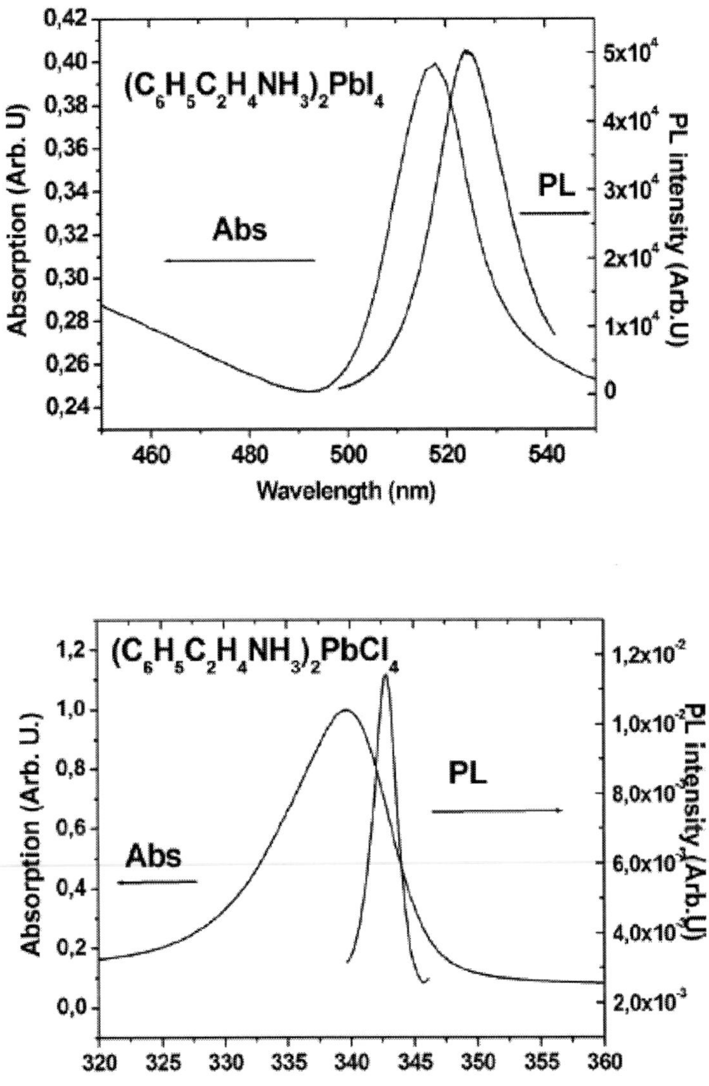

Figure 5. Absorption and PL spectra of $(C_6H_5C_2H_4NH_3)_2PbI_4$ and $(C_6H_5C_2H_4NH_3)_2PbCl_4$ perovskites.

B. Ellipsometric Measurements

From the results presented here, the potential of the halide based perovskites for application in the field of different hybrid organic-inorganic optical devices is clearly seen. A crucial point during the fabrication of such devices, as light emitting diodes and lasers, is the knowledge of the extinction coefficient *k* and the refractive index *n* of the materials. Ellipsometry data were acquired with a Sopra model GESP5 ellipsometer using UV-VIS-IR light sources. The incident angle was 75°, and the polarizer was set at 45°. For the ellipsometric measurements, perovskites were deposited on silicon substrates.

In table 2 are presented the results from the ellipsometric measurements of the studied perovskite thin films. The results were obtained by measuring the change in the polarization state of light upon reflection from the sample surface. For the ellipsometric measurements, perovskite thin films were deposited on Si substrate. The measurements were done in the range from 2 to 5 eV.

Table 2. Refractive index values at maximal absorption and 630 nm, extinction coefficient at maximal absorption, and optical energy bang gap of the studied perovskites

	n	k	n	Eg
Perovskite	λ_{max_Abs}	λ_{max_Abs}	630nm	(eV)
$(C_6H_{13}NH_3)_2(CH_3NH_3)Pb_2I_7$	1.851	0.073	1.889	2.134
$(C_6H_{13}NH_3)_2PbI_4$	2.117	1.337	2.115	2.376
$(C_6H_5C_2H_4NH_3)_2PbI_4$	2.459	1.690	2.298	2.376
$(C_6H_{13}NH_3)_2(CH_3NH_3)_2Pb_3Br_{10}$	2.334	0.275	1.608	2.672
$(C_6H_{13}NH_3)_2(CH_3NH_3)Pb_2Br_7$	2.088	0.490	1.558	2.812
$(C_6H_5C_2H_4NH_3)_2PbBr_4$	1.775	0.687	1.601	2.988
$(C_6H_{13}NH_3)_2PbBr_4$	1.816	1.345	1.965	3.100
$(C_6H_5C_2H_4NH_3)_2PbCl_4$	1.730	1.560	1.713	3.605

Optical constants were determined by fitting experimental data using the Cauchy's law and the least square method [34].

From table 2 we can see the relatively high refractive indexes between 1.7 and 2.5 of all perovskite structures at the maximum absorption wavelength with respect to most of the organic materials. High refractive indexes are characteristic for perovskite thin films even in the transparency region at 630 nm. The highest extinction coefficients were observed in the case of $(RNH_3)_2PbX_4$ (X = Cl, Br, I). The extinction coefficients of $(RNH_3)_2(CH_3NH_3)_{m-1}Pb_mI_{m+3}$ (X = Br, I) (m = 2, 3) perovskites thin films were about two times lower than those of the $(RNH_3)_2PbX_4$ thin films. These results were in good agreement with those obtained from the absorption measurements. They can be explained by the weaker quantum (size) confinement effect in the multilayered perovskite structures, which have normally double or triple monolayer well thickness [36].

In general, electronic states in covalently bonded inorganic semiconductors are significantly more delocalized than those of organic materials. Consequently, carrier screening is more efficient in covalently bonded inorganic semiconductors and they have larger dielectric constants. In turn, polarization energies are smaller and the exciton binding energy in a material such as GaAs is only a few meV. The minimum energy of formation of a separated free electron and hole, i.e. the transport gap, E_t is therefore very close to absorption edge, i.e. the optical gap, E_{opt} [35]. Thus, in the case of the perovskite hybrid organic-inorganic semiconductors (Wanier excitons [27-29]) the transport gap can be accurately calculated from the absorption spectrum. The optical band gap energy of each film in this work was deduced from the spectral dependence of the extinction coefficient k. The decrease of the band gaps of the perovskite thin films with structure $(RNH_3)_2(CH_3NH_3)_{m-1}Pb_mI_{m+3}$ (X = Cl, Br, I) (m = 1, 2, 3) in direction Cl : Br : I , and 1 : 2 : 3 is related to the decrease of the exciton energies, which has been discussed in details above.

C. Atomic Force Microscopy (AFM)

We studied the surface morphology of the investigated thin films with AFM. To determine the effect of the spincoating process on the surface morphology of the perovskite thin films, the surface profiles of the $(RNH_3)_2PbX_4$ (X = Cl, Br, I) perovskites with highest extinction coefficients were measured. The room-temperature AFM images of the various thin films were recorded on VEECO Explorer microscope. Non-contact mode tip was used for all measurements. The scan area was typically 2 x 2 micrometers [figure 6.].

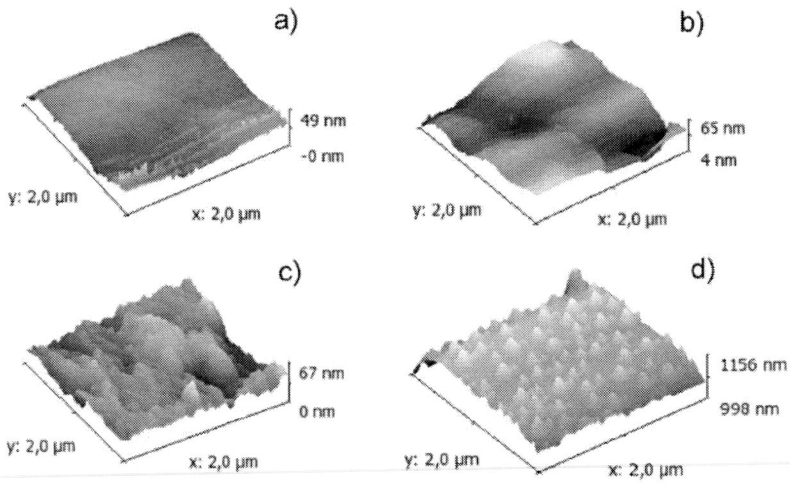

Figure 6. Surface topography of perovskite thin films: a) $(C_6H_5C_2H_4NH_3)_2PbI_4$; b) $(C_6H_{13}NH_3)_2PbI_4$; c) $(C_6H_5C_2H_4NH_3)_2PbBr_4$; d) $(C_6H_5C_2H_4NH_3)_2PbCl_4$.

The average surface roughness (Ra), of the thin films was calculated from the AFM surface topographic data in a scanning area of 2 µm x 2 µm by the equation:

$$Ra = \frac{1}{MN}\sum_{i=1}^{M}\sum_{j=1}^{N}\left|Z(x_i, y_j)\right| \tag{1}$$

where Z (x_i, y_j) denotes the height of a surface point (x_i, y_j) relative to the mean plane; MN is the number of points in the analyzed area. Figure 6 shows the measurement results. The surface roughness on the spincoated surfaces varied from around 8 nm for iodine and bromine based perovskites to 80 nm for the chlorine based perovskites. The obtained Ra values were similar to those obtained by profilometry. Despite the observable grain-like structures due to the polycrystalline structure of the films, studied thin-films had the advantage of being relatively flat, which can increase the opportunity for application of these materials in various optical and electronic devices.

D. Strong Exciton-Photon Coupling in a Microcavity Containing Perovskite Thin Films

The characterization of the emitters of light such as excitons inside the one-dimensional microcavities is very interesting. Just as the exciton energy is controlled by the quantum-well thickness, so can the absorption energy and the coupling coefficient be tuned by the microcavity design. These techniques may be employed to understand fundamental physics of photon-matter interaction, and may be applicable to energy-effective, high-speed light emitters, nonlinear optical switches, and polariton lasers [11, 38]. In this chapter we investigate the strong coupling of microcavities containing $(C_6H_5C_2H_4NH_3)_2PbI_4$ and $(C_6H_5C_2H_4NH_3)_2PbCl_4$ perovskite structures. The general structure of the Perot-Fabry microcavities used in this work is shown in figure 7. In the case of cavities containing lead iodide based perovskites, the bottom mirror was a commercial dielectric mirror (LayerTec corp.), centered at 2.4 eV at normal incidence, the reflectivity at 2.4 eV is 98%, the stop-band extends from 1.98 eV to 2.95 eV. For the lead chloride containing microcavities we used as a bottom mirror a dielectric mirror having a stop band centered at 3.49 eV at normal incidence. The dielectric mirrors in this case were prepared by thermal evaporation of four pairs of successive TiO_2 and SiO_2 layers on a fused silica substrate. The reflectivity at normal incidence at the centre of the stop band is 85%, the stop band extends from 3.1 eV to 4 eV.

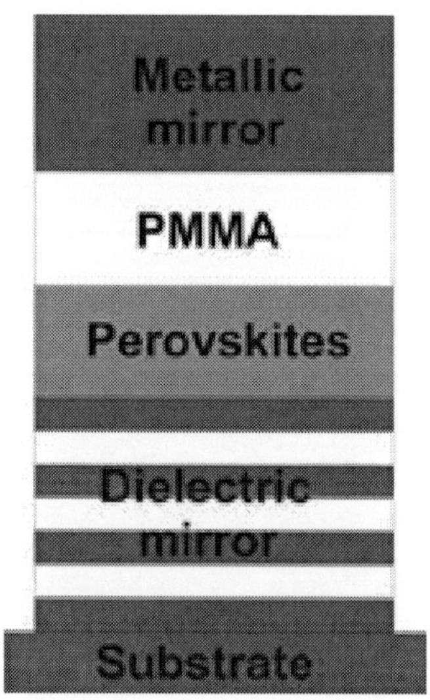

Figure 7. Scheme of the Fabry-Perot microcavity structure used in this study.

A perovskite thin film was deposited on top of the dielectric mirrors by spincoating a 1.5 wt% solution of $C_6H_5C_2H_4NH_3X$ and PbX_2 (X = Cl, I) dissolved in stoichiometric amounts in DMF. Then a polymethylmetacrylate (PMMA) layer acting as a spacer layer was spincoated in order to tune the cavity photon mode close to the exciton wavelength. The top mirror of the microcavity was then produced by thermal evaporation of silver or aluminum with thickness of 30 nm on the PMMA layer (for the lead iodide and lead chloride based microcavities, respectively). All layer thicknesses were determined by profilometry. We studied the strong coupling regime in the microcavity by varying the angle of incidence θ relative to the surface normal, which allows to tune the relative energy separation (and hence the degree of interaction) between the perovskite exciton energy (which is dispersionless, and so angle-independent) and the cavity photon mode energy. Angle-resolved reflectivity (R) measurements of the microcavities were performed using a Xenon lamp as the excitation source, between angles of incidence of 0° and 80°, at room temperature. Because of the high reflectivity of the dielectric mirror, transmission (T) measurements were not possible. A scheme of the reflectivity measurements setup is shown in figure 8. The cavity photon mode energy E_{ph} is related to θ by:

$$E_{ph}(\theta) = \frac{E_0}{\cos(\theta_{int})} \tag{2}$$

where E_0 is the photon mode energy at normal incidence ($\theta = 0^{o}$), and $\cos(\theta_{int}) = \sqrt{1 - \frac{\sin^2(\theta)}{n_{eff}^2}}$, where n_{eff} is the effective refractive index of the entire cavity [37]. In figure 9a and 10a are shown series of reflectivity spectra performed at room temperature of the microcavity containing a thin film of $(C_6H_5C_2H_2NH_3)_2PbI_4$ and $(C_6H_5C_2H_2NH_3)_2PbCl_4$, respectively. Two modes, whose energy position, intensity and linewidth are angle-dependant were observed in both figures. In the case of $(C_6H_5C_2H_2NH_3)_2PbI_4$ perovskites for $\theta = 10^{\circ}$, the sharp dip at 2.22 eV is identified as a photonlike mode, and the second weaker dip at 2.43 eV as an excitonlike mode (figure 9a.). Similar is the case of $(C_6H_5C_2H_2NH_3)_2PbCl_4$ perovskites: for $\theta = 10^{\circ}$ the sharp dip at 3.57 eV is identified as a photonlike mode, and the second weaker dip at 3.77 eV as an excitonlike mode (figure 10a.). The difference in the intensity of the modes at this angle can be attributed to their large difference in energy. At progressively larger viewing angles, the photonlike mode approaches the excitonic peak (excitons measured in absorption spectra at 2.41 eV and 3.65 eV, respectively), and the coupling between the exciton and photon modes increases, and mixing of the modes can be observed. Such mixed modes are called cavity polaritons. The minimum value of the energy separation between the two modes is at 40° for $(C_6H_5C_2H_2NH_3)_2PbI_4$, and 20° for $(C_6H_5C_2H_2NH_3)_2PbCl_4$, and at this point each of the modes contains approximately equal admixtures of a photon and an exciton. We say that in this case our system is at resonance. At viewing angles $\geq 50^{\circ}$ for $(C_6H_5C_2H_2NH_3)_2PbI_4$, and $\geq 30^{\circ}$ for $(C_6H_5C_2H_2NH_3)_2PbCl_4$, the higher-energy mode moves to higher energies. In figure 9a and 10a, two lines are used as a guide to the eye to identify both the upper (UB) and lower (LB) polariton branches. In the angular resolved reflectivity spectra (figure 9a) we observed, that for angles superior to 40°, the upper energy dip becomes the stronger one, while the lower energy dip intensity decreases. Above 70°, the intensity of the low energy transition is too low to be measured. In the case of $(C_6H_5C_2H_2NH_3)_2PbCl_4$ perovskites thin film containing microcavity, there is one feature in the angular dependent reflectivity spectra that is not commonly observed in experiments on inorganic and organic microcavities. At positive detuning (corresponding to viewing angles >30°) the upper branch does not increase significantly in visibility (figure 10a). This occurs because the DBR mirror has a limited reflectivity at ~4 eV. Therefore at large detunings, the DBR no longer acts as an effective mirror for the upper branch. The order of magnitude of the observed anti-crossing is quite coherent with polariton absorption results obtained by Ishihara et al [38] in a distributed feedback microcavity containing perovskites.

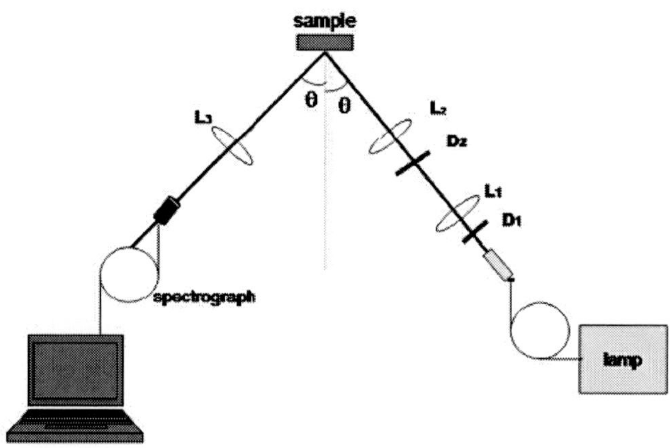

Figure 8. Angle resolved reflectivity measurements set-up.

The energies of the two minima observed in figure 9a and 10a, which correspond to the UB and the LB, are reported as a function of the angle of incidence θ in figure 9b and 10b, respectively. The experimental results are fitted to the dispersion using a standard two-level model [37] (solid lines in figure 9b and figure 10b):

$$E_{UB,LB} = \frac{E_{ph}(\theta) + Ex}{2} \pm \sqrt{V^2 + \frac{E_{ph}(\theta) - Ex}{4}} \quad (3)$$

This relation is the same as the relation obtained for two coupled oscillators with a coupling energy of V. V is a fit parameter, assumed to be constant at all angles. $E_{ph}(\theta)$ is the angular dispersion of the photon mode. The horizontal dash-dotted line (figure 9b and 10b at Ex = 2.41 eV and Ex = 3.65 eV, respectively) corresponds to the perovskite exciton transition energy measured by absorption spectroscopy and ellipsometry (see table 1 and 2). The very good agreement between the experimental and calculated results shows unambiguously that strong coupling leads to a Rabi splitting equal to 2V = 150 meV in the case of cavity containing $(C_6H_5C_2H_2NH_3)_2PbI_4$ thin film and 2V = 250 meV in the cavity with $(C_6H_5C_2H_2NH_3)_2PbCl_4$ thin film, respectively. The splitting is due to the Rabi splitting of the transition between the exciton and the molecule, attributed to the reversible changes in the structure of the film during the strong coupling [39].

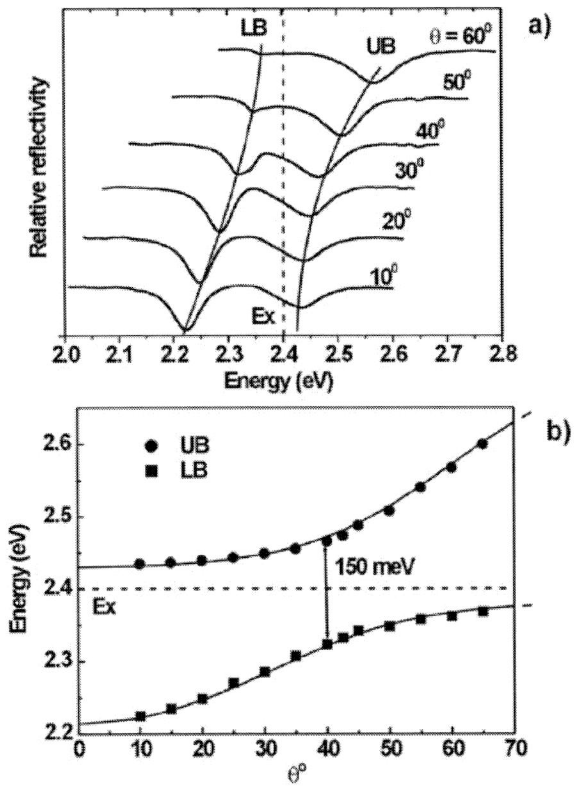

Figure 9. a) Reflectivity spectra from the cavity containing a $(C_6H_5C_2H_4NH_3)_2PbI_4$ thin film as a function of external viewing angle. The energy of the absorption exciton Ex is marked by a vertical dashed line. The solid lines through the data mark the upper branch and lower branch using UB, and LB, respectively are guide to the eye. b) Angular dependent energy dispersion of the two cavity polariton branches UB, and LB, respectively. The solid lines are best fits to the data (see text for details). The horizontal dashed line is the exciton absorption energy of $(C_6H_5C_2H_4NH_3)_2PbI_4$ perovskite within the cavity.

Far from the resonance, the LB branch has a strong cavity photon mode character. In the case of $(C_6H_5C_2H_2NH_3)_2PbI_4$: the cavity modes linewidth is around 100 meV, leading to a typical quality factor at 2.4 eV of about 25. From the absorption measurements of perovskite in table 1, the linewidth of the perovskite exciton is deduced to be 90 meV, despite these relative large linewidth it was possible to observe anti-crossing because of its high value of 150 meV. The high value of the anti-crossing is due to the increased interaction strength between perovskite excitons and photons. We evaluated the oscillator strength from the spectroscopic ellipsometry measurement of the complex refraction index of a 210 nm thick layer of $(C_6H_5C_2H_4NH_3)_2PbI_4$ deposited on a silicon substrate. The fit of this measurement with a Lorentzian oscillator provides an oscillator strength of $5x10^{15}$ cm^2 which can be compared to the oscillator strength of J-aggregates [16]. As our perovskite layer has a multiple QWs structure, it is possible to define an oscillator strength per QW: f_{QW}. Knowing the thickness of the PbI_4 layer : 6 Å; and the thickness of the organic chain : 10 Å; from the crystalline structure [44] , we evaluate $f_{QW} = 3.6x10^{13}$ cm^2 per QW. It is remarkable that this value is quite larger than in an InGaAs structure [45] : $4.2x10^{12}$ cm^2 per QW. In the case of $(C_6H_5C_2H_2NH_3)_2PbCl_4$ containing cavities, the cavity modes FWHM is around 70 meV, leading to a typical quality factor at 3.65 eV of about 50. The oscillator strength of the material was not evaluated because of the lack of crystallographic data. The stronger Rabi splitting observed in the cavities containing lead chloride based perovskites can be attributed to the higher quality factor of these microcavities in respect to the lead iodide containing ones.

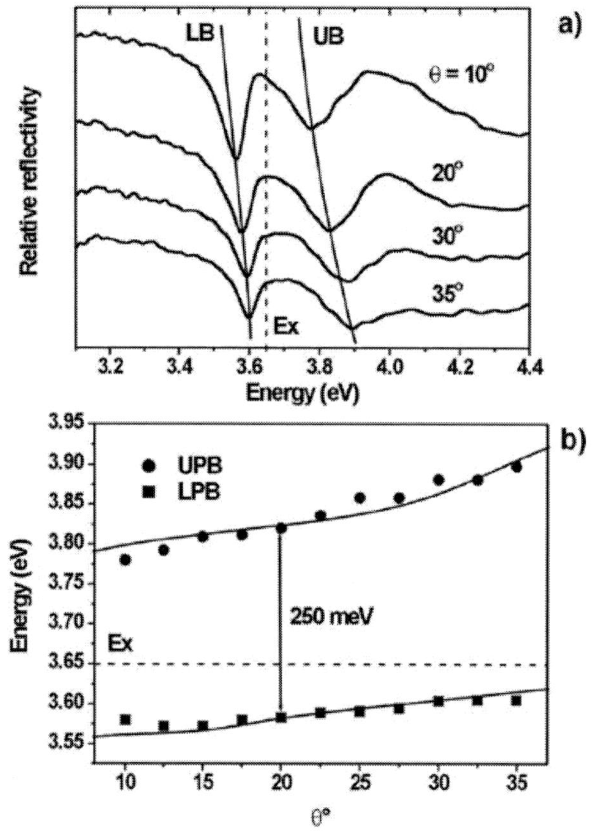

Figure 10. a) Reflectivity spectra from the cavity containing a $(C_6H_5C_2H_4NH_3)_2PbCl_4$ thin film as a function of external viewing angle. The energy of the absorption exciton Ex is marked by a vertical dashed line. The solid lines through the data mark the upper branch and lower branch using UB, and LB, respectively are guide to the eye. b) Angular dependent energy dispersion of the two cavity polariton branches UB, and LB, respectively. The solid lines are best fits to the data (see text for details). The horizontal dashed line is the exciton absorption energy of $(C_6H_5C_2H_4NH_3)_2PbCl_4$ perovskite within the cavity.

IV. Conclusion

The organic light emitting diodes (OLEDs) and organic field effect transistors (OFETs) technologies are fast approaching commercial applications. Organic materials are very attractive because of the possibility of low-cost, low-temperature processing and the range of interesting physical properties observed. Although providing wide range important opportunities in their own right, organic and inorganic materials have limitations imposed by the chemistry and physics underlying each materials class. The combination of the characteristics of organic and inorganic materials opens an interesting opportunity. It can offer an important means of acquiring more functionality within the materials used to create new electronic and optical systems. The organic–inorganic perovskite thin films are one important class of crystalline hybrid materials that have recently received substantial attention.

In this chapter we have investigated the optical properties of lead halide based perovskite thin films. They were chosen because of their high stability at ambient conditions. The inorganic type excitons in these hybrid materials provide the potential for high electrical mobility, better thermal stability, magnetic and dielectric transitions, and templating for the organic cations. A crucial point during the fabrication of devices, such as light emitting diodes and lasers is the knowledge of the extinction coefficient *k* and the refractive index *n* of the materials, which are presented in this chapter. A number of important applications of perovskites in the field of hybrid electronic devices have already been demonstrated. The field effect transistors with the field effect mobilities comparable to that of amorphous silicon have been produced by using perovskites semiconducting layer [6]. Light emitting diodes based on hybrid perovskites thin films have also been demonstrated. By combining the efficient fluorescence of organic molecules with the templating and electronic band structure tenability of metal halide sheets, relatively efficient organic-inorganic light emitting diodes have been built [4,5].

In this chapter we also demonstrated the strong coupling regime between the two-dimensional perovskite exciton and the confined photon modes in a Perot-Fabry microcavity at room temperature. It opens the way to more complex structures where the perovskite excitons can be coupled via the cavity photon [40] to different excitons. In particular, following the theoretical work of Agranovich et al [41], the possibility of coupling a perovskite exciton and an inorganic semiconductor exciton in the same Perot-Fabry microcavity will be explored in our laboratory. These experiments show the opportunity of using the perovskites in many applications as high-speed light emitters, nonlinear optical switches, and polariton lasers [11, 38].

Moreover, given the possibility of combining charge separation at an organic–inorganic interface with the high electrical mobility of an inorganic framework and the photosensitivity of an organic component, photoconductor and solar cell applications are worth considering [42, 43]. We have shown that it is possible to extend the field of optical applications of the perovskites to the UV range.

ACKNOWLEDGEMENTS

The authors thank Arnaud Brosseau for performing the atomic force microscopy measurements and Pierre Audebert for the fruitful discussions. This work is supported by the European Commission via the 5th Framework RTN 'HYTEC'(NoHPRN-CT-2002-00315).

REFERENCES

[1] Tsumura, A.; Koezuka, H.; Ando T. *Appl. Phys. Lett.* 1986, 49, 1210-1212.
[2] Tang, C. W.; VanSlyke, S. A.; Chen, C. H. *J. Appl. Phys.* 1989, 65, 3610-3616.

[3] Ishihara, T.; Takahashi, J.; Goto, T. *Phys. Rev. B* 1990, 42, 11099-11107.

[4] Chemla, D. S.; Miller, D. A. B.; Smith, P. W. Semiconductors and Semimetals, Academic Press, New York, 1987 chap. 5

[5] Ishihara, T.; Takahashi, J.; Goto, T. *Solid State Commun.* 1989, 69, 933-936.

[6] Kagan, C. R. ; Mitzi, D. B. ; Dimitrakopoulos C. D. *Science* 1999, 286, 945-947.

[7] Mitzi, D. B.; Feild, C. A.; Schleisinger, Z.; Laibowitz, R. B. *J. Solid State Chem.* 1995, 114, 159-163.

[8] Poglitsch, A.; Weber, D. *J. Chem. Phys.* 1987, 87, 6373-6378.

[9] Mitzi, D. B. *J. Chem. Soc., Dalton Trans.* 2001, 1-12.

[10] Shannon, R. D. *Acta Crystallogr., Sect. A* 1976, 32, 751-767.

[11] Bulovic, V.B. ; Khalfin, G. Gu. ; Burrows, P.E. ; Garbuizov, D.Z. ; Forrest, S.R. *Phys. Rev. B* 1998, 58, 3730-3740.

[12] Flechter, R.B.; Lidzey, D.G.; Bradley, D.D.C.; Bernius, M.; Walker, S. *Appl. Phys. Lett.* 2000, 77, 1262-1264.

[13] Weisbuch, C.; Nishioka, M.; Ashikawa, A.; Arakawa, Y. *Phys. Rev. Lett.* 1992, 69, 3314-3317.

[14] Stevenson, R.M.; Astratov, V.N.; Skolnick, M.S.; Whittaker, D.M.; Emam-Isamail, M; Tartakovskii, A.I.; Savvidis, P.G.; Baumberg, J.J; Roberts, J.S. *Phys. Rev. Lett.* 2000, 85, 3680-3683.

[15] Saba, M.; Ciuti, C.; Bloch, J.; Thierry-Mieg, V.; Andre, R.; Dang, L. S.; Kundermann, S.; Mura, A.; Bongiovanni, G.; Staehli, J. L.; Deveaud, B. *Nature* 2001, 414, 731-735.

[16] Lidzey, D. G.; Bradley, D. D. C.; Skolnick, M. S.; Virgili, T.; Walker S.; Whittaker, D.M. *Nature* 1998, 395, 53-55.

[17] Hobson, P. A.; Barnes, W. L.; Lidzey, D. G.; Gehring, G .A.; Whittaker, D. M.; Skolnick M. S.; Walker, S. *Appl. Phys. Lett.* 2002, 81, 3519-3521.

[18] Tabuchi, Y.; Asai, K.; Rikukawa, M.; Sanui, K.; Ishigure, K. *J. Phys. Chem. Solids* 2000, 61, 837-845.

[19] Tanaka, K.; Kondo, T. *Sci. Techn. Adv. Mater.* 2003, 4, 599-604.

[20] Doni, E; Grosso, G; Harbeke, G; Meier, E; Tosstti, E. *Phys. Status Solidi* 1975, B68, 569-574.

[21] Fujita, M ; Nakagawa, H ; Fukui, K ; Matsumoto, H ; Miyanaga, T ; Watanabe, M. *J. Phys. Soc. Jpn.* 1991, 60, 4393-4394.

[22] Schluter, I. Ch ; Schluter, M. *Phys. Rev. B* 1972, 9, 1652-1663.

[23] Kitazawa, N. *Jpn. J. Appl. Phys.* 1996, 35, 6202-6207.

[24] Hirasawa, M; Ishihara, T.; Goto, T. *J. Phys. Soc. Jpn.* 1994, 63 , 3870-3879.

[25] Muljarov, E. A.; Tikhodeev, S. G.; Gippius, N. A.; Ishihara, T. *Phys. Rew. B* 1995, 51, 14370-14378.

[26] Tanaka, K.; Sano, F.; Takahashi, T.; Kondo, T.; Ito, R.; Ema, K. *Solid State Commun.* 2002, 122, 249-252.

[27] Koutselast, I. B.; Ducasse, L.; Papavassiliou, G. C. *J. Phys. Condens. Matter.* 1996, 8,1217-1227.

[28] Tanaka, K.; Takahashi, T.; Kondo, T.; Umeda, K.; Ema, K.; Umebayashi, T.; Asai, K.; Uchida, K.; Miura, N. *Jpn. J. Appl. Phys.* 2005, 44, 5923-5932.

[29] Shimizu, M.; Fujisawa, J.-I.; Ishi-Hayase, J. *Phys. Rew. B* 2005, 71, 205306-1-205306-9.

[30] Ishihara, T. *J. Luminesc.* 1994, 60, 269-274.

[31] Kanzaki, H.; Sakuragi, S.; Sakamoto, K. *Solid State Commun.* 1971, 9, 999-1002.

[32] Takahei, K.; Kobayashi, K. *J. Phys. Soc. Jpn.* 1978, 44, 1850-1860.

[33] Shinozuka, Y.; Toyozawa, Y. *J. Phys. Soc. Jpn.* 1979, 46, 505-514.

[34] Press, W. H.; Flannery, B. P.; Teukolski, S. A.; Vetterling, W. T. *Numerical Recipes*, 523-526, Cambridge, 1986.

[35] Hill, I. G.; Kahn, A.; Soos, Z. G; Pascal, R. A. *Chem. Phys. Lett.* 2000, 327, 181-188.

[36] Hong, X.; Ishihara, T.; Nurmikko, A. V. Phys. Rev. B 1992, 45, 6961-6964.

[37] Lidzey, D. G.; Fox, A. M.; Rahn, M. D.; Skolnik, M. S.; Agranovich, V. M.; Walker, S. *Phys. Rev. B* 2002, 65, 195312-1-195312-10.

[38] Fujita, T; Sato, Y; Kuitani, T; Ishihara, T. *Phys. Rev. B* 1998, 57, 12428-12434.

[39] Shimizu, M. *Phys. Rev. B* 2005, 71, 033316-1-033316-4.

[40] Lidzey, D. G.; Bradley, D. D. C; Armitage, A.; Walker, S.; Skolnick M.S. *Science* 2000, 288, 1620-1623.

[41] Agranovich, V.; Benisty, H.; Weisbuch C. *Solid State Comm.* 1997, 102, 631-636.

[42] Takada, J.; Awaji, H.; Koshioka, M.; Nakajima, A.; Nevin, W. A. *Appl. Phys. Lett.* 1992, 61, 2184-2186.

[43] Takada, J. *Jpn. J. Appl. Phys.* 1995, 34, 3864-3870.

[44] Calabrese, J.; Jones, N. L.; Harlow, R. L.; Herron, N.; Thorn D. L.; Wang, Y. *J. Am. Chem.Soc.* 1991, 113, 2328-2330.

[45] Skolnick, M.S.; Fisher, T.A.; Whittaker, D.M. *Semicond. Sci. Technol.* 1998, 13, 645-669.

In: Progress in Advanced Materials Research
Editor: Nicolas H. Voler, pp. 139-159
ISBN: 978-1-60021-893-4

Chapter 5

FABRICATION AND APPLICATION OF SURFACE RELIEF DUE TO MASS MIGRATION

Takashi Ubukata
Department of Advanced Materials Chemistry,
Graduate School of Engineering, Yokohama National University

ABSTRACT

Photochromic compounds are very interesting and potentially useful for optical data storage, optical switching, optical filter, and so on. In recent years, unique photoinduced surface relief formation caused by lateral material migration has been reported and has attracted great attention. Large amplitude surface relief structures can be inscribed on thin films containing the azobenzene moiety upon irradiation with patterned light. This article focuses to selected research topics on possible applications proposed in our group using reversible dynamic lateral motion in the formation and erasure process of surface relief structure. They involve i) new methodology for optical patterning and alignment of various light-inert materials, ii) organic distributed feedback laser with oscillation wavelength programmability. In addition to these topics on surface relief structures formed on azobenzene-containing materials, new class of surface relief formation using a photochromic compound other than the azobenzene derivatives is presented. Promotions of these research areas are anticipated to provide wide and new opportunities for material processing and creation of new class of photonic devices.

INTRODUCTION

1. Photoinduced Surface Relief

In 1995, a new phenomenon was reported by two research groups, Natansohn and Rochon group and Tripathy and Kumar group, independently.[1, 2] When thin films of azobenzene-containing polymers were exposed to an interference of coherent light, the surfaces of the films were changed to regular topographical undulation with the same period as its interference pattern due to material migration (figure 1). The structure was named as "surface relief grating, SRG". The shape of SRG was stable at temperature below glass transition temperature (Tg) of polymers, but could be erased by heating the polymers above their Tgs and returned to initial flat films. Erasure of SRG could also be attained by irradiating a single beam of laser light with the appropriate condition.[3] Since then, formation of SRG has been one of the most attracting targets in the current research of photoinduced motion of azobenzene-containing polymer films.

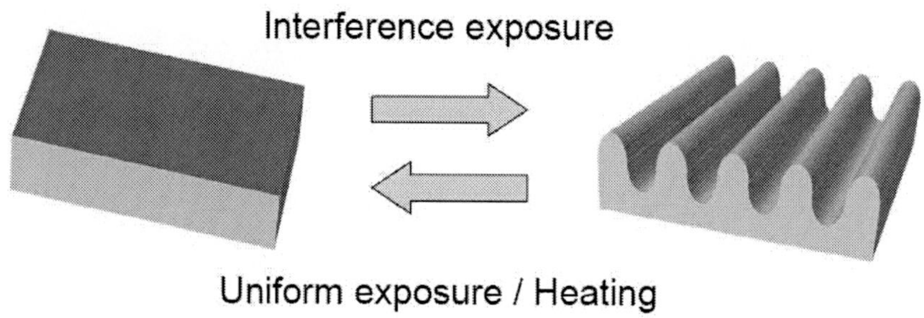

Figure 1. Schematic illustration of photoinduced surface relief gratings.

2. Optical Conditions to Generate SRG

Most SRGs are frequently inscribed on films of azobenzene-containing polymers by exposure to interference of argon ion continuous laser beams with 488 or 514 nm. The power of light required for SRG formation is not so large. SRGs have usually been formed with laser power of a few to several hundred mW cm^{-2}. Sinusoidal undulations are often constructed using a Lloyd interferometer (figure 2).[4, 5] A collimated argon ion laser beam is split into two: one is directly incident onto the film and the other is reflected by a mirror rigidly held perpendicular to the film. By simply rotating the film held to the mirror, a wide-range grating period (from sub to several micrometers) can be obtained according to

$$\Lambda = \lambda_w / 2\sin\theta \quad (1)$$

where λ_w is wavelength of the writing beams, θ the incidence angle of the writing beams onto the film, and Λ the period of SRG. The formation process of the SRG is monitored from the change in the first-order diffraction intensity of a helium-neon (He-Ne) laser beam on the transmission side using a photodetector. The height of the inscriptions can be actually controlled by demanding conditions, such as the exposure energy and polarization combinations of writing beams. Inscribed feature can be on the order of initial thickness of the film, up to 1 μm.

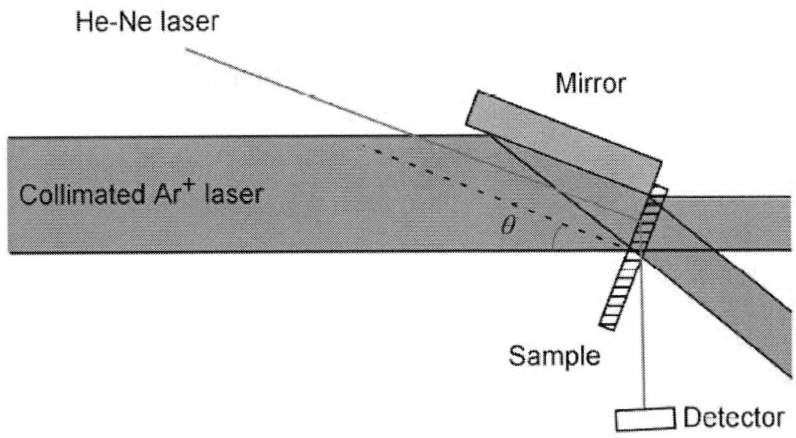

Figure 2. Experimental setup for the fabricating the surface relief gratings (Lloyd interferometer).

SRG inscription experiments using interfering laser beams with different polarizations demonstrated that spatial variations of both magnitude and alternation of resultant electric field vector are an essential condition to induce surface modulations. [6] Further detailed investigation of polarization dependence revealed that efficient SRG formation requires photoinduced anisotropy and a component of optical field gradient in the direction of the grating vector.[7] This result strongly supported the field gradient force model, which is one of the proposed models to explain the driving force of SRG formation.[8]

Surface relief structures could also be fabricated on azobenzene-containing films by various patterns of light irradiation, such as diffraction of an edge,[1, 9] a focused single Gaussian laser beam,[10, 11] exposure thorough a transmission mask,[12, 13] and near-field illumination.[14-16] Resultant surface relief structures were closely correlated with intensity distributions, as SRGs were formed by interference exposure. Therefore, the above various topological surface undulations are sometimes called "Photoinduced surface relief, PSR" in recent years.[16] Most of the cases exhibited that relief are π-shifted to the light distribution, with the light intensity maxima corresponding to surface relief minima, that is, material migration from high-light intensity to low intensity regions. In any cases, strong polarization-dependence was admitted, indicating that PSR was formed by photonic process, not due to the thermal effects.

3. Material Properties to Form SRG

The isomerization of azobenzene is a key to the formation of SRGs. Typically, the absorption of an ultra-violet (UV) light transforms an elongated rod-like form (*trans* isomer) into a bent form (*cis* isomer), and backward transformation of *cis* isomers into *trans* ones is induced by visible light (figure 3). As the *trans* isomer of azobenzene is thermodynamically more stable than the *cis* isomer, thermal backward isomerization also occurs. Rau classified azobenzenes into three types; azobenzene, aminoazobenzene, and pseudo-stilbene.[17] Most experiments have been performed using azobenzenes of pseudo-stilbene type with substitution of electron donor-electron acceptor groups in the para ring positions though it is not a critical factor for formation of SRG.[18] Since the π–π* band of *trans* isomer of this type of azobenzene is shifted to longer wavelengths, both the *trans* and *cis* isomers can be effectively photoexcited with the single wavelength around 500 nm.

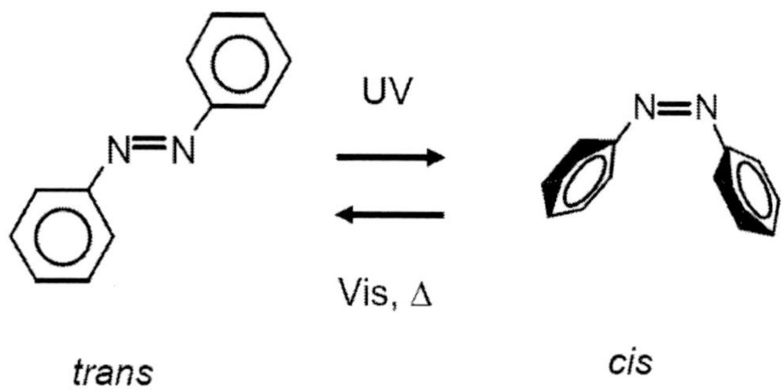

Figure 3. Photoisomerization of azobenzene.

A large variety of azobenzene pendant amorphous polymers with different main chain have been extensively investigated since first two reports using acrylate-based[1] and epoxy-based[2] polymers.[19, 20] Though most of the SRGs were prepared on side chain modified polymer films, main chain modified polymer, where the azobenzene are assembled in the polymer backbones, were also investigated to afford thermal stability. The SRG formation due to large-scale mass transport occurred readily at room temperature in amorphous polyurea with azobenzene groups in the main chain with high Tg.[21] Erasure could be achieved by heating above its Tg or by exposure to a single beam in the manner similar to side-chain azobenzene polymers. As to ultimate thermal stability, the SRG formed on azocarbazole-based polyimide film showed no deformation after heating 1 h at 240 °C.[22]

Materials for inscription of SRG are not limited to amorphous polymers. Ramanujam and Hvilsted group have made efforts to study the SRG formation in liquid crystalline polymers with low Tgs around room temperature, and clearly demonstrated the different characteristics between amorphous and liquid crystalline materials. A series of polyesters having azobenzene with different substitutes through an alkyl spacer between the polymer backbone and the azobenzene chromophore have been examined. Focusing on the influence of their Tgs, efficient SRG formation was observed in a narrow temperature range around Tg.[23] The most efficient migration occurred exactly at or slightly above Tg. Ikeda and the coworkers have also shown similar results, that SRG formation was associated with photochemical phase transition.[24] Experiments using transmission mask as well as single focused beam demonstrated that the direction of material migration is dependent on substituents and backbones. [25] A film of amorphous polyester E1(CN)P, which has the same azobenzene chromophore as P6(CN)12 but a shorter spacer and an aromatic main chain, was irradiated through a transmission mask for comparison. This material moved out of illuminated areas, forming valleys, in contrast to liquid crystalline polymer P6(CN)12. In an attempt to explain this discrepancy, Helgert and the coworkers have conducted an experiment in which they illuminated linearly polarized light onto E1(CN)P and P6(CN)12 films floating on water. Thin films of these polymers underwent opposite changes of shape under the influence of polarized light, indicating that a liquid crystalline polymers behaved differently from the amorphous ones with response to the polarized light.[26]

Apart form the grafted system in which azobenzene chromophore is covalently attached to the polymer chain, attempts to create SRGs with guest-host systems in which low molecular weight azobenzene is doped in polymer matrix were also conducted. [13, 27] However, only a much weaker undulation (about 20 nm) was measured in the guest-host systems, indicating that azobenzene grafting to the polymer chain should be an important factor for effective SRG formation. In some copolymer cases, ability of SRG formation is dependent on the azobenzene concentration, showing sigmoidal curve with threshold.[28] Considering this result, rather weak ability of SRG formation in guest-host systems may be ascribed to the low azobenzene concentration in homogeneous films to prevent azobenzene-containing polymer segregation. More recently, SRG formation has been explored in different classes of azobenzene films, such as amorphous molecular materials [29] and single crystal.[30]

4. Photo-Triggered Surface Relief

Seki group introduced a novel approach combining pre-irradiation with UV light and liquid crystalline property (figure 4). A markedly rapid material migration took place in a liquid crystalline azobenzene polymer hybridized with low molecular mass liquid crystal molecules (figure 4a).[31, 32] The total energy required for the full SRG inscriptions was reduced by a level of approximately three orders of magnitude, resulting in the reduction of the exposure time to as short as only a few seconds. The irradiation with UV light made the film a *cis*-rich state, which resulted in the photochemical phase transition of the liquid crystalline phase into isotropic phase.[33] Thus, this highly sensitive SRG formation should have been triggered by the spatially patterned *cis*-to-*trans* back photoisomerization during a photochemical phase transition from the isotropic liquid to liquid crystalline phase. In this context, we sometimes call this highly sensitive relief formation as "Photo-triggered surface relief" rather than "Photoinduced surface relief".[34] Similar sensitivity enhancement was achieved in a single component liquid crystalline azobenzene copolymer, possessing a flexible oligo(ethylene oxide) group in the side chain as a plasticizer (figure 4b).[35, 36] In this case, stability of resultant SRG could be improved by the chemical cross-linking by a mixed vapor of hydrogen chloride and formaldehyde. The dependence of SRG formation on the polarization characteristics in these highly sensitive systems showed marked contrast with those widely observed amorphous azobenzene polymer films.[37]

Since the discovery of SRG, the application of the SRG is often limited to attaining static functions based on the surface shape, such as holographic optical recording, waveguide formation, and liquid crystal alignment perhaps due to the requirement of high light doses. On the other hand, we have recently discovered the highly sensitive SRG formation system as described in this section. Then our effort has been devoted to dynamic functions in the process of SRG formation.

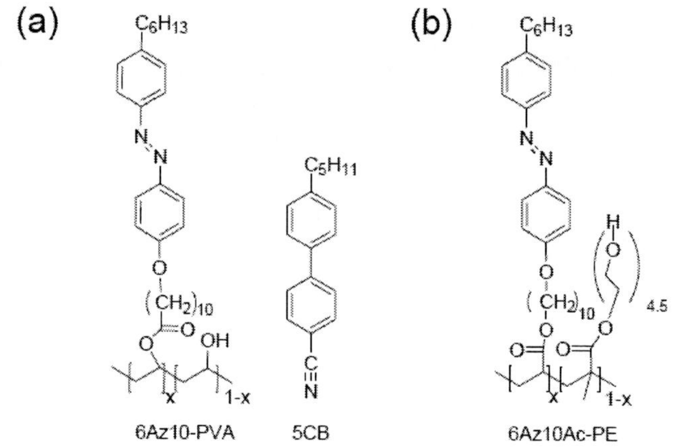

Figure 4. Chemical structures used in photo-triggered surface relief. a) Binary system. b) Single system.

Results

1. New Methodology for Optical Patterning and Alignment

We report herein the anomalous self-transport motions in this material system, and further propose a new optical patterning and manipulation procedure for "light-inert" materials utilizing these motions. Schematic illustrations of the process are presented in figure 5. This is of great technological advantage for desired manipulations without using a complex set-up and expensive optical components.

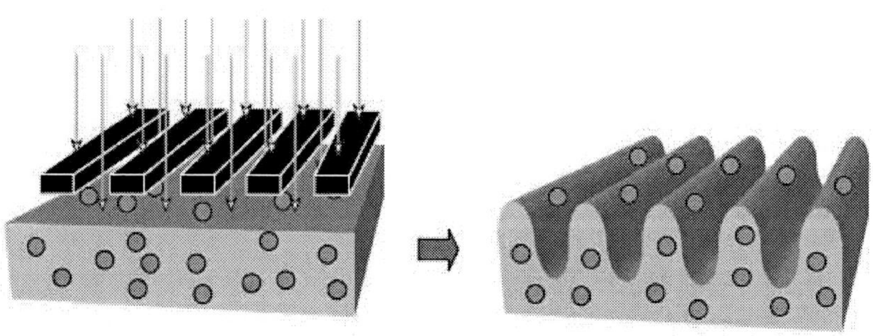

Figure 5. Schematic illustration of conveyer system.

1.1.DICHROIC DYE

As a guest molecule, we firstly selected dichroic dye (figure 6a).[38] A blue rodlike diaminoanthraquinone-type dye (LCD118) was further incorporated into the hybrid film of azobenzene polymer (6Az10-PVA) and liquid crystal molecule (5CB) (figure 4a). The ternary hybrid film (ca. 150 nm thickness) was prepared from a chloroform solution dissolving 6Az10-PVA, 5CB and LCD118 (47/41/12 wt%). Freshly prepared films were transparent and showed no in-plane anisotropy, as confirmed by polarized transmission microscopy and polarized UV-vis spectroscopy. The entire film was first exposed to 365 nm light from a high-pressure Hg lamp and then patterned irradiation was performed with argon ion laser (488 nm). As shown in the right handed picture of figure 6b, the irradiated and shaded areas were observed as bright and dark parts, respectively. The difference was clearly indicated by microscopic absorption spectra observed in a 10-μm-diameter area (figure 6c). The deeply colored region corresponds to the protruding parts shown in the height profile obtained by surface profilometry (figure 6d). Thus, the hybrid material migrated from the irradiated regions to the shaded ones. It is stressed that the material is almost perfectly expelled from the irradiated area, as judged from the flat profile, with the substrate surface most probably being flat. In fact, after the hybrid film was scratched with an atomic force microscope (AFM) tip, the cross-sectional height profile clearly indicated the complete removal of the film material from the exposed regions.

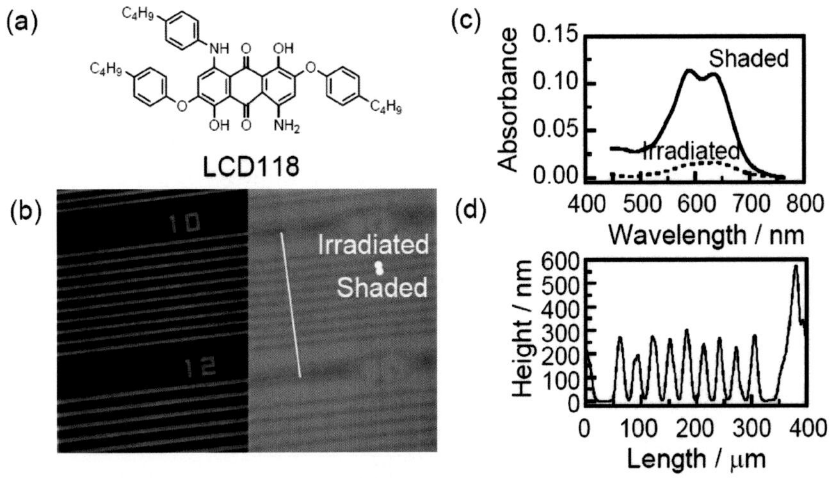

Figure 6. Photo-triggered mass migration in the hybrid film visualized by dichroic dye LCD118. a) Chemical structure of LCD118. b) Optical-microscope images of a photomask (left) and the hybrid film (right) after irradiation with an argon ion laser through the photomask. c) Microscopic UV-vis absorption spectra in the shaded (solid line) and irradiated (dashed line) regions shown as white spots in (b). d) Cross-sectional height profile along the white line in (b).

When the dye content was increased, the UV irradiation followed by dark adaptation at room temperature for a day led to the growth of dendritic crystals of the dye (figure 7a). Most probably, the rodlike dye molecules showed self-aggregation in the fluid film matrix due to the *cis*-rich state of 6Az10-PVA. When the crystallization was performed after photoinscription of the lined relief (18-μm width in this case), the direction of dye organization was restrained and aligned along the relief structure (figure 7b). After sufficient organization at room temperature, the protruding line showed high optical anisotropy. The polarized optical-microscope image revealed that the dichroic dye was aligned in the direction of the SRG line produced by photoirradiation through the mask. Figure 7c represents microscopic polarized absorption spectra of the self-assembled projection. The absorption moment of the dichroic dye was highly aligned parallel to the direction of the inscribed line, and the dichroic ratio at the maximum absorption wavelength (588 nm) in this example was as high as 6.6.

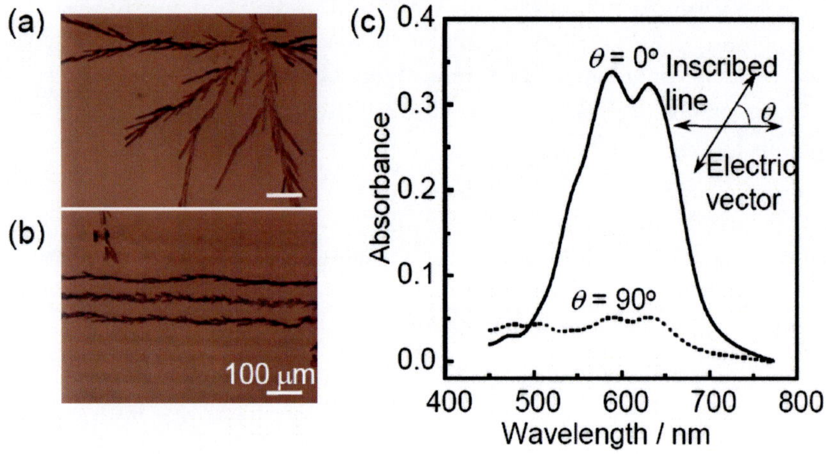

Figure 7. Patterned organization of LCD118 using the material conveyance system. a and b) Photographs taken at areas without irradiation and with photofabricated surface relief, respectively. c) Microscopic polarized absorption spectra of the crystallized spot (10-μm diameter). The electric vectors of probing light are parallel (solid line) and perpendicular (dashed line) to the inscribed line.

Without surface relief inscription, the dendrite grew in random directions in the flat surface of the hybrid film (figure 7a), however, the long axis of the dye was oriented parallel to the direction of crystal growth. Most probably, the orientational anisotropy is induced by the uniaxial growth in the diffusion-limited aggregation process, since the dye molecules should diffuse preferentially along the molecular long axis. To date, the photoalignment of molecules has been achieved by modifying the light irradiation using linearly polarized light or oblique incidence.[39] In the present system, the uniaxial alignment is achieved by anisotropic molecular diffusion that is restricted in the light-inscribed geometry.

1.2.π-CONJUGATED POLYMER

Imaging of π-conjugated polymers is considered as essential technology because these polymers are promising as active materials in many application areas including microelectronics and photonics. In this section, we focus on the possibility for two-dimensional patterning of π-conjugated polymers using a photo-triggered mass transfer system.[40]

A soft liquid crystalline polyacrylate having azobenzene and oligo(ethylene oxide) side chains (6Az10Ac-PE4.5) was used as the host polymer (figure 4b).[35] We adopted a π-conjugated polymer, poly(3-dodecylthiphene) (PDT) as the guest polymer (figure 8a). The weight averaged molecular mass of the synthesized PDT (random head-tail sequence) was 8.2 x 10^4 as determined by gel permeation chromatography with polystyrene standard calibration. Binary polymer blend films (ca. 30 nm thickness) were prepared by spin coating method from a toluene solution dissolving 6Az10AcPE4.5/PDT (99/1 wt%). Discernible phase separation was not observed as revealed by optical microscopy.

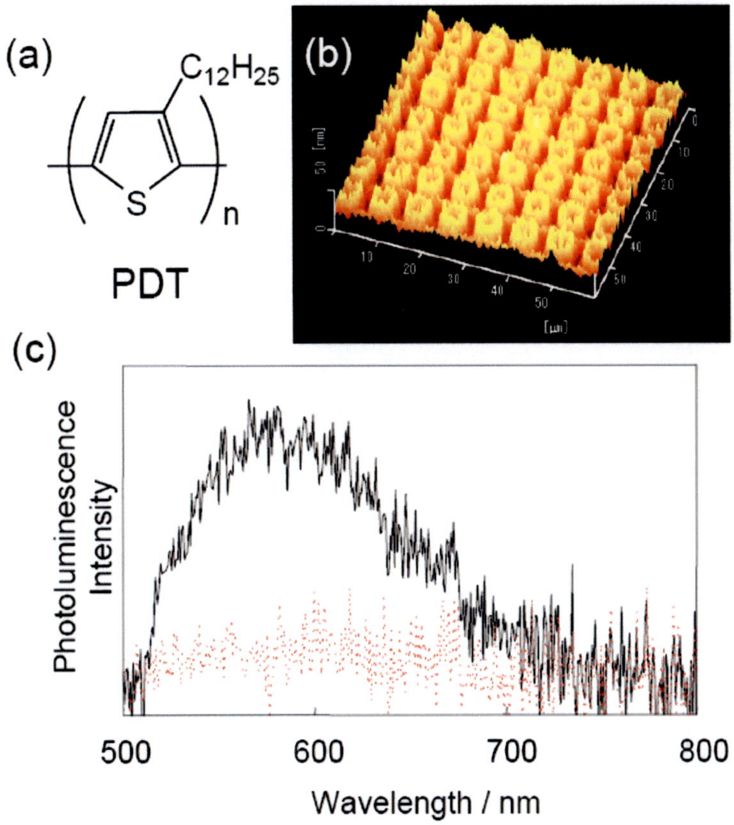

Figure 8. Two-dimensional patterning of PDT using the material conveyance system. a) Chemical structure of PDT. b) AFM image of the resulting inscription after irradiation with 436 nm light from a mercury lamp (4 mW cm^{-2} for 30 s). c) Microscopic fluorescence spectra in the shaded (solid line) and illuminated (dashed line) regions.

The photo-triggered mass migration of PDT was achieved by patterned incoherent blue light from a mercury lamp using a grid photomask. An inscribed relief structure resulting from the irradiation through a photomask is shown in figure 8b. The tetragonal features of the surface topology evidently indicated that the mass migration of azobenzene polymer occurred from the bright regions to the dark ones within the plane. Moreover, the height difference from valley to top of the tetragonal structure was excess of or equal to the initial flat film. This result suggests that a large modulation of light intensity produced by the photomask leads to the perfect removal of materials from the bright regions.

It was possible to detect the two dimensional distribution of PDT in the SRG film by microscopic fluorescence spectroscopy. As shown in figure 8c, fluorescence intensity at higher parts of relief structure was ca. 5 times stronger than that at lower parts. As shown from above results, PDT should be transferred together with the migration of azobenzene polymer; in other words, PDT was "conveyed" by the azobenzene polymer.

1.3. SEMICONDUCTOR NANOCRYSTALS

Application of the mass conveying principle is not limited to organic materials. We herein apply semiconductor nanocrystals (NCs) known as quantum dots to this methodology.[38] We adopted small cadmium selenide (CdSe) clusters whose surface is surrounded by a layer of organic ligands (figure 9a). Due to surrounded layer, the CdSe NCs readily dissolves in common organic solvents. The averaged size of the NCs was 6 nm in diameter and the averaged NCs mass was 6 x 10^5. They were incorporated into the hybrid film for the experimental advantage that NCs migration could be readily visualized by fluorescence microscopy because of their light emitting properties.

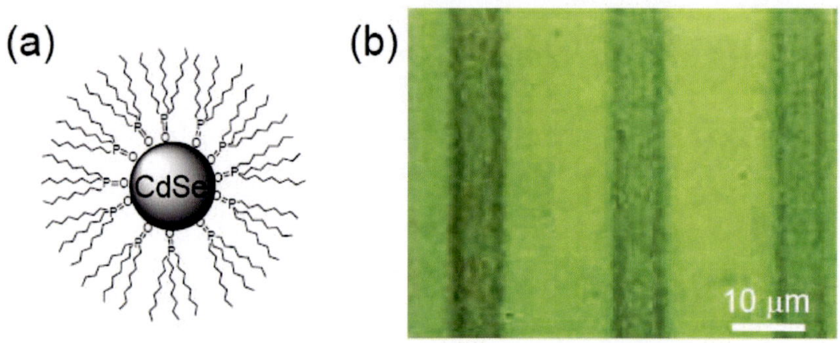

Figure 9. Patterned organization of CdSe NCs using the material conveyance system. a) Schematic illustration of CdSe NCs. b) Fluorescence-microscope image of CdSe NCs -containing (5 % by weight) hybrid film after photoinscription of surface relief structure with a photomask.

Spin-cast films were prepared from mixed solution composed of 6Az10-PVA, 5CB, and CdSe NCs (50/45/5 wt%). After a short light exposure of 436 nm line from a mercury lamp through a grating photomask, surface relief undulations were clearly formed. Observation of luminescence from NCs by fluorescence microscopy (figure 9b) revealed that NCs were transported following the patterned irradiation. These results indicate that the photomigrating film carried the CdSe NCs.

We newly developed a patterning technique of functional materials using the instant mass migration system. In this method, the processing steps can be reduced with less complexity. According to designed light intensity patterning, functional materials can be conveyed in any two-dimensional directions together with the lateral motion of azobenzene polymer. Our results reveal important aspects of material migration assisted by the azobenzene polymer film. Once the materials are compatible with the polymer film, target materials from molecules to NCs can be manipulated. We emphasize that the photo-triggered conveyance system is not limited to the three examples shown here, but should be widely applicable to functional materials that are themselves inert to light. The present photo-triggered hybrid system is effective at a low light exposure dose, therefore the target materials are hardly damaged by light. The instant mass migration technique would be useful method in the two-dimensional patterning of a great number of functional materials.

2. Organic Distributed Feedback Laser with Oscillation Wavelength Programmability

The diffraction of light by the periodic structure of SRGs can be applied to various optical elements such as a polarization discriminator, a light coupler, and a narrow-band waveguide filter.[19, 20] SRGs can also be employed as a waveguide distributed feedback (DFB) laser, where optical feedback allowing laser oscillation is provided by diffraction on a permanent index grating due to the periodic modulation based on the SRGs.[41, 42] However, SRG was used just as a template of the grating in these reports. In this section, we demonstrate organic DFB lasers with wavelength programmability, i.e., erase-and-rewritability, utilizing the advantages of SRG.[43, 44]

In the design of organic DFB waveguide lasers, single lateral-mode propagation, sufficient optical gain, and Bragg diffraction are necessary in obtaining a single-peak laser output spectrum. We firstly attempted to realize a DFB laser using a dye-doped azobenzene polymer film with SRG. However, the fluorescence of all the laser dyes investigated was significantly quenched, so that the optical gain of the doped film was insufficient for lasing. This quenching was attributed to electron transfer from the azobenzene to the laser dye.[45] To prevent this, we designed a double-layered DFB laser composed of a diffraction layer of an azopolymer (pDR1M) and an active layer of laser-dye-doped poly(vinyl alcohol) (PVA) on a synthetic quartz substrate. We employed malachite green (MG) and sulforhodamine (SR) as laser dyes. The cross section of the double-layered structure and the chemical structure of the compounds used in this work are shown in figure 10. Scanning electron microscopy (SEM) confirmed the double-layered structure composed of a diffraction layer and an active layer. By placing the azopolymer layer on the top of the active layer, the azopolymer layer has a free surface which enables us to construct, erase, and rewrite SRG.

The fluorescence properties of both the MG-PVA/pDR1M and SR-PVA/pDR1M double-layered films were investigated. Sufficient photoluminescences were observed for both films. The fluorescence maxima were 673 and 613 nm, and the full width at half maxima (FWHM) of the fluorescence band were 73 and 32 nm, respectively.

The lasing properties were investigated by photoexciting the double-layered film transversally from the substrate side with a pulse laser. The wavelength of the excitation beam was set to 605 nm, which was longer than the π–π^* band of the azobenzene chromophores, to prevent film flattening by photoisomerization. Thus, another important factor for selecting MG and SR as the laser dyes was their efficient fluorescence with excitation at 605 nm.

Figure 11(a) shows the emission intensity of a surface-relief-structured (Λ = 713.4 nm, Δh = 72 nm) MG-PVA/pDR1M double-layered film as a function of excitation pulse energy. When the excitation energy exceeded the threshold of 70 nJ pulse^{-1} corresponding to a fluence of 0.4 mJ cm^{-2} pulse^{-1}, the emitted light with a very sharp spectral line was detected at 714 nm. On the other hand, only broadband emission (λ_{max} = 733 nm, FWHM = 19 nm) assigned to amplified spontaneous emission (ASE) was observed from the flat double-layered structure without inscribing SRG. Moreover, the laser emission was found to be linearly polarized normal to the film plane, i.e., the TM mode. We also found that predominance of the TM mode does not depend on pump beam polarization and laser dyes.

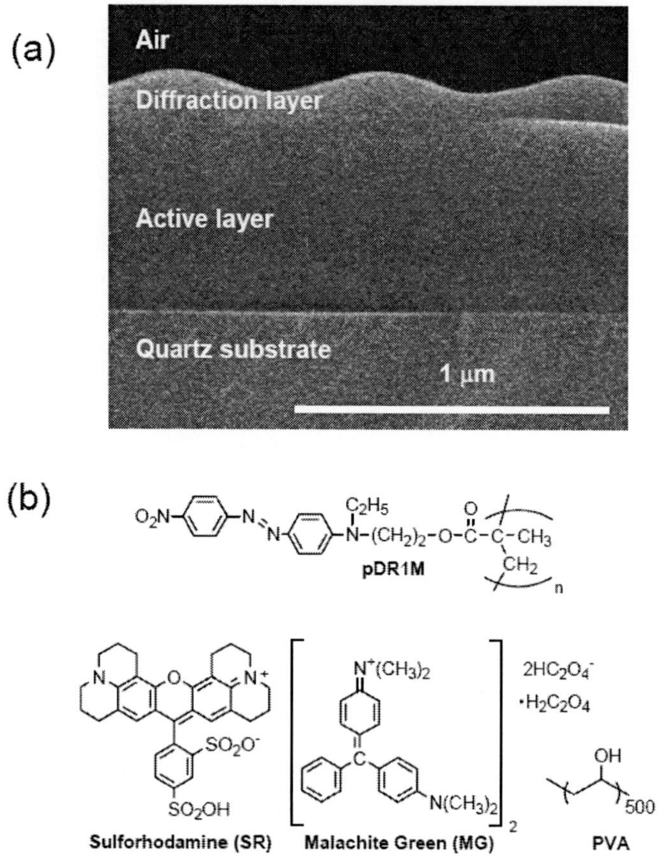

Figure 10. Double-layered DFB laser. a) Cross-sectional SEM image of double-layered polymer film. b) Chemical structures of the components of the DFB laser.

To confirm that this is DFB lasing, we calculated the Bragg diffraction wavelength λ_{Bragg}, which approximates the lasing wavelength of a DFB laser.[46] λ_{Bragg} is expressed as

$$\lambda_{Bragg} = 2n_e\Lambda / m, \quad (2)$$

where m is an integer representing the diffraction order and n_e the effective refractive index of the waveguide mode. We calculated n_e based on a three-layer planar waveguide model, in which the surface undulation of an azopolymer is approximated by a flat layer of an azopolymer/air mixture. We obtained $n_e = 1.514$ at 730 nm and $\lambda_{Bragg} = 720$ nm at $m = 3$. This was in good agreement with the observed lasing wavelength, within measurement errors of the refractive indices and the thicknesses of the layers. This confirmed that the lasing results from the distributed feedback induced by the SRG structure. Furthermore, as shown in figure 11b, we succeeded in changing the lasing wavelength systematically from 714 to 749 nm by adjusting the grating period according to eq. (2). This result shows that an MG-doped double-layered polymer laser can be tuned precisely in a wide wavelength range of ASE (710-750 nm).

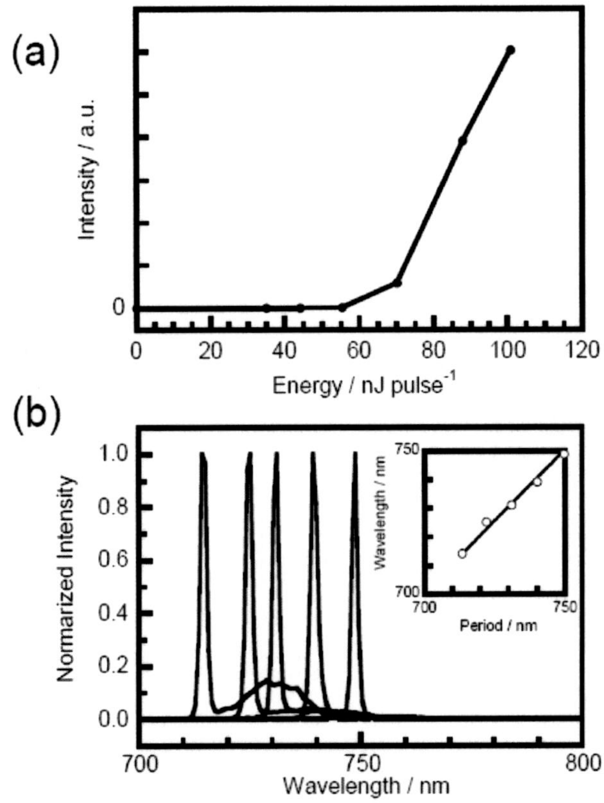

Figure 11. DFB laser emission from double-layered films doped with MG dye. a) Pump energy dependence of peak intensity at 714 nm for a double-layered film with surface relief period of 713.4 nm. b) Tuning of emission wavelength by changing the period of SRG. The inset represents the oscillation wavelength as a function of SRG period.

The most important advantage of our DFB laser based on a photoinduced SRG system is that the wavelength of DFB lasing is reprogrammable by erasing and rewriting SRG with another period as shown in figure 12. A double-layered film of an SR-PVA/pDR1M was photoinscribed with interfering beams (θ= 21.75°). An SRG structure (Λ = 658.5 nm, Δh = 35 nm) was formed on the azopolymer layer as observed by AFM (figure 12a). When the film was photoexcited transversally with a pulse laser exceeding the threshold, a very sharp line spectrum was detected at 663 nm due to the third-order DFB lasing (figure 12d).

By irradiating a uniform argon ion laser beam of 20 mW cm^{-2} for 2 hours, the original surface relief structure disappeared almost completely (figure 12b). DFB lasing was not detected from the film even if photoexcitation energy was further increased up to 280 nJ pulse^{-1}, although slight undulation (Δh = 3 nm) remained.

The film was again exposed to the interfering argon ion laser beams, aiming to prepare SRG with another period using a different incident angle (θ= 21.50°), so that the SRG structure with the corresponding period (Λ = 665.8 nm, Δh = 72 nm) was reformed on the azopolymer layer, as shown in figure 12c. By photoexcitation, DFB lasing at the corresponding wavelength of 675 nm was detected, as shown in figure 12d.

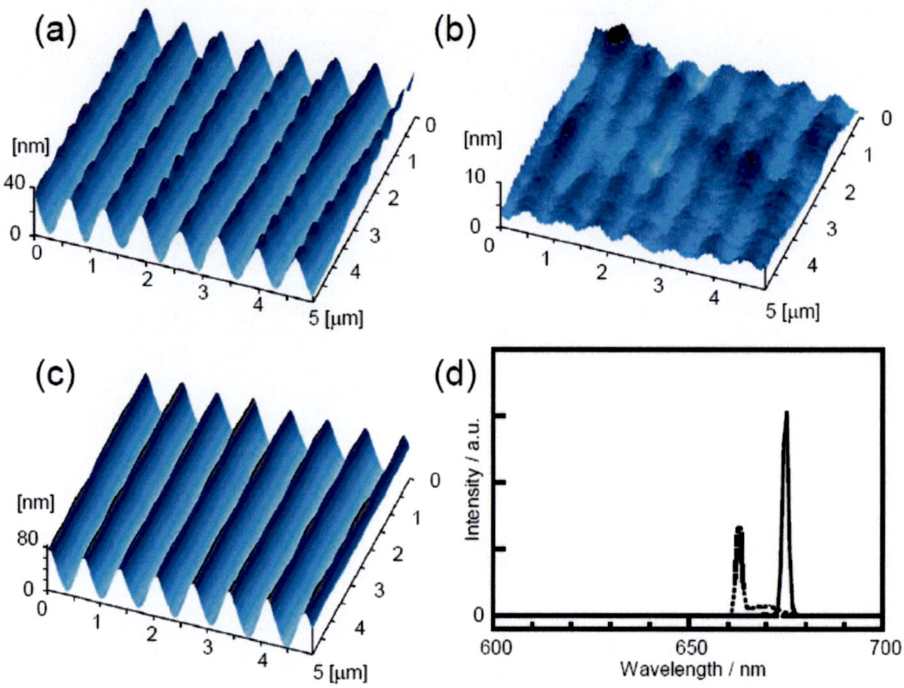

Figure 12. Control of DFB laser emission from a double-layered film doped with SR dye by writing, erasing, and rewriting SRG. a-c) Topographical AFM images of double-layered film at each stage, after the first writing of SRG (a), erasing by uniform exposure (b), and rewriting of SRG with another period (c). d) Emission spectra of double-layered film at each stage, dotted line (a), broken line (b), and solid line (c), respectively.

These results indicate the advantages of this DFB laser based on photoinduced polymer migration system, in terms of not only the programmability of lasing wavelength by tuning the period of SRG but also the programmability of threshold and differential efficiency by tuning the depth of SRG. These advantages originate in the dynamic function in the formation process of SRG, which are properties that no organic and semiconductor DFB laser reported so far can realize.

In conclusion of this section, a double-layered structure composed of an azobenzene polymer layer that can form SRG assisted by light and a dye-doped polymer layer that provides sufficient optical gain, enabled us to fabricate a wavelength-programmable organic DFB laser. This structure provides lasing at the desired wavelength simply by adjusting the interference angle of the writing beams, and the lasing wavelength is reprogrammable by erasing and rewriting SRG.

3. New Class of Surface Relief Formation

Although a few researchers have recently reported on PSRs formed on the materials other than azobenzene-containing materials, they have been limited to the PSRs formed by the thermal effect caused by irradiation of the pulsed laser[47] and the PSRs formed on the cross-linkable materials.[48-50] There is no example of reversible PSR formation by the continuous wave light irradiation except for azobenzene-containing materials. When the reversible PSR formation occurs on the films composed of the compounds other than azobenzene, a number of ways of application will be expected as well as the important knowledge to be obtained as to the PSRs formation mechanism. Therefore we have started the research to find the novel agents that can induce PSRs. In this section, we describe PSRs formed on a polymer film containing a photochromic compound that is not an azobenzene-related compound.

3.1. SPIROPYRAN

In the beginning, we employed spiropyrans (SPs) because they are representative thermally reversible photochromic compounds like azobenzenes.[51] SPs show a large structural change upon photoirradiation; an approximately 90° rotation of one half of the molecule during the ring opening and closing reactions between the spiro-form (SP-form) and the merocyanine form (MC form) (figure 13). In this section, we report the formation of the photoinduced surface relief structure using SP1 and other SPs.

A series of SPs shown in figure 13 were synthesized by the condensation of a nitrosalicylaldehyde with an N-alkylindolenium bromide.[52] The spin-coated films were prepared from the chloroform solutions of each SPs and poly(methyl methacrylate) (PMMA) on a cleaned glass substrate. The thickness of films was 0.3 μm, when the film was composed of 60 % of SP and 40 % of PMMA in weight.

The wavelength of the light for irradiation longer than 350 nm was taken out from a high-pressure mercury lamp by a long pass filter and water filter. The light was collimated to obtain a homogeneous light intensity (0.1 W cm^{-2}) over the entire area to be exposed. Patterned light irradiation to form surface relief structures was performed through a photomask attached to the film. The photomask was composed of stripes with the same widths of the transparent and opaque lines.[53] After the film was exposed to the irradiation light for 2 h, the surface of the film was observed through a differential interference microscopy and AFM.

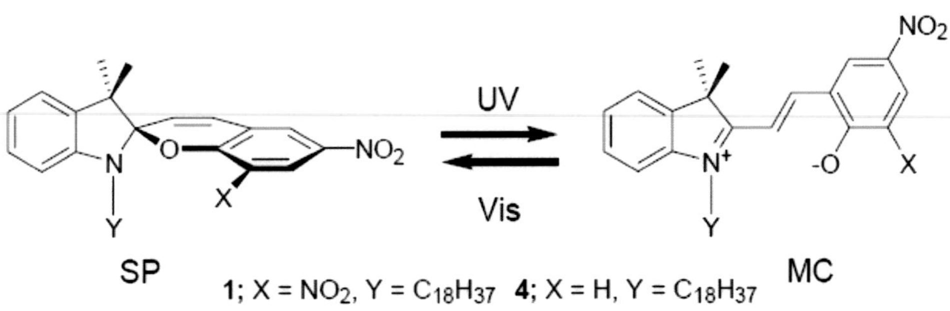

SP MC

1; X = NO_2, Y = $C_{18}H_{37}$ **4**; X = H, Y = $C_{18}H_{37}$

2; X = NO_2, Y = C_8H_{17} **5**; X = H, Y = C_8H_{17}

3; X = NO_2, Y = CH_3 **6**; X = H, Y = CH_3

Figure 13. Interconversion of spiropyran (SP) and merocyanine (MC) isomers of 1-6.

Figure 14 shows the surface relief structure formed on the SP1-doped PMMA film. When the light was shaded completely, only a flat surface was observed as shown on the left-hand side of figure 14. On the other hand, when the film was irradiated with patterned light, regular surface modulation was produced, and its spatial period was coincident with that of the photomask as shown on the right-hand side of figure 14. When we examined the cross-sectional topography, we found that the top of the convex is higher than the initial surface level and base of depression is lower than the initial level. This result clearly shows that the lateral material motion occurred at the film surface. In order to investigate the direction of material migration, irradiation experiments through a photomask possessing a grating-like slits with 150 μm, and each slit has 6.25-μm width, were performed. In the case of the SP1-doped PMMA film, irradiation produced a significant topographic maximum, indicating that polymer migration occurred from the shaded areas to the exposed areas. The direction of migration was the same for all of our experiments, irrespective of the materials.

To induce surface relief structures, we have to select the proper light of energy. It means that the back-and-forth photoisomerization between MC and SP forms is inevitable to form PSRs. In azobenzene polymers, it is believed that the *trans-cis-trans* isomerization cycles to induce the decrease in the viscosity of the film[54] have played an important role in the formation of PSR. Therefore, as for the case of SP1, increase in the photoinduced flexibility of the film caused by the large motion of the SP1 molecules during the iterative photoisomerization must be one of the requirements to induce the material transfer.

It was reported that the low-molecular-mass azobenzene-dye-doped systems exhibited rather small ability to form PSR, while the covalently bonded azobenzene polymer systems usually showed several hundred nm height differences.[13, 27] On the other hand, we observed PSRs with large surface modulation in SP dye-doped polymer systems. This difference was probably due to the high solubility of SP1 to the polymer matrix, showing no dye-polymer segregation at even the relatively high concentration.

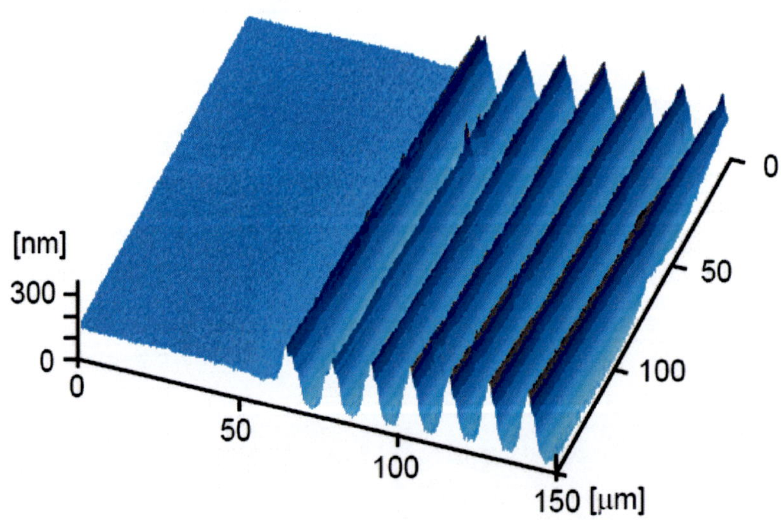

Figure 14. Atomic force microscopy image of an SP1 doped PMMA film after exposed to high-pressure mercury lamp (> 350 nm) with 0.1 W cm^{-2} for 2 h through a photomask.

PSR formation experiments on the PMMA films doped with other SPs were also investigated as listed in table 1. Though SP2 and SP3 formed no PSRs because of their low compatibility with PMMA, PSRs were obtained on the films of SP4-SP6 as well, similar to the case of SP1. However, the modulation depth was shallower than that of SP1 under the same conditions. SP1, possessing two electron-withdrawing nitro groups on the chromophore to stabilize the colored zwitterionic open-ring form, shows negative photochromism. To the contrary, SP4 shows the positive photochromism because the colored form is less stable than the colorless form. These differences may have affected to the photoreactivity and the ability of PSR formation. More importantly, long alkyl chain on the nitrogen atom plays the decisive role to improve the ability of PSR formation, by comparing the modulation depth obtained from electronically equivalent SP4-SP6. This result indicates, by taking the results of the PSR formation in the azobenzene dye systems into account, that the large structural change during the photoisomerization is the essential factor. Twining of the polymer chain with the long alkyl chain of SPs may be important because of the host-guest system in our case.

Table 1. Modulation depth and shape of surface relief in SP doped PMMA films

Spiropyran	Type	Approximate height / nm	Shape
1	N	200	Convex
2	N	-	-
3	N	-	-
4	P	80	Convex
5	P	30	Convex
6	P	10	Complex

N: negative photochromism
P: positive photochromism

3.2. DIARYLETHENE

Azobenzene and spiropyran can be classified as the thermally reversible organic photochromic compounds that exhibit structural change thermally as well as photochemically. However, there are no reports on PSR formation with thermally irreversible photochromic compounds, classified into the other counterpart. In this section, we report on PSR formation on polymer films containing diarylethene (DAE) known as one of the representatives of the thermally irreversible photochromic compounds. Since DAEs can undergo photochromic reactions between two isomers (open and closed ring forms) upon photoirradiation with significant photo-fatigue resistance (figure 15),[56] improvement of rewritability of PSR can also be expected.

Syntheses of DAE was carried out with standard procedures reported.[56] Optically transparent films, about 0.4 μm thick, were prepared by spin-coating method onto a cleaned glass plate, from the chloroform solution of each DAE and PMMA, 60 wt% and 40 wt%, respectively.

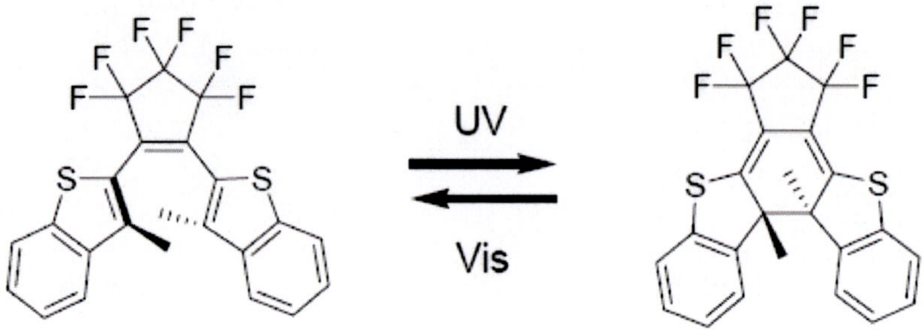

Figure 15. Interconversion of diarylethene.

Figure 16a shows the surface relief structure produced by the patterned light irradiation (> 350 nm) to the PMMA film doped with DAE through a photomask. The regular surface modulation was produced, and its spatial period was coincident with that of the photomask. In order to investigate the direction of material transfer, irradiation experiments through a photomask possessing a slit with 4-μm width were performed. In the case of the film doped with DAE, irradiation produced a significant topographic maximum, indicating that material migration occurred from the shaded areas to the exposed areas (figure 16b). The ability of PSR formation of DAE was not so large when it is compared with the azobenzene and spiropyran compounds, probably due to its small structural change during photoisomerization.

In conclusion of this section, PSRs were recorded on the surface of a PMMA film doped with photochromic compound such as SPs and DAE. As far as we know, this is the first PSR formation using a photochromic compound other than the azobenzene derivatives. With this section, we believe that we have broken the superstition that PSRs can be formed only with the azobenzene derivatives.[20]

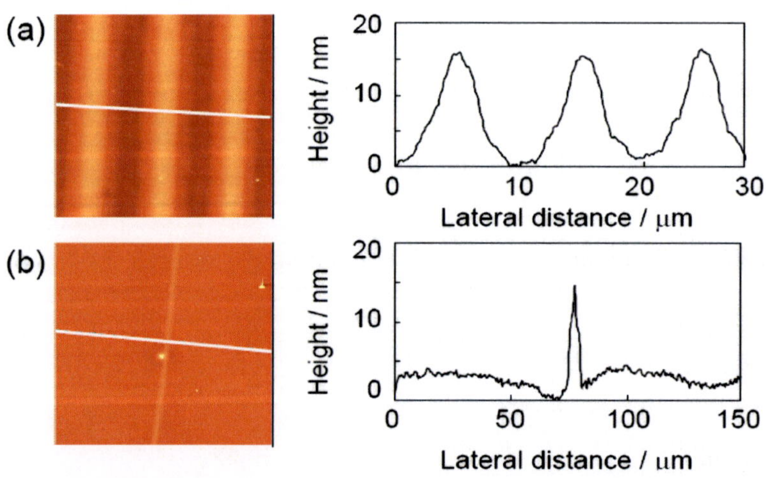

Figure 16. AFM images of surface relief structure produced by the patterned light irradiation (> 350 nm, 0.1 W cm^{-2}). The photomask was composed of stripes with the same widths (4 μm) of the transparent and opaque lines (a), a slit line with 4-μm width (b). The figures on the right side indicate the cross-sectional height profiles for the solid lines shown in the AFM images.

Conclusion

Decade years have passed since first discovery of photoinduced surface relief formation due to unusual material migration. Intensive studies as to photoinduced surface relief have been developed in view of both fundamental science and practical applications, though they are not introduced here. This chapter focuses to selected research topics on possible applications proposed in our group using reversible dynamic lateral migration in the formation and erasure process of surface relief structure. We developed a patterning technique of functional materials using the photo-triggered material migration system. According to designed light intensity patterning, functional materials from organic molecules to inorganic nanoparticles can be conveyed in any two-dimensional directions together with the lateral migration of azobenzene polymer. Erasable and rewritable surface relief grating was employed as an organic distributed feedback laser with oscillation wavelength programmability. A double-layered structure composed of an azobenzene polymer layer and a dye-doped polymer layer, enabled us to obtain a wavelength-programmable organic distributed feedback laser. In addition to these topics on surface relief structures formed on azobenzene polymers, new class of surface relief formations using a photochromic compound other than the azobenzene derivatives are presented. Promotions of these research areas are anticipated to provide wide and new opportunities for material processing and creation of new class of photonic devices.

Acknowledgement

The author is grateful to Prof. T. Seki for long encouragement, guidance, and discussions. The author thanks Prof. K. Ichimura, Prof. M. Hara, Dr. T. Isoshima, and Prof. Y. Yokoyama for collaboration and helpful discussions. The author also thanks co-workers, Dr. N. Zettsu, Mr. T. Higuchi, Mr. K. Takahashi, Mr. S. Yamaguchi, Mr. Y. Moriya for their insightful contributions and efforts. This work was supported, in part, by a Grant-in-Aid for Scientific Research <KAKENHI (14740391)> from the Ministry of Education, Culture, Sports, Science, and Technology, Industrial Technology Research Grant Program from New Energy and Industrial Technology Development Organization (NEDO) of Japan, the fellowship 'Special Postdoctoral Researchers Program' provided by RIKEN, and the Iketani science and technology foundation.

References

[1] P. Rochon, E. Batalla, and A. Natansohn, *Appl. Phys. Lett.,* 66, 136 (1995).
[2] D. Y. Kim, S. K. Tripathy, L. Li, and J. Kumar, *Appl. Phys. Lett.,* 66, 1166 (1995).
[3] X. L. Jiang, L. Li, J. Kumar, D. Y. Kim, and S. K. Tripathy, *Appl. Phys. Lett.,* 72, 2502 (1998).
[4] D. Y. Kim, L. Li, X. L. Jiang, V. Shivshankar, J. Kumar, and S. K. Tripathy, *Macromolecules,* 28, 8835 (1995).
[5] C. J. Barrett, A. L. Natansohn, and P. L. Rochon, *J. Phys. Chem.,* 100, 8836 (1996).

[6] X. L. Jiang, L. Li, J. Kumar, D. Y. Kim, V. Shivshankar, and S. K. Tripathy, *Appl. Phys. Lett.,* 68, 2618 (1996).

[7] N. K. Viswanathan, S. Balasubramanian, L. Li, S. K. Tripathy, and J. Kumar, *Jpn. J. Appl. Phys.,* 38, 5928 (1999).

[8] K. G. Yager, and C. J. Barrett, *Curr. Opin. Solid State Mater. Sci.,* 5, 487 (2001).

[9] J. Kumar, L. Li, X. L. Jiang, D.-Y. Kim, T. S. Lee, and S. Tripathy, *Appl. Phys. Lett.,* 72, 2096 (1998).

[10] S. Bian, J. M. Williams, D. Y. Kim, S. Balasubramanian, J. Kumar, and S. Tripathy, *J. Appl. Phys.,* 86, 4498 (1999)

[11] S. Bian, L. Li, J. Kumar, D. Y. Kim, J. Williams, and S. K. Tripathy, *Appl. Phys. Lett.,* 73, 1817 (1998).

[12] N. C. R. Holme, L. Nikolova, S. Hvilsted, P. H. Rasmussen, R. H. Berg, and P. S. Ramanujam, *Appl. Phys. Lett.,* 74, 519 (1999).

[13] C. Fiorini, N. Prudhomme, G. de Veyrac, I. Maurin, P. Raimond, and J.-M. Nunzi, *Synth. Met.,* 115, 121 (2000).

[14] T. Ikawa, T. Mitsuoka, M. Hasegawa, M. Tsuchimori, O. Watanabe, Y. Kawata, C. Egami, O. Sugihara, and N. Okamoto, *J. Phys. Chem. B,* 104, 9055 (2000).

[15] N. Landraud, J. Peretti, F. Chaput, G. Lampel, J.-P. Boilot, K. Lahlil, and V. I. Safarov, *Appl. Phys. Lett.,* 79, 4562 (2001).

[16] T. Fukuda, K. Sumaru, T. Kimura, H. Matsuda, Y. Narita, T. Inoue, and F. Sato, *Jpn. J. Appl. Phys.,* 40, L900 (2001).

[17] H. Rau, in Photochromism; Molecules and Systems, edited H. Dürr and H. Bouas-Laurent, *Elsevier,* Amsterdam (1990), p. 165.

[18] M. J. Kim, C. Chun, T. Nakayama, and D. Y. Kim, *Jpn. J. Appl. Phys.,* 45, L169 (2006).

[19] Natansohn and P. Rochon, *Adv. Mater.,* 11, 1387 (1999).

[20] N. K. Viswanathan, D. Y. Kim, S. Bian, J. Williams, W. Liu, L. Li, L. Samuelson, J. Kumar, and S. K. Tripathy, *J. Mater. Chem.,* 9, 1941 (1999).

[21] T. S. Lee, D.-Y. Kim, X. L. Jiang, L. Li, J. Kumar, and S. Tripathy, *J. Polym. Sci. A,* 36, 283 (1998).

[22] J. P. Chen, F. L. Labarthet, A. Natansohn, and P. Rochon, *Macromolecules,* 32, 8572 (1999).

[23] M. Pedersen, S. Hvilsted, N. C. R. Holme, and P. S. Ramanujam, *Macromol. Symp.,* 137, 115 (1999).

[24] T. Yamamoto, M. Hasegawa, A. Kanazawa, T. Shiono, and T. Ikeda, *J. Mater. Chem.,* 10, 337 (2000).

[25] M. Helgert, L. Wenke, S. Hvilsted, and P. S. Ramanujam, *Appl. Phys.* B, 72, 429 (2001).

[26] D. Bublitz, M. Helgert, B. Fleck, L. Wenke, S. Hvilsted, and P. S. Ramanujam, *Appl. Phys.* B, 70, 863 (2000).

[27] F. L. Labarthet, T. Buffeteau, and C. Sourisseau, *J. Phys. Chem.* B, 102, 2654 (1998).

[28] T. Fukuda, H. Matsuda, T. Shiraga, T. Kimura, M. Kato, N. K. Viswanathan, J. Kumar, and S. K. Tripathy, *Macromolecules,* 33, 4220 (2000).

[29] H. Nakano, T. Takahashi, T. Kadota, and Y. Shirota, *Adv. Mater.,* 14, 1157 (2002).

[30] H. Nakano, T. Tanino, and Y. Shirota, *Appl. Phys. Lett.,* 87, 061910 (2005).

[31] T. Ubukata, T. Seki, and K. Ichimura, *Adv. Mater.,* 12, 1675 (2000).

[32] T. Ubukata, T. Seki, and K. Ichimura, *Colloids and Surfaces A,* 198-200, 113 (2002).
[33] T. Ikeda, *J. Mater. Chem.,* 13, 2037 (2003).
[34] T. Seki, *Polym. J.,* 36, 435 (2004).
[35] N. Zettsu, T. Ubukata, T. Seki, and K. Ichimura, *Adv. Mater.,* 13, 1693 (2001).
[36] N. Zettsu and T. Seki, *Macromolecules,* 37, 8692 (2004).
[37] N. Zettsu, T. Fukuda, H. Matsuda, and T. Seki, *Appl. Phys. Lett.,* 83, 4960 (2003).
[38] T. Ubukata, M. Hara, K. Ichimura, and T. Seki, *Adv. Mater.,* 16, 220 (2004).
[39] K. Ichimura, *Chem. Rev.,* 100, 1847(2000).
[40] N. Zettsu, T. Ubukata, and T. Seki, *Jpn. J. Appl. Phys.,* 43, L1169 (2004).
[41] L. Rocha, V. Dumarcher, C. Denis, P. Raimond, C. Fiorini, and J.-M. Nunzi, *J. Appl. Phys.,* 89, 3067 (2001).
[42] T. Matsui, M. Ozaki, K. Yoshino, and F. Kajzar, *Jpn. J. Appl. Phys.,* 41, L1386 (2002).
[43] T. Ubukata, T. Isoshima, and M. Hara, *Adv. Mater.,* 17, 1630 (2005).
[44] T. Ubukata, T. Isoshima, and M. Hara, *Mol. Cryst. Liq. Cryst.,* 445, 269 (2006).
[45] T. Isoshima, E. Ito, T. Ubukata, and M. Hara, *Mol. Cryst. Liq. Cryst.,* 444, 81(2006).
[46] H. Kogelnik, and C. V. Shank, *Appl. Phys. Lett.,* 18, 152 (1971).
[47] O. Baldus, A. Leopold, R. Hagen, T. Bieringer, and S. J. Zilker, *J. Chem. Phys.* 114, 1344 (2001).
[48] S. Yamaki, M. Nakagawa, S. Morino, and K. Ichimura, *Appl. Phys. Lett.,*76, 2520 (2000).
[49] H. Ono, A. Emoto, N. Kawatsuki, and T. Hasegawa, *Appl. Phys. Lett.,* 82, 1359 (2003).
[50] P. S. Ramanujam and R. H. Berg, *Appl. Phys. Lett.,*85, 1665 (2004).
[51] T. Ubukata, T. Takahashi, and Y. Yokoyama, submitted.
[52] Y. Yokoyama, W. Hara, T. Inoue, T. Ubukata, M. Sakomura, and H. Tukada, *Chem. Lett.,* 34, 1622 (2005).
[53] T. Ubukata, T. Higuchi, N. Zettsu, T. Seki, and M. Hara, *Colloids and Surface A,* 257-258, 123 (2005).
[54] T. Srikhirin, A. Laschitsch, D. Neher, and D. Johannsmann, *Appl. Phys. Lett.,* 77, 963 (2000).
[55] T. Ubukata, S. Yamaguchi, and Y. Yokoyama, in preparation.
[56] M. Irie, *Chem. Rev.,* 100, 1685 (2000).

In: Progress in Advanced Materials Research
Editor: Nicolas H. Voler, pp. 161-184
ISBN: 978-1-60021-893-4

Chapter 6

ALIGNED-CRYSTALLINE SILICON FILMS ON POLYCRYSTALLINE METAL-ALLOY SUBSTRATES

Alp T. Findikoglu[1], Woong Choi[1], Marilyn Hawley[2], Manuel J. Romero[3], Kim M. Jones[3] and Mowafak M. Al-Jassim[3]

[1] Materials Physics and Applications Division,
Los Alamos National Laboratory, Los Alamos, New Mexico 87545
[2] Materials Science and Technology Division,
Los Alamos National Laboratory, Los Alamos, New Mexico 87545
[3] National Renewable Energy Laboratory, Golden, Colorado 80401

ABSTRACT

This chapter summarizes recent progress in growth and characterization of aligned-crystalline silicon (ACSi) films on polycrystalline metal-alloy substrates. The ACSi deposition process uses, as a key technique, ion-beam-assisted deposition (IBAD) texturing to achieve *near-single-crystalline* silicon films with good electrical properties on low-cost, large-area, non-single-crystalline substrates. First, we describe the general aspects of IBAD texturing technique, and the motivation for applying it to semiconductors. This is followed by a detailed description of multilayer buffer structures implemented in this work. We have performed materials characterization using scanning electron microscopy, atomic force microscopy, electron backscattering diffraction, transmission electron microscopy, and scanning tunneling microscopy, and combined them with electrical measurements, such as carrier mobility, to understand the materials-property relationships in these structures. Optimized ACSi films on metal-alloy substrates with insulating IBAD buffer layer stacks demonstrate out-of-plane and in-plane mosaic spreads of 0.8° and 1.3°, respectively, and a room-temperature Hall mobility of 90 $cm^2/V.s$ (~50% of what is achievable with epitaxial Si films, and ~1000 times that of amorphous Si films) for a p-type doping concentration of ~$4x10^{16}$ cm^{-3}. By using various microscopic techniques, we have confirmed the underlying near-single-crystalline order and the superior characteristics of low-angle (<5°) grain boundaries in ACSi films. Also, forming gas anneal experiments indicated that Si films with low-angle grain boundaries do not need to be passivated to demonstrate improved majority carrier transport properties. Moreover, we have demonstrated a new conducting IBAD material, TiN, in addition to the standard insulating IBAD MgO, that can be integrated with Si. This has led to demonstration of a conducting nitride buffer stack that could be especially suitable for photovoltaic applications of ACSi films. As a further demonstration of the photovoltaic potential, we have fabricated thick (~1-5-micron-thick) ACSi films with well-defined doping profiles on these buffer layers. In the electrical characterization area, in addition to having extensive studies on the majority carrier transport using Hall measurements, we have performed surface potential measurements using AFM and transport measurements based on STM that showed that carrier transport is not significantly impeded by low-angle grain boundaries in optimized ACSi films. Also, with an eye for thin film transistor (TFT) applications, we have prepared metal-insulator-semiconductor structures and measured C-V curves. These measurements yielded near-electronic-grade surface properties and low surface defect densities in the ACSi films. Currently, no other monolithic technology combines the high performance of epitaxial silicon films with the low cost of non-single-crystalline substrates. As such, the ACSi film technology holds promise to make a major impact in the semiconductor industry by providing a new means to improve the performance (or, reduce material and fabrication costs) of silicon-based devices, such as solar cells and flat-panel displays.

INTRODUCTION

High performance and low cost are the two main drivers for large scale applicability of semiconductor electronics and sensors. Thus, the challenge facing the semiconductor industry is to reduce production cost while maintaining or improving device performance. In many thin film semiconductor applications, including electronics, photovoltaics, and sensors, a major technical difficulty is the lack of a suitable epitaxial template for the growth of well-oriented films. Favorable device characteristics are generally better defined and more pronounced in well-oriented (i.e., single-crystalline or near-single-crystalline) thin films, but conventional epitaxial film-growth techniques require single crystal templates that are either expensive or of limited availability or both.

The ion-beam-assisted deposition (IBAD) texturing technique can produce thin films of certain materials with biaxially-oriented grains (i.e., with preferred out-of-plane and in-plane crystallographic orientations) by using an ion beam to establish a preferred orientation for buffer layer growth on a non-single-crystalline substrate.[1,2,3] Once established, this IBAD-textured buffer layer serves as a biaxially-oriented template for the epitaxial growth of subsequent layers.[4] By careful selection of subsequent layers, we can achieve the grain-aligned, matched, and robust surface necessary to grow biaxially-oriented Si films.[5] As will be demonstrated in the following sections, with optimized processes, these biaxially-textured films of Si, which we will refer to as Aligned-Crystalline Silicon (ACSi), demonstrate near-single-crystalline characteristics both microstructurally and electrically.

For the rest of our discussion in this chapter, all layers of the multilayer architecture we use are electron-beam evaporated (or sputtered) *in situ* on a flexible, polycrystalline nickel-alloy tape, a corrosion-resistant Hastelloy C-276. This commercially obtained 1-cm-wide and 100-μm-thick tape is cleaned with solvents and polished, if necessary, to achieve a root mean square surface roughness of <10 nm, for a $5 \times 5\ \mu m^2$ area.

OVERVIEW OF IBAD TEXTURING PROCESS AND ACSI FILMS

The implementation of the IBAD texturing technique to the growth of ACSi is the focus of this chapter. The IBAD texturing technique begins with any common polycrystalline or amorphous substrate, such as a flexible polycrystalline metal tape or glass. On this substrate, we deposit a special buffer-layer stack that is either insulating or conductive, depending on the end-product's requirements. This buffer-layer stack, which includes several distinct layers of material, allows the transition from a substrate surface with poor crystalline order to the highly-ordered and lattice-matched crystalline texture of conventional epitaxial systems. Figure 1 shows transmission electron microscopy (TEM) image of an ACSi sample that was fabricated on a metal-alloy substrate.[5] Within the buffer-layer stack, the key transition from a polycrystalline or amorphous surface to the highly-oriented layer is achieved with a 3-to-5-nm-thick IBAD layer (marked by red dashed line). The optimization of the growth conditions for this *IBAD nanolayer* is the single-most important parameter in determining the final properties of the ACSi films. There have been several publications on IBAD texturing, mostly from experimental and phenomenological points of view. However, the microscopic basis of IBAD texturing is still poorly understood, although some simulation work has provided verifiable predictions.[6,7,8]

We have previously published the details of our IBAD texturing process, and its use for achieving ACSi films.[5] In figure 1, the strain contrast indicates misfit dislocations and stacking faults in the ACSi film: most of the misfit dislocations are concentrated at the Si/buffer-layer interface, while stacking faults propagate further into the ACSi film.[5]

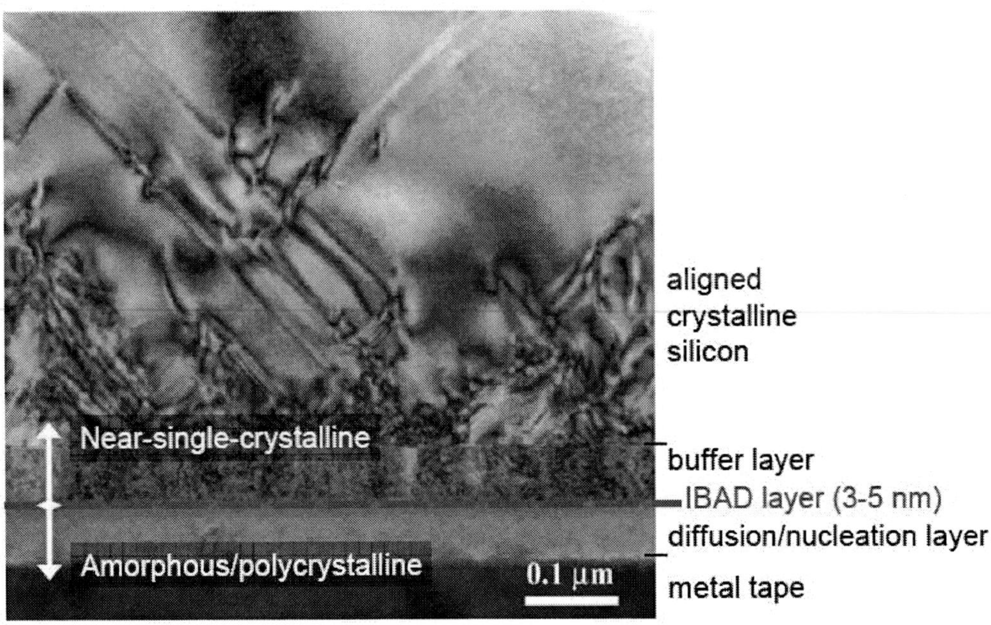

Figure 1. Tranmission electron microscopy (TEM) image of an Aligned-Crystalline Si (ACSi) film on buffer stack on metal tape, showing 3-to-5-nm-thick IBAD MgO layer that facilitates transition from a non-single-crystalline (amorphous/polycrystalline) surface to a biaxially-oriented, near-single-crystalline texture (ref. [5]).

Figure 2 shows that the microstructure of the ACSi film on metal tape, where the buffer layer termination is the cubic form of Al_2O_3 (γ-Al_2O_3), is very similar to that of standard Si film on single crystal α-Al_2O_3 (sapphire).[9]

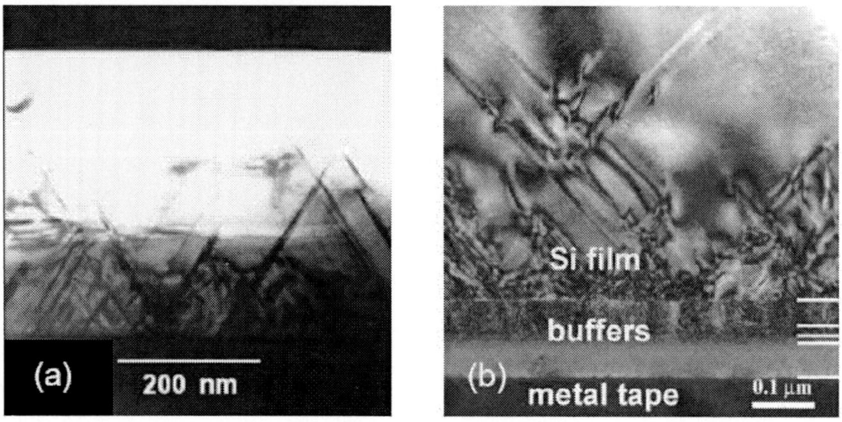

Figure 2. Tranmission electron microscopy image of (a) Si film on single-crystal sapphire vs (b) our IBAD-buffered ACSi film on metal tape, showing similar microstructure (refs. [5,9]).

Diffusion Barrier / Nucleation Layer

In most applications, it is necessary to have a total buffer stack thickness of 0.1 - 0.5 μm to act as an effective diffusion barrier because epitaxial growth of ACSi films at high rates (>1 nm/sec) requires relatively high temperatures (>700 °C), and at such high temperatures diffusion of certain elements from the substrate into the ACSi films could have detrimental effects on electrical performance.

The nucleation layer is a layer with amorphous or nano-crystalline surface where the IBAD texturing process could be initiated. Whereas the diffusion barrier needs to be relatively thick, the nucleation layer could be only several monolayers (~3 nm) thick. Also, the nucleation layer can be a deposited film, or could be obtained by treatment of an existing material surface, such as ion bombardment of the surface of a metal-alloy substrate.

Our early work used a 50-100 nm thick amorphous Al_2O_3 layer on the metal tape as a diffusion barrier,[10] followed by a 7-nm-thick Y_2O_3 nucleation layer, both deposited at room temperature. We have extensively used reflection high energy electron diffraction (RHEED) to monitor film surface texture during and after such film growth. For example, the diffuse RHEED pattern in figure 3 indicates that the top ~3 nm of the Y_2O_3 nucleation layer is amorphous/nanocrystalline. Our experience has been that such a diffuse RHEED pattern is a necessary condition for obtaining good IBAD texturing.

We have demonstrated good IBAD texturing using a variety of diffusion-barrier/nucleation-layer combinations. For example, nucleation layers can be formed directly on metal-alloy substrates by using Ar^+ and N^+ ion bombardment. Alternatively, films of various oxides and nitrides that are deposited in amorphous/nanocrystalline form can serve the purpose of nucleation layers.

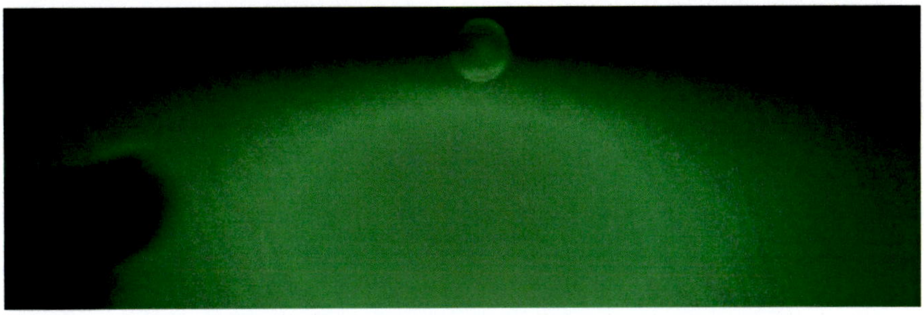

Figure 3. Diffuse reflection high energy electron diffraction (RHEED) image, indicating amorphous/nanocrystalline surface, of a 7-nm thick Y2O3 nucleation layer on 50 nm thick amorphous Al2O3 layer on metal-alloy tape.

IBAD LAYER

The IBAD texturing process relies on deposition of a face-centered cubic material on an appropriate nucleation layer under collimated ion bombardment. In the example schematically shown in figure 4, a 3-to-5-nm-thick MgO film is deposited via IBAD texturing while directing a 750-eV Ar^+ ion beam at the substrate at ambient temperature and with a 45° angle between the ion beam and the substrate—forming a *biaxially oriented crystalline layer*. Figure 5 shows the RHEED image on such an IBAD surface. The RHEED spots indicate that MgO grains are well aligned both in-the-plane and out-of-the-plane of the substrate. Also, from the separation of the spots we calculate an in-plane lattice constant of ~ 0.42 nm, which is close to the single-crystal value. Once established, this IBAD-textured layer serves as an epitaxial template for the growth of the subsequent layers that have a high degree of crystallographic orientation and grain alignment.

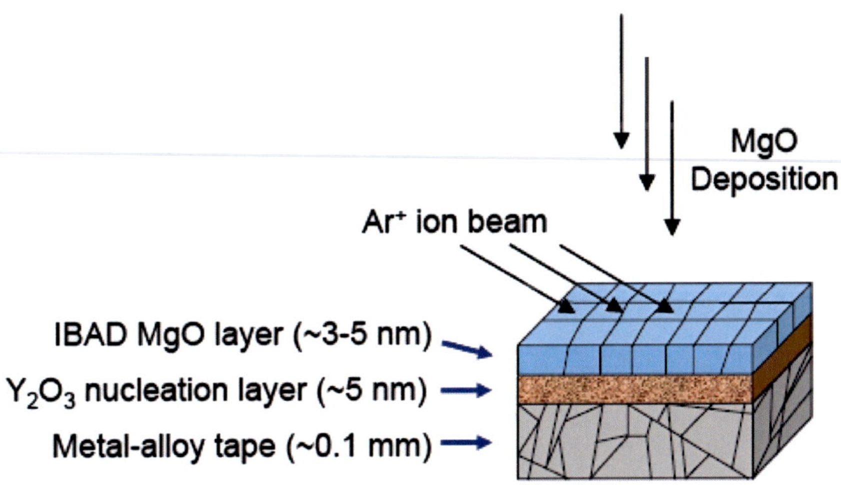

Figure 4. Schematic showing a representative IBAD texturing process that yields a 3-5-nm-thick biaxially-oriented MgO film on a 5-nm-thick Y_2O_3 nucleation layer on a polycrystalline metal-alloy tape.

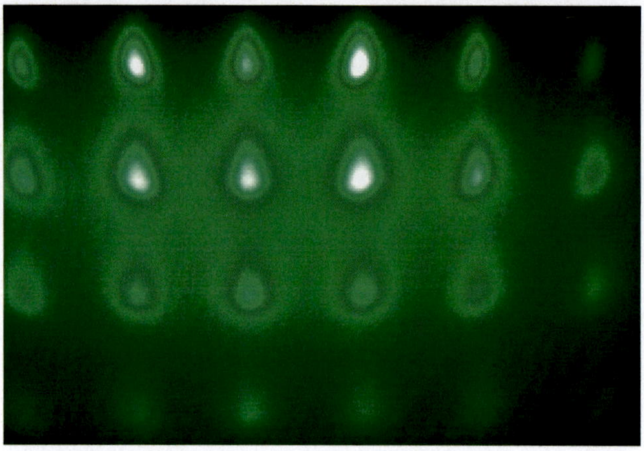

Figure 5. RHEED image on an IBAD MgO layer, indicating biaxially-oriented crystalline texture and a lattice constant close to that of bulk MgO.

We have previously published a detailed study of the effect of process conditions on the biaxial texture of MgO films grown by the IBAD texturing process.[4] The biaxial texture shows a strong dependence on the Ar^+/MgO flux ratio, but a weak dependence on the divergence of the Ar^+ beam. These effects are not fully explained by existing analytical models.

Recently, we have demonstrated IBAD texturing using other face-centered cubic materials. Figure 6 shows a RHEED image on IBAD TiN deposited directly on a metal-alloy tape. In this case, the nucleation layer is formed on the tape in a thin surface layer that is treated by an ion beam. From the separation of spots in figure 6 we calculate an in-plane lattice constant of ~0.42 nm, which is within the range of published data.

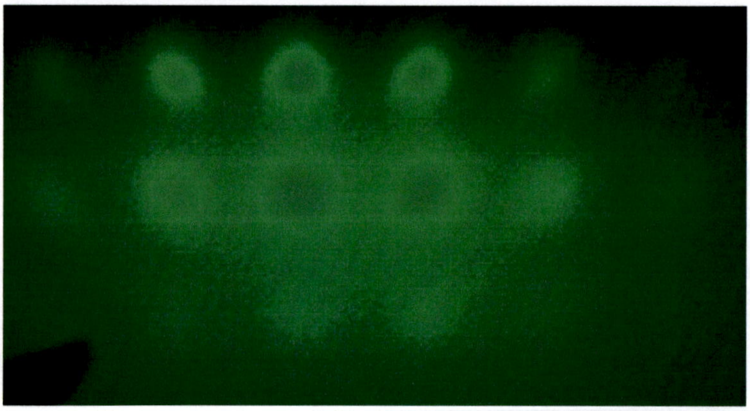

Figure 6. RHEED image on IBAD TiN layer, indicating biaxially-oriented crystalline texture and a lattice constant close to that of MgO.

Buffer Layer

The transition from the thin metal oxide insulating layer (IBAD MgO) or the thin metal nitride conductive layer (IBAD TiN) to the non-oxide or -nitride semiconductor layer (Si) requires careful selection of a buffer layer that acts not only as a diffusion barrier between the semiconductor film and the underlying metal substrate, but also as a robust, non-reacting, lattice-matched template layer for the epitaxial semiconductor film growth at elevated temperatures (>700 °C).

In the case of the IBAD MgO structure, the texture and surface coverage improve with the deposition of a ~0.1-to-0.3-μm-thick homoepitaxial MgO layer on the IBAD MgO layer. Figure 7(a) shows the RHEED image of the homoepitaxial MgO layer. The smaller spot sizes are indicative of larger grains and/or better alignment of the grains, whereas the vertical streaks that are connecting the bright spots are due to enhanced 2-D character of the reflection, i.e., the presence of a smooth surface.

The silicon-on-sapphire (SOS) is a well-known technology that combines Si with a single-crystal oxide substrate. To complete the buffer layer stack, we borrow from earlier SOS work and grow (at ~700-800 °C) the cubic form of alumina, γ-Al_2O_3, that acts as a transitional epitaxial layer between the MgO layer and the Si film. Figure 7(b) shows a RHEED image on the heteroepitaxially deposited γ-Al_2O_3 on the MgO layer.

(a)
(b)
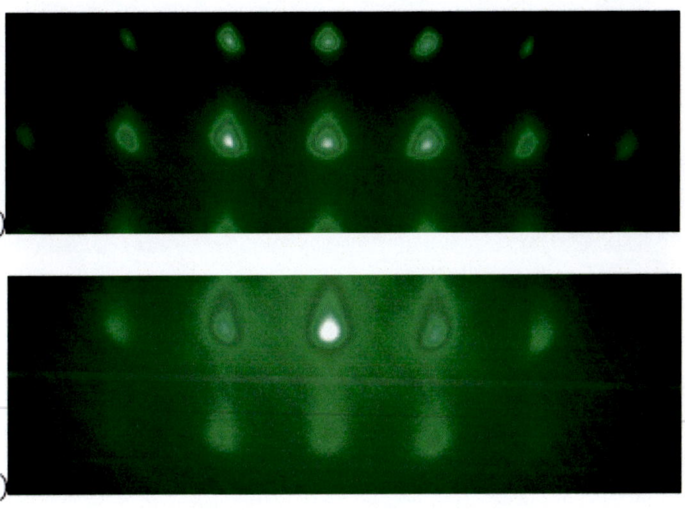

Figure 7. RHEED image for: (a) homoepitaxially-grown 0.1-μm-thick MgO layer, (b) heteroepitaxially-grown 0.15-μm-thick γ-Al_2O_3 layer, indicating preservation of biaxially-oriented crystalline texture.

Under our process conditions, we need between 0.1 to 0.4 μm total thickness in the buffer stack for it to act as an effective diffusion barrier. In the case of IBAD TiN, only a homoepitaxial TiN buffer layer (grown at ~500 - 700 °C) is sufficient to complete the buffer layer stack because TiN is a good diffusion barrier and it allows hetero-epitaxial growth of Si. Furthermore, the TiN buffer stack, unlike the oxide buffer stack, is a conducting structure that could be especially useful for certain applications, such as photovoltaics. Figure 8 shows a RHEED image for a homoepitaxially-grown TiN film, with a thickness of ~0.2 μm.

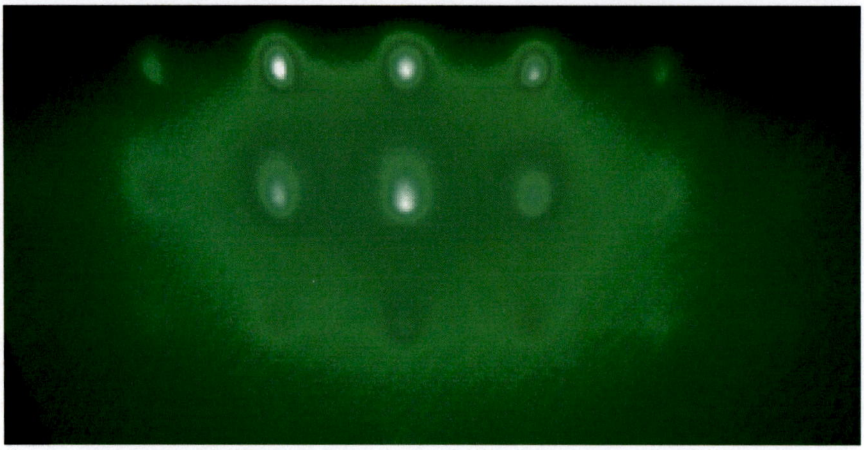

Figure 8. RHEED image for the homoepitaxially-grown 0.2-μm-thick TiN on the IBAD TiN layer, indicating biaxially-oriented crystalline texture.

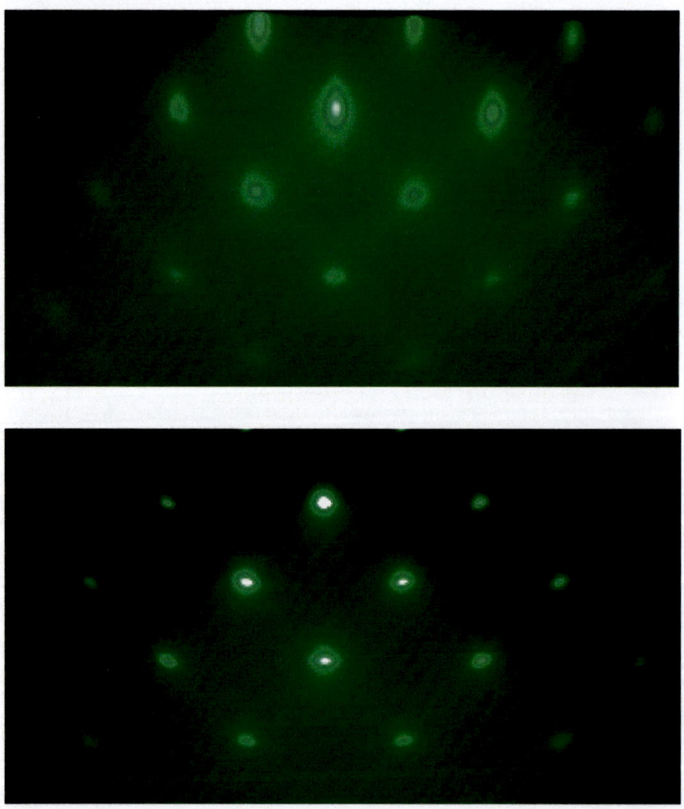

(a)

(b)

Figure 9. RHEED image for: (a) heteroepitaxially-grown 4-μm-thick ACSi film on an oxide buffer stack, (b) heteroepitaxially-grown 4-μm-thick ACSi film on a nitride buffer stack, indicating a high level of biaxially-oriented crystalline texture.

ACSi Film

The carefully grown top surface of the buffer stack approaches single-crystalline order and provides the ordered and robust surface necessary to grow biaxially oriented grains (i.e., with preferred out-of-plane and in-plane crystallographic orientations) of near-single-crystalline ACSi films. Since the starting IBAD layers are oriented <001> with respect to the substrate, all the subsequent layers, including the ACSi films, that are grown epitaxially follow the same orientational relationship with respect to the substrate. Using e-beam evaporation, we have been able to grow epitaxial ACSi films at deposition rates of up to ~2 nm/s, and at approximate substrate temperatures of 700 – 800 °C. The thickness of silicon films that we have deposited in situ on these buffer stacks ranges from 0.1 to 5 μm. Also, we have created various doping profiles (in situ during film growth, or ex situ by diffusion) depending on the device application. Figure 9 shows RHEED images of 4-□m-thick ACSi films deposited on both insulating oxide and conductive nitride buffer stacks on metal-alloy tapes. These images show good crystalline order in the ACSi films. Also, they indicate an in-plane lattice constant of 0.54 nm, which is close to bulk values.

Microstructure of ACSi Films

We studied the microstructure, crystallographic orientation, and epitaxial relationship of the ACSi films and buffer layers by x-ray diffraction (XRD), transmission electron microscopy (TEM), selective area diffraction (SAD), atomic force microscopy (AFM), scanning electron microscopy (SEM), and electron backscattering diffraction (EBSD). These measurements confirm our *in situ* RHEED results that all biaxially oriented buffer layers and ACSi films grow predominantly with a <001> orientation perpendicular to the substrate surface with a high degree of biaxial grain alignment.

Figure 10 shows XRD θ-2θ scans, indicating <00n> growth for all epitaxial layers of MgO, γ-Al_2O_3, TiN, and ACSi for multilayer structures of ACSi/γ-Al_2O_3/epi-MgO/IBAD-MgO/Y_2O_3//tape and ACSi/epi-TiN/IBAD-TiN//tape. In the rest of this chapter, we will refer to these structures as ACSi/oxide-stack//tape and ACSi/nitride-stack//tape, respectively. In both cases, the 4-μm-thick ACSi films had B doping profiles of ~10^{19} 1/cm^3 at the bottom 1 μm (substrate side) and 10^{16}-10^{17} 1/cm^3 at the top 3 μm. We achieved these doping profiles by using Si evaporation sources with different levels of B addition. As indicated in the boxes in figure 10 (a) and (b), both ACSi films had similar out-of-plane ($\Delta\omega_{Si}$ = 1.5 and 1.6 °C) and in-plane ($\Delta\phi_{Si}$ = 3.3 and 4.1 °C) mosaic spreads.

The cross-sectional TEM images shown in figures 1 and 2 indicate sharp interfaces between the buffer stacks and the 0.4-μm-thick ACSi films. They also show that the contrast due to stacking faults and misfit dislocations in ACSi anneal out to some degree with increasing thickness. Figure 11 shows similar features for a 4-μm-thick ACSi film on an oxide-stack on metal tape. It illustrates a dense ACSi film with a smooth surface and a columnar structure of grains with low-contrast, i.e., low-defect density, grain boundaries.

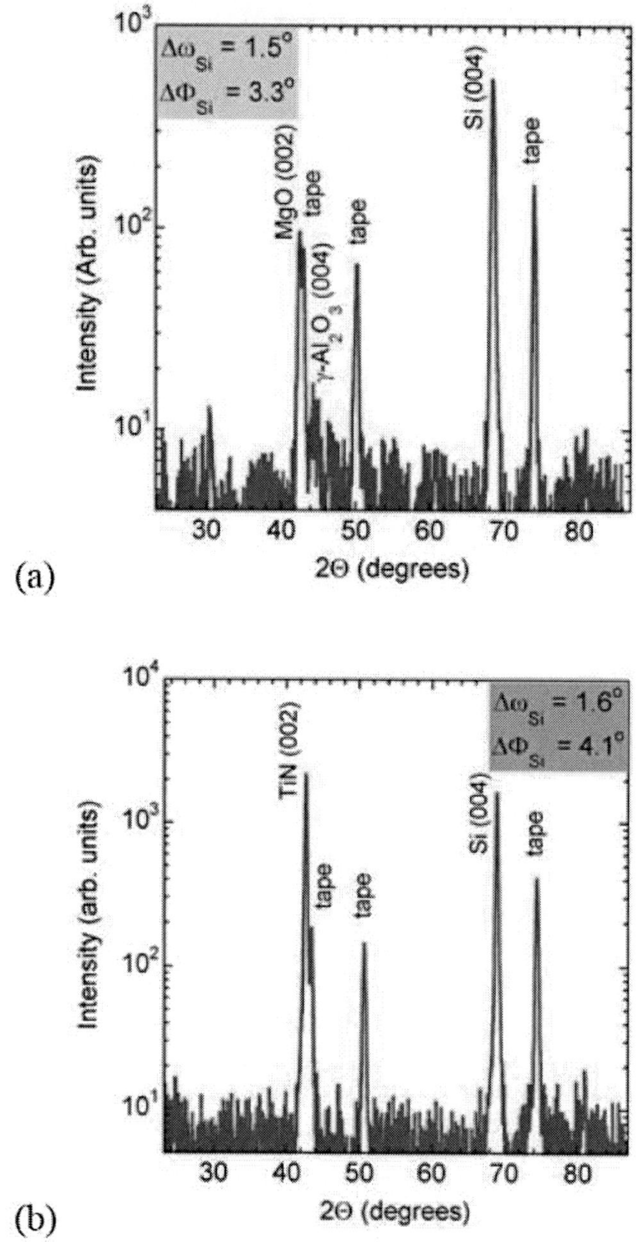

Figure 10. XRD θ-2θ scans for 4-μm-thick ACSi films on: (a) γ-Al_2O_3/epi-MgO/IBAD-MgO/Y_2O_3//tape (ACSi/oxide-stack//tape), and (b) epi-TiN/IBAD-TiN//tape (ACSi/nitride-stack//tape), indicating <00n> growth for all epitaxial layers. Also indicated are out-of-plane and in-plane mosaic spreads $\Delta\omega_{Si}$ and $\Delta\phi_{Si}$, respectively.

Figure 12 shows plan-view TEM micrographs and SAD images on 4-μm-thick ACSi films on both oxide and nitride stacks on metal tape. The bright field TEM images indicate planar grain sizes of order 0.5 μm, and SAD spots reveal highly oriented ACSi films, with preferred <001> orientation.

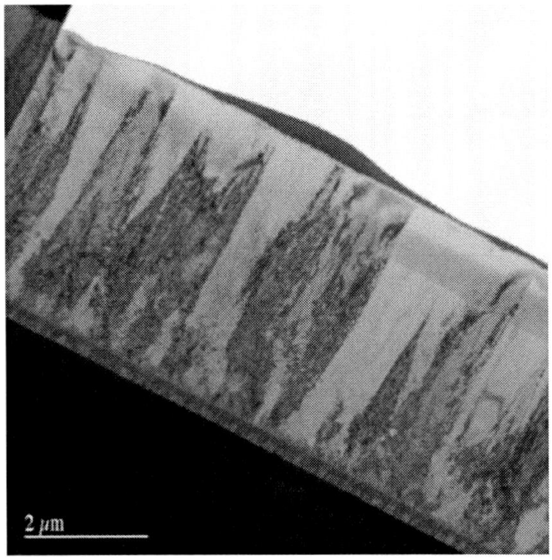

Figure 11. Tranmission electron microscopy (TEM) image of a 4-μm-thick ACSi film on an oxide buffer stack on metal tape.

Figure 12. (a) Plan-view TEM and (b) selective area diffraction (SAD) image of an ACSi film on an oxide stack on metal tape; (c) Plan-view TEM and (d) SAD image of an ACSi film on a nitride stack on metal tape.

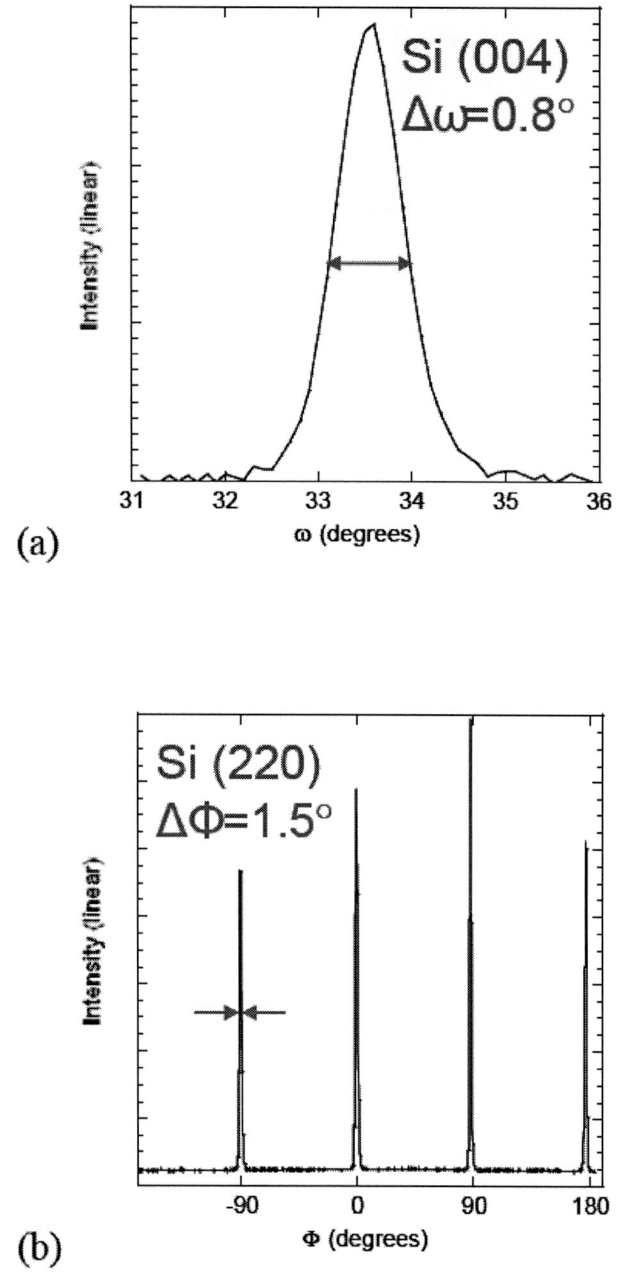

Figure 13. XRD (a) rocking curve and (b) phi scan for an optimized 0.4-μm-thick ACSi film on an oxide-stack on metal tape, yielding out-of-plane and in-plane mosaic spreads of $\Delta\omega_{Si} = 0.8°$ and $\Delta\phi_{Si} = 1.5°$, respectively (ref. [5]).

With optimized process conditions, we have achieved ACSi films with very low mosaic spreads: $\Delta\omega_{Si} = 0.8$ ° and $\Delta\phi_{Si} = 1.5$ ° (see figure 13). Such ACSi films also show surface microstructure reminiscent of epitaxial films grown on single-crystal substrates, as shown in figure 14. Figure 14 is a composite where the left side shows the amplitude of the AFM signal, which is sensitive to surface height, whereas the right side shows the phase of the AFM signal, which is sensitive to change in height. Figure 14 reveals a faceted surface in this optimized ACSi film with an approximately 24 nm step height. It also shows a relatively smooth surface with a root mean square (rms) surface roughness of ~16.7 nm (measured on a 5 μm x 5 μm area) for a total ACSi film thickness of ~4 μm. In addition, in such ACSi films with low mosaic spreads, the grain boundaries become almost indistinguishable by AFM.

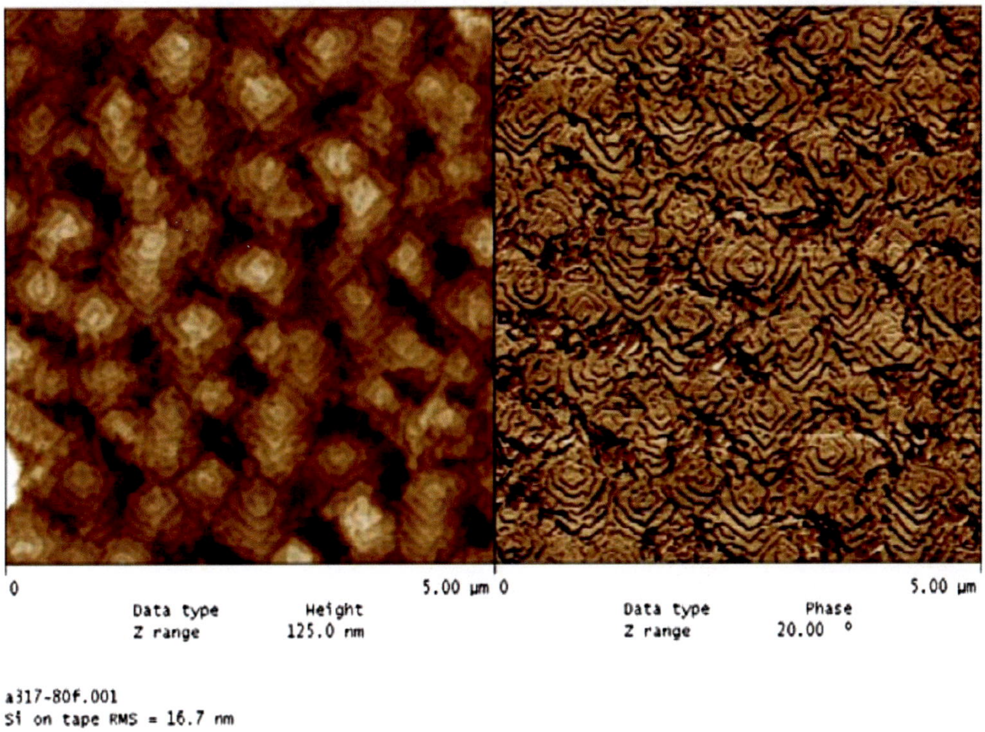

Figure 14. Atomic force microscopy (AFM) image composite (amplitude on the left, phase on the right) of an optimized 4-μm-thick ACSi film on an oxide stack on metal tape, showing a faceted surface and smooth grain boundaries.

We have also performed EBSD measurements to gain more detailed information, at the grain level, about grain alignment in ACSi films. We have observed a strong correlation between XRD results and EBSD results. Figure 15 is a composite, for an optimized 4-μm-thick ACSi film, that combines an SEM image with a {100} pole figure (PF) perpendicular to the sample surface and plots of grain misorientation distributions (i.e., mosaic spread) relative to the Z axis (tilt) and X axis (rotation). The PF shows that ACSi grains are predominantly oriented towards <001>. Plots of tilt and rotation distributions also indicate a high degree of grain alignment: the tilt (Z) scatters over 2-3° and the rotation (X) scatters over 4-6°. We do not observe a strong correlation between tilt and rotation, suggesting that the misorientation follows a Gaussian distribution.

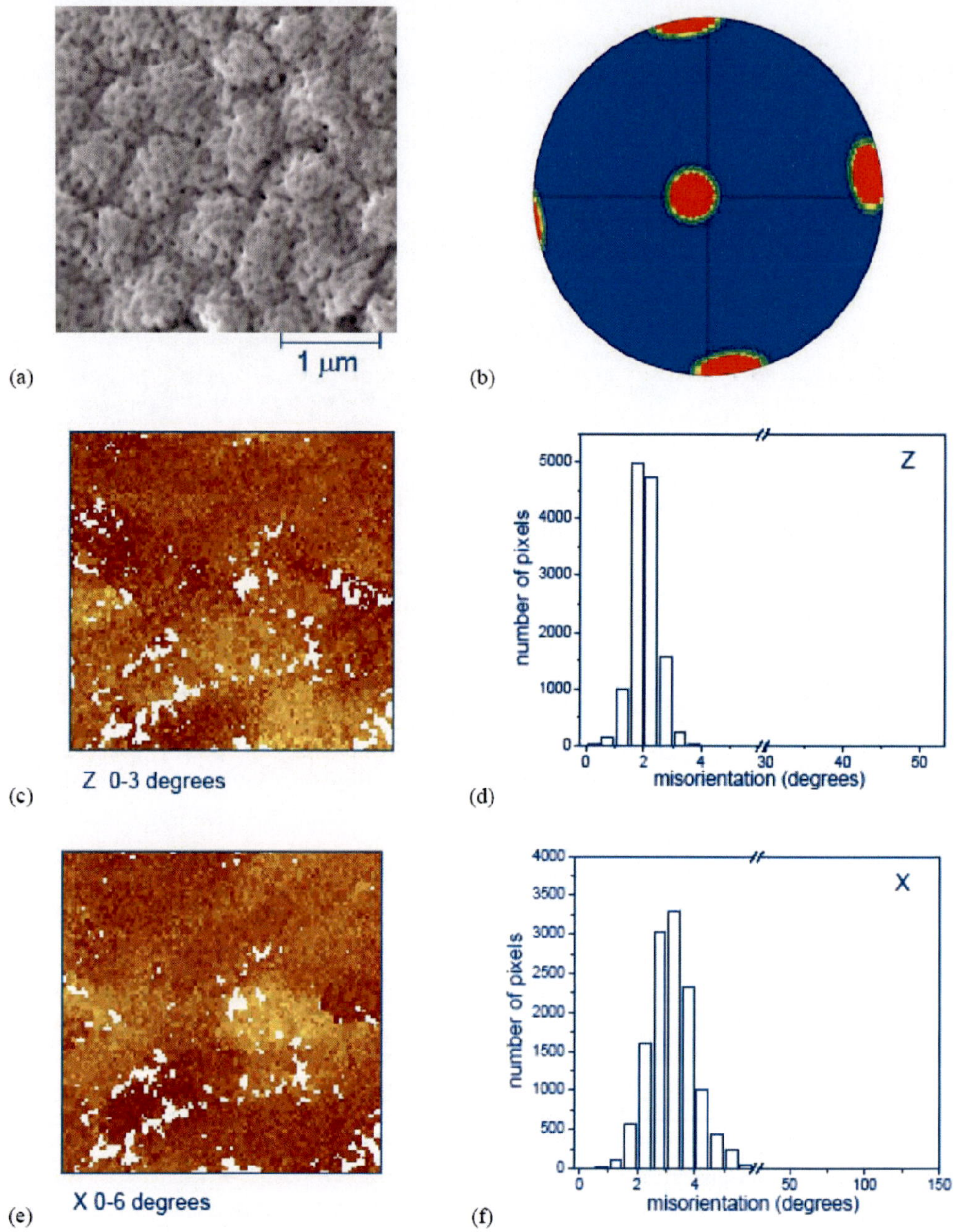

Figure 15. Electron backscattering diffraction (EBSD) results for a 4-μm-thick ACSi film on an oxide stack on metal tape: (a) SEM image that shows a smooth surface and ~0.5 μm grain size, (b) {100} pole figure (PF) that shows that the ACSi grains are predominantly oriented towards <001>, (c) map for grain misorientation distribution (i.e., grain mosaic spread) relative to the Z axis (tilt), (d) bar graph of grain misorientation distribution relative to the Z axis, (e) map of grain misorientation distribution relative to the X axis (rotation), (f) bar graph of grain misorientation distribution relative to the X axis.

ELECTRICAL PROPERTIES OF ACSI FILMS

To study the electrical properties of ACSi films, we have prepared ACSi films with varying grain alignment, thickness and doping concentration, and used both quantitative and qualitative measurement tools. For quantitative analysis of the effect of microstructure on majority carrier transport, we have photolithographically patterned the ACSi films for Hall measurements. For qualitative analysis of electronic properties at the nanometer scale, we have used surface voltage measurements based on AFM and current transport measurements based on STM.

For the Hall measurements, we used bridge patterns with channel widths of 400 μm and channel lengths of between 0.4 and 1.2 mm. The contact electrodes were made with sputtered Al followed by forming gas (94% Ar, 6% H_2) anneal at 350 °C for 30 mins. For comparison purposes, we also patterned a lightly B-doped 400-nm-thick <001>-oriented SOS film from a commercial vendor (Applied Technologies Enterprises, Inc., Irmo, South Carolina, US) and measured Hall properties on this sample following the same processing procedures. We measured room-temperature Hall mobility of about 90 $cm^2/V{\cdot}s$ for a doping concentration of about $4x10^{16}$ cm^{-3} for our optimized p-type samples, whereas the commercial p-type SOS sample yielded 135 $cm^2/V{\cdot}s$ at a doping concentration of $8.1x10^{14}$ cm^{-3}. This mobility value for optimized ACSi films is within a factor of two of the best values reported for similarly thin single-crystalline Si films on single-crystal substrates and is about one thousand times better than what is achievable with standard amorphous Si film technology.[5]

Previously, we defined a total mosaic spread, *TMS*, from three orthogonal mosaic spreads (tilt along x, $\Delta\omega_x$; tilt along y, $\Delta\omega_y$; and, rotation along z, $\Delta\phi_z$) as:[5]

$$TMS = (\Delta\omega_x^2 + \Delta\omega_y^2 + \Delta\phi_z^2)^{1/2} \quad (1)$$

Also, to be able to compare the carrier mobility of samples with different doping concentrations, we have normalized the carrier mobility of the films with respect to bulk single-crystal Si (without taking into account the film thickness effects), and used a Hall factor of 0.8 to convert drift mobilities to Hall mobilities.[11,12] Figure 16 shows the normalized carrier mobility versus total mosaic spread for ACSi films with varying thickness and doping concentration. The lines are a guide to the eye. We observe that the lower the carrier concentration and the thinner the Si film, the more sensitively the mobility depends on mosaic spread. For the 0.2-μm-thick ACSi samples (triangles) with doping concentrations of 10^{16}-10^{17} cm^{-3}, the normalized mobility increases by more than one order of magnitude for a decrease in total mosaic spread of only a few degrees, from 6.5° to 3.5°. For 0.4-μm-thick samples (circles) with similar doping concentrations, the effect is slightly reduced. On the other hand, 2-to-4-μm-thick, highly-doped (10^{18}-10^{19} cm^{-3}) ACSi films (squares) show only marginal improvement in normalized carrier mobility with decreasing mosaic spread. We note that the normalized carrier mobility values are already quite high even at large mosaic spread values for such films.

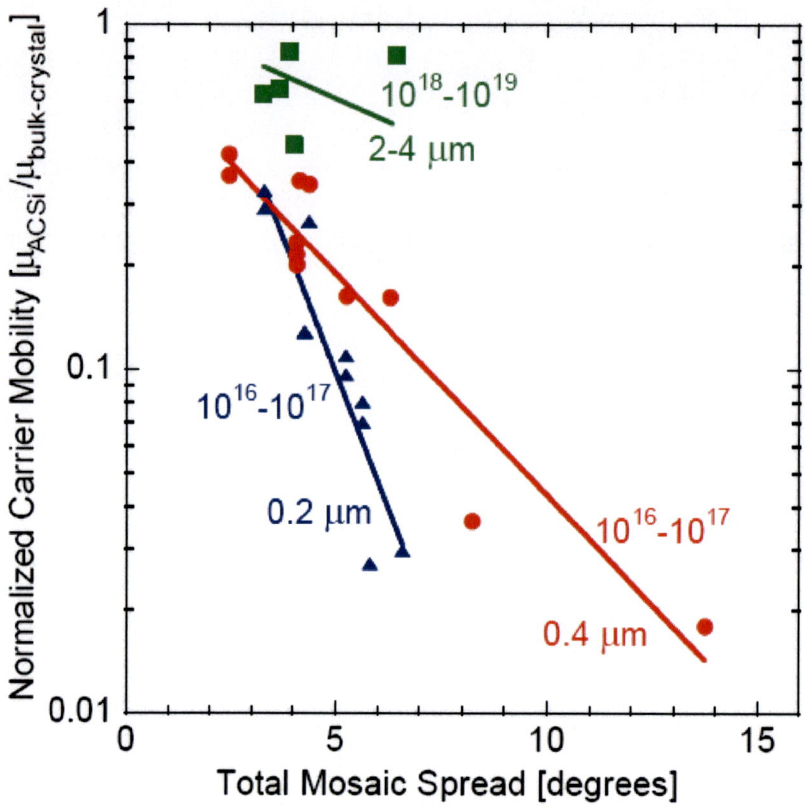

Figure 16. Normalized carrier mobility vs total mosaic spread for ACSi films with varying thickness and carrier concentration (cm^{-3}) on insulating oxide stack on metal tape.

To further study the effects of inter-grain alignment on the majority carrier transport across ACSi films, we use a simple model that combines carrier scattering in the grains and at the grain boundaries, with associated definitions of intra-grain mobility (μ_G) and inter-grain mobility (μ_{GB}), respectively.[13] The details of this model are given elsewhere.[14] The total mobility μ according to this model is given by:

$$\mu^{-1} = \mu_G^{-1} + \mu_{GB}^{-1}, \tag{2}$$

where $\mu_G = C\mu_B$ (C: constant)[15] and $\mu_{GB} = C'qL(2\pi m^* kT)^{-1/2} e^{-E_B/kT}$ (C': constant, q: unit charge, L: grain size, m^*: effective mass, k: Boltzmann constant, T: temperature, and E_B: energy barrier height at the grain boundary).[14,16]

In the model, we assume intra-grain mobility is similar to bulk mobility of single-crystalline Si. The enhanced scattering, and thus reduced mobility, at the grain boundary can be represented by an energy barrier height E_B, as shown in figure 17.

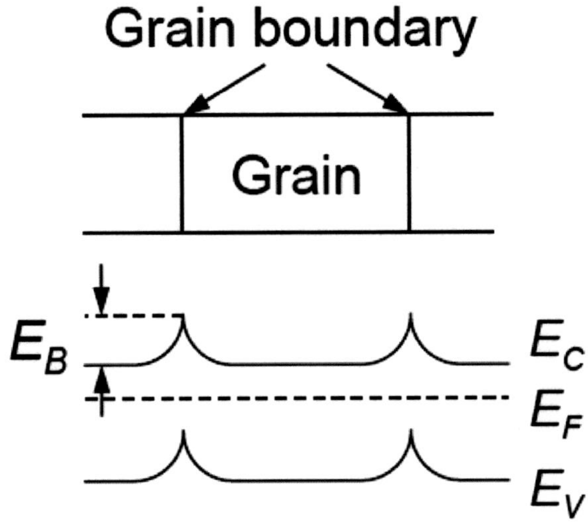

Figure 17. Model used for the effect of grain boundaries, with associated energy barrier height E_B, on majority carrier transport, where E_C, E_F, and E_V are the conduction band energy level, the Fermi level, and the valence band energy level, respectively.

Figure 18 shows the effect of temperature as well as of a hydrogen passivation process on the carrier mobility of ACSi films with different *TMS*. The details of this study have been published elsewhere.[17] As a reference, we have also included results on a similarly processed SOS sample. We observe that ACSi films with low *TMS* values show increasing mobility with decreasing temperature, similar to what is observed in single-crystalline Si films on single-crystal substrates (for example, SOS). This behavior is indicative of carrier scattering that is dominated by processes within the grains. On the other hand, ACSi films with large-angle grain boundaries (i.e., with large *TMS*) show behavior that indicates carrier scattering that is dominated by grain boundary processes. In addition, ACSi films with low *TMS* and the SOS reference sample do not show any change in behavior with a hydrogen passivation treatment, whereas ACSi films with large *TMS* show appreciable improvement. This again supports our earlier results that ACSi films with predominantly low-angle grain boundaries have a low density of defects, and they don't need grain boundary modification (i.e., hydrogen passivation and/or grain recrystallization) to exhibit good transport properties. In fact, a detailed study we have performed show that as the *TMS* decreases from ~15° to ~3°, the energy barrier height associated with the grain boundary scattering decreases monotonically from ~150 meV to less than 1 meV, respectively.[17]

We have also used surface voltage measurements combined with AFM, and local excitation of electron-hole (e-h) pairs combined with STM to study the effect of grain alignment on electrical properties in more microscopic detail. Figure 19 shows an AFM image (on the left) of an ACSi film with a *TMS* of ~4°, and a corresponding surface potential map (on the right). We have used the surface potential measurement as a qualitative tool here to determine variations in the surface potential of ACSi films due to changes in work function and/or resistivity. The surface potential follows the granular structure of the ACSi film, with a relatively uniform modulation across the whole scan area.

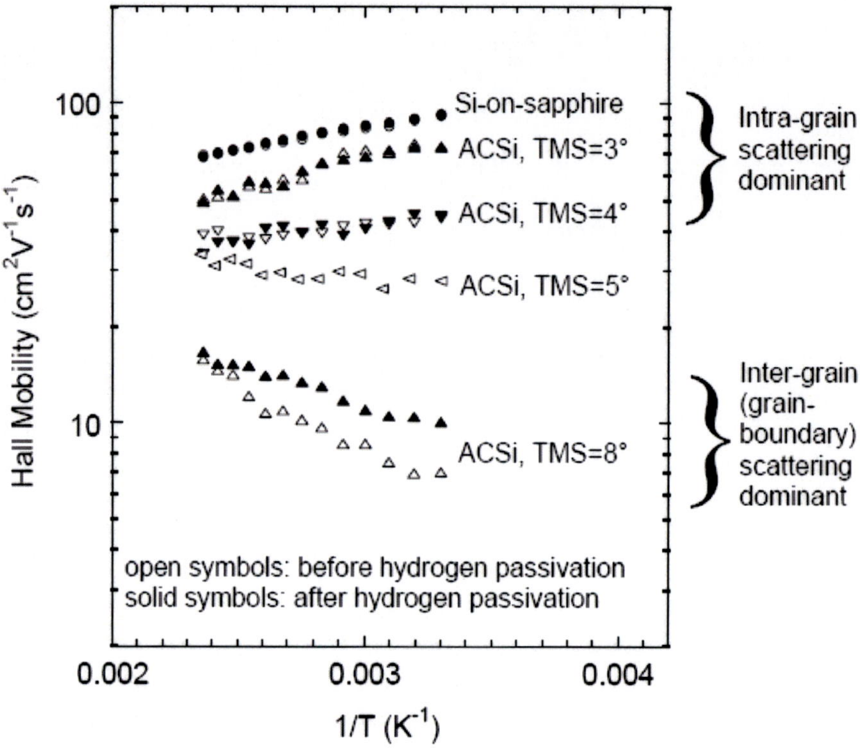

Figure 18. Hall mobility vs inverse temperature for ACSi films with varying total mosaic spread *TMS* and a Si-on-sapphire sample, indicating that ACSi films with low *TMS* behave like single-crystalline Si films in terms of both temperature dependence and hydrogen passivation behavior (ref. [17]).

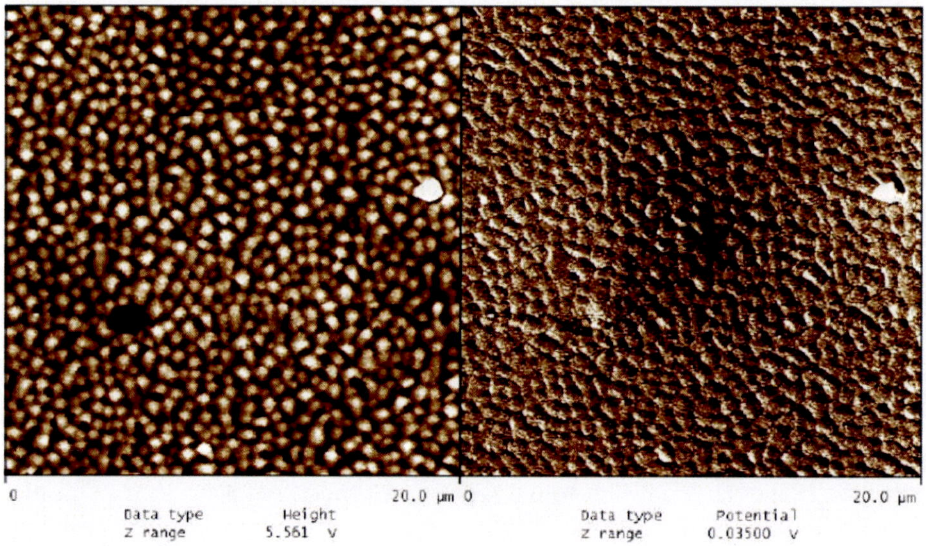

Figure 19. AFM surface height (on the left) and surface potential (on the right) map of a 4-µm-thick ACSi film with a TMS of ~4°, showing low-intensity uniform modulation of surface potential in connection with grain boundaries.

A combination of STM and the local excitation of e-h pairs (i.e., electro-assisted STM) also allows us to investigate the effect of grain boundaries on the electrical properties of ACSi films at the grain level.[18] The details of these experiments are published elsewhere.[17] Figure 20 shows an STM constant-current image (on the left) of a 0.2-µm-thick p-type ACSi film with relatively large *TMS* of ~8°, and a corresponding map (on the right) of the contribution of the electron excitation to the tunneling current (I_{cc}). In figure 20, we observe domains where the collected current sensed by the STM tip decreases substantially (indicated by arrows), revealing those grain interiors delimited by an active grain boundary acting as a barrier for electron transport.

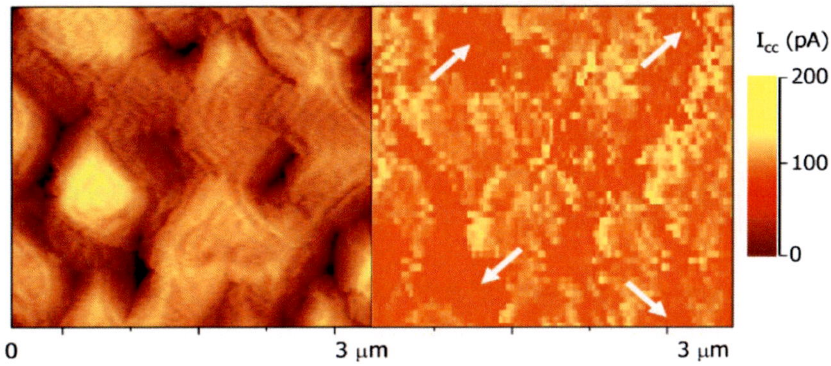

Figure 20. A constant current mode STM image (on the left) of a 0.2-µm-thick ACSi film with *TMS* of ~8°, and a corresponding map (on the right) of the electro-assisted tunneling current (I_{cc}) (arrows indicate the domains where the collected current sensed by the STM tip decreases substantially) (ref.[17]).

Figure 21 shows an STM constant-current image (on the left) of a 1-µm-thick p-type ACSi film with a relatively small *TMS* of ~4°, and a corresponding map (on the right) of the contribution of the electron excitation to the tunneling current. In figure 21, we observe a relatively uniform collected current, indicating that the electron transport is not impeded significantly by the presence of low-angle grain boundaries.

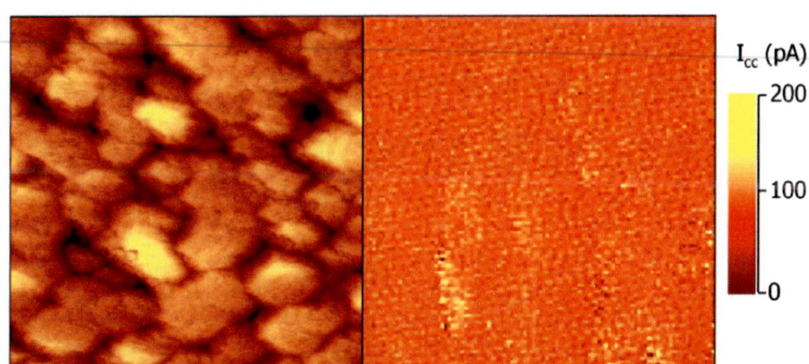

Figure 21. A constant current mode STM image (on the left) of a 1-µm-thick ACSi film with *TMS* of ~4°, and corresponding map (on the right) of the electro-assisted tunneling current (I_{cc}), indicating relatively uniform collected current throughout the area under investigation (ref.[17]).

We have also studied the effect of grain alignment on the surface electronic properties of ACSi films using metal-oxide-semiconductor capacitor structures. For this work, we have fabricated MOS capacitors based on ACSi films with thermally-grown oxides on them. The details of the fabrication of and the capacitance-voltage (C-V) measurement on these structures are given elsewhere.[19] In figure 22, C-V results show well defined regions of carrier accumulation, depletion, and inversion for a 4-μm-thick ACSi film with a light (p) to heavy (p+) doping profile from ACSi surface to substrate side. An accumulation capacitance appears at negative voltages indicating that the top layer of the ACSi film does not have a high density of defects. It also shows typical high- and low- frequency effects at 100 kHz and 1 kHz, respectively. The low frequency effect at 1 kHz under positive voltage bias is indicative of carrier inversion from p-type to n-type at the surface of the ACSi film. This inversion effect is crucial for operation of thin-film transistor (TFT) structures such as metal-oxide-semiconductor field-effect-transistors (MOSFETs).

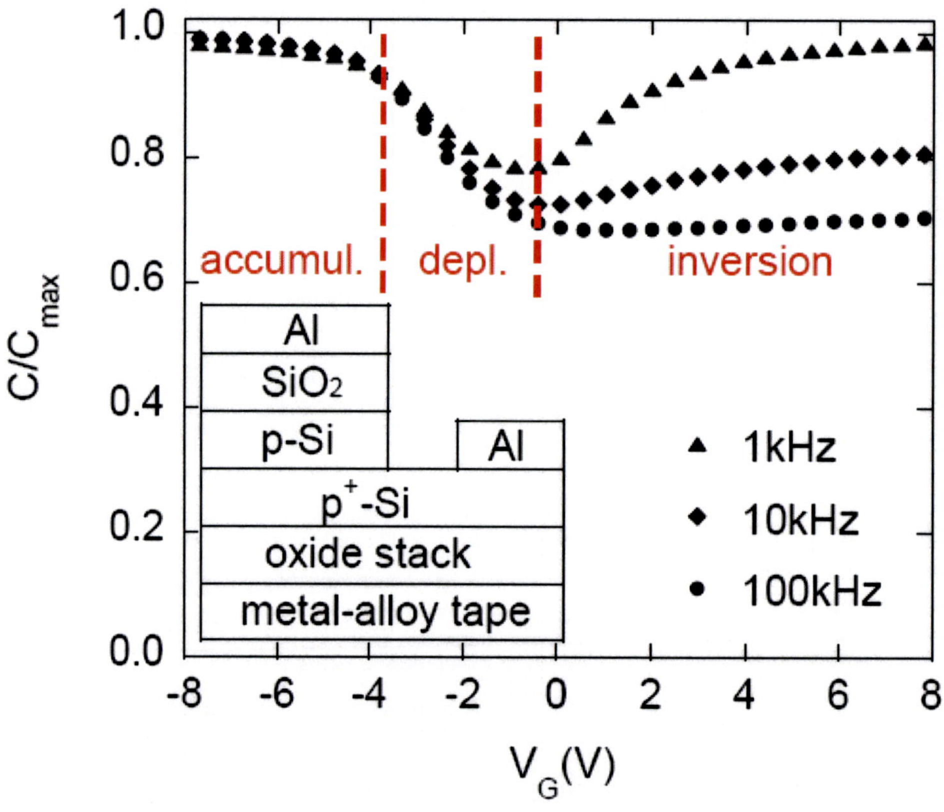

Figure 22. Normalized capacitance (C/C_{max}) vs gate voltage (V_G) at three different measurement frequencies (1kHz, 10 kHz, 100 kHz), showing carrier accumulation, depletion and inversion in the ACSi film. Also shown is a schematic cross-section of the metal-oxide-semiconductor (MOS) capacitor structure (ref. [19]).

Figure 23 (a) shows high frequency (1 MHz) C-V curves of MOS capacitors based on ACSi films with *TMS* of 6.5°, 8°, 12°, and 14°. As *TMS* increases, the C-V curve stretches out further along the voltage axis indicating a progressively higher density of surface/interface traps.[20] By using a model based on the high-frequency method,[21] we calculate the interface trap density at mid-gap (D_{it}) for ACSi films with these varying *TMS* values.[19] The results in figure 23(b) show that with improved alignment of Si grains (i.e., with decreasing *TMS*) ACSi films monotonically approach the low defect levels observed in single-crystal bulk Si and near-single-crystalline film Si (i.e., SOS).

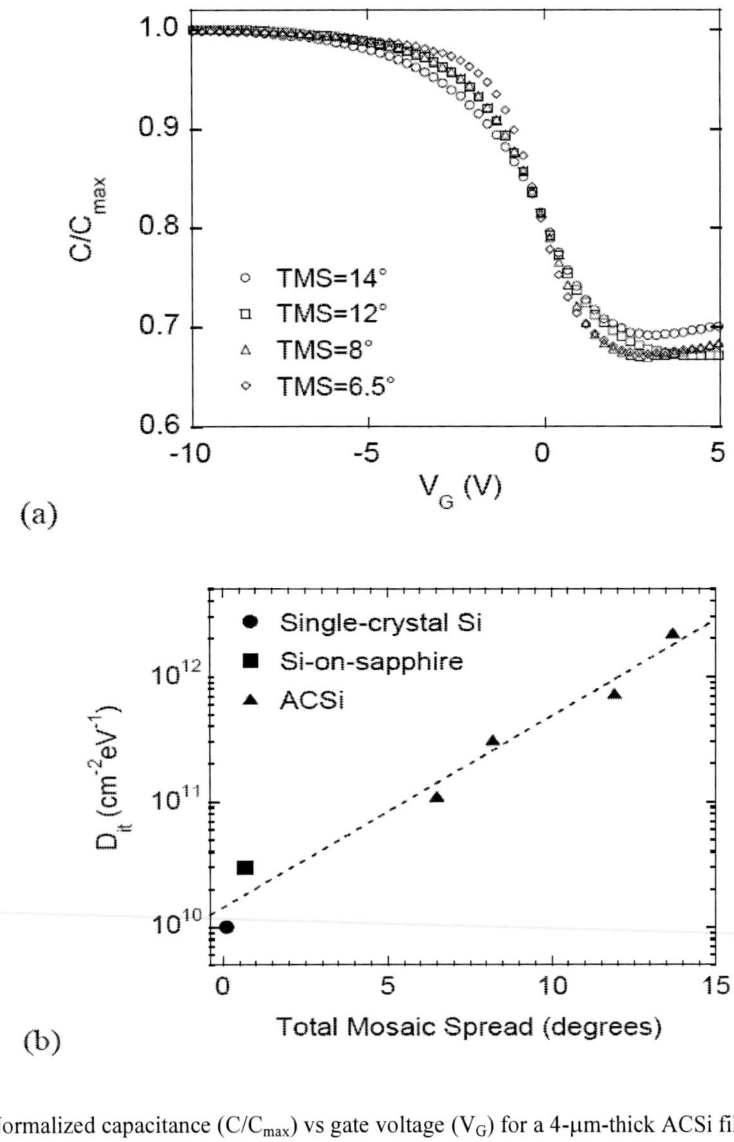

Figure 23. (a) Normalized capacitance (C/C_{max}) vs gate voltage (V_G) for a 4-μm-thick ACSi film with four different *TMS* values, showing increased "stretch-out" with increasing *TMS*. (b) Interface trap density at mid-gap (D_{it}) vs TMS, showing decreasing trap density with improved grain alignment (dashed line is a guide to the eye). Data on single-crystal Si and SOS are also included for reference (ref. [19]).

TECHNOLOGICAL POTENTIAL

The optimized ACSi films demonstrate improved electrical performance due to improved alignment and connectivity of the grains in the films. Our microstructural and electrical measurement results demonstrate that improving the intergrain alignment of sub-micron-sized grains in ACSi films is an effective alternative to increasing grain size (for example, by laser annealing) in order to approach single-crystal-like electrical performance. Currently, no other technology combines the high-performance aspects of single-crystalline silicon wafers with the low-cost aspects of amorphous and polycrystalline silicon films, which use inexpensive, non-single-crystalline substrates. As such, the ACSi technology that we have described here could allow the Si industry to lower its costs while maintaining or improving the performance of its products.

The current electronic display market is dominated by active-matrix liquid crystal displays (LCDs), which are mostly based on amorphous Si thin-film transistors (TFTs) on glass substrates.[22] However, amorphous Si has very low carrier mobility that causes limited circuit speed and large pixel size, which, in turn, lead to low-resolution displays with limited refresh rates. Also, glass, an inflexible material, cannot deliver many of the potential benefits that flexible substrates offer, such as non-flat displays. The ACSi technology has commercial potential in flexible display applications, because the carrier mobility in ACSi silicon films is about one thousand times better than in amorphous Si films, and within a factor of two of single-crystalline Si films. The improved mobility we obtain in ACSi films on flexible substrates could allow manufacture of displays with improved display resolution, refresh rate, and system reliability, combined with reduced manufacturing costs.

Si is by far the most-widely used material (more than 98% of the world market in 2005) in the photovoltaics industry.[23] However, with increasing Si feedstock and wafer costs, and competition from other renewable energy resources, the solar industry is seeking alternative technologies, mostly in the form of continuous film coating technologies that will reduce materials and fabrication costs while retaining high solar cell conversion efficiency. The emerging amorphous and polycrystalline Si film technologies aim to achieve large reductions in materials and fabrication costs by using 1-to-30 micron-thick Si films on inexpensive substrates, while tolerating some reduction in efficiency. Despite recent advances, current Si film photovoltaics technologies force buyers to sacrifice efficiency for cost savings. The ACSi technology, on the other hand, holds promise for combining the cost advantage of Si films with the high efficiency of bulk Si wafers for the development of cost-effective, high-performance Si solar cells.

CONCLUSION

The ACSi film technology we have described above provides near-single-crystalline Si films on polycrystalline metal-alloy tape substrates. Once optimized, these ACSi films show in- and out-of-plane mosaic spreads as low as 1.3° and 0.8°, respectively. We have observed similar crystalline quality for ACSi films deposited on both insulating oxide and conducting nitride IBAD buffer stacks, demonstrating the flexibility of the technique. We have observed a strong correlation between good grain alignment, i.e., low total mosaic spread (<5°), and improved electrical performance in these films. ACSi films with good grain alignment also showed high carrier mobilities (~50% that of single-crystal Si film, ~1000 times that of amorphous Si film), in addition to low potential barrier heights at grain boundaries (<1 meV) and low trap densities at the film surface ($\leq 10^{11}$ $cm^{-2}eV^{-1}$). We believe the ACSi technology that we have outlined above, once fully developed, could make a significant impact in electronics, especially in the areas of photovoltaics and flat panel displays.

Acknowledgements

This work was funded by Laboratory Directed Research and Development – Exploratory Research (LDRD-ER), Technology Maturation Fund, and MST Division Office at Los Alamos National Laboratory (LANL), and by DOE-EERE at National Renewable Energy Laboratory (NREL). We thank Jung-Kun Lee, Vlad Matias, Terry Holesinger, Quanxi Jia, Paul Arendt, Brady Gibbons, Randy Groves, Ray DePaula, Liliana Stan, Ian Campbell, and Darryl Smith of LANL for their various technical contributions to the work at LANL, and Jeffrey O. Willis of LANL for his useful comments on the manuscript.

References

[1] Reade, R. P.; Berdahl, P.; Russo, R. E.; Garrison, S. M. *Appl. Phys. Lett.* 1992, 61, 2231.

[2] Arendt, P. N.; Foltyn, S. R. *Mater. Res. Soc. Bull.* 2004, *29*, 543.

[3] Iijima, Y.; Kakimoto, K.; Sutoh, Y.; Ajimura, S.; Saitoh, T. *Supercond. Sci. Tech.* 2004, *17*, 264.

[4] Findikoglu, A. T.; Kreiskott, S.; te Riele, P. M.; Matias, V. *J. Mater. Res.* 2004, *19*, 501.

[5] Findikoglu, A. T.; Choi, W.; Matias, V.; Holesinger, T. G.; Jia, Q. X.; Peterson, D. E. *Adv. Mater.* 2005, *17*, 1527.

[6] Dong, L.; Zepeda-Ruiz, L. A.; Srolovitz, D. J. *J. Appl. Phys.* 2001, *89*, 4105.

[7] Zepeda-Ruiz, L. A.; Srolovitz, D. J. *J. Appl. Phys.* 2002, *91*, 10169.

[8] Huhne, R.; Fahler, S.; Holzapfel, B.; Oertel, C. G.; Schultz, L.; Skrotzki, W. *Physica C* 2002, *372*, 825.

[9] Cho, H. Y.; Park, C. J. *Physica E* 2003, *16*, 489.

[10] Usov, I.; Arendt, P.; Stan, L.; DePaula, R.; Wang, H.; Foltyn, S.; Dowden, P. *J. Mater. Res.* 2004, *19*, 1175.

[11] Lin, J. F.; Li, S. S.; Linares, L. C.; Teng, K. W. *Solid State Electron.* 1981, *24*, 827.

[12] Sze, S. M. *Physics of Semiconductor Devices*, Wiley-Interscience, New York 1981, pp. 30-35.

[13] Choi, W.; Matias, V.; Lee, J. K.; Findikoglu, A. T. *Appl. Phys. Lett.* 2005, *87*, 152104.

[14] Levinson, J.; Shepherd, F. R.; Scanlon, P. J.; Westwood, W. D.; Estes, G.; Rider, M. *J.Appl. Phys.* 1982, *53*, 1193.

[15] Kazmerski, L. L.; Berry, W. B.; Allen, C. W. *J. Appl. Phys.* 1972, *43*, 3515.

[16] Seto, J. Y. W. *J.Appl. Phys.* 1975, *46*, 5247.

[17] Choi, W.; Findikoglu, A. T.; Romero, M. J.; Al-Jassim, M. *J. Mater. Res.*(in press).

[18] Romero, M. J.; Jiang, C. S.; Noufi, R.; Al-Jassim, M. *Appl. Phys. Lett.* 2005, *87*, 172106.

[19] Choi, W.; Lee, J. K.; Findikoglu, A. T. *Appl. Phys. Lett.* (in press).

[20] Sze, S. M. *Physics of Semiconductor Devices*, Wiley, New York, 1981, pp. 362-407.

[21] Termin, L. M. *Solid-State Electron.* 1962, *5*, 285.

[22] Toshiharu, S. *J. Appl. Phys.* 2006, *99*, 11.

[23] Green, M. A.; Emery, K.; King, D. L.; Igari, S.; Warta, W. *Progress in Photovoltaics: Research and Applications*, 2005, *13*, 49.

In: Progress in Advanced Materials Research
Editor: Nicolas H. Voler, pp. 185-219
ISBN: 978-1-60021-893-4

Chapter 7

FABRICATION AND CHARACTERIZATION OF ONE-DIMENSIONAL NANOSCALE HETEROSTRUCTURES

Jinhua Zhan
Department of Chemistry, Shandong University, Jinan 250100, China and National Institute for Materials Science, Namiki 1-1, Tsukuba, Ibaraki 305-0044, Japan

ABSTRACT

One-dimensional (1D) nanomaterials (such as nanowires, nanotubes, nanobelts and nanorods) have attracted considerable attention due to their potential applications as building blocks in nanoscale circuits, optoelectronic, electromechanical and sensing devices. 1D nanostuctures with controlled diameters can act as a model system for systematic studies on the size-dependent electrical, optical, and thermal transport phenomena. Precision engineering or tailoring of materials is always the goal of much of the effort driving synthetic technology of 1D nanomaterials. The synthesis and fabrication of 1D nanostuctures nanostructures can be approached from one of two directions. One is the "top down" approach in which nanostructures are generated using a number of nanolithographic techniques. The other is the "bottom up" approach in which the nanostructures are built up from individual atoms or molecules. This is the basis of most "cluster science" as well as synthesis of crystal materials, usually via chemical means. In contrast with the nanolithographic techniques, the chemical method provides an attractive route to produce 1D nanostuctures of various materials at large scale and reasonably low cost.

This chapter will focus on our current research activities on the fabrication and characterization of one-dimensional nanowire heterostructures. The authors' endeavors over the recent years with respect to the chemical synthesis of novel inorganic 1D nanostructures via simple thermal evaporation, their detailed structural analysis using state-of-the-art electron microscopy techniques, and electrical property evaluation are presented. These nanostructures including carbon nanotube or silicon-based heterostructures, namely, gallium-filled carbon nanotubes, Ga_2O_3-C core-shell 1D nanostructures, Si-ZnS side-by-side composite nanowires, Si-Cd_4SiS_6 composite nanowires and Si-In end-to-end nanowire heterostructures. These materials are thoroughly analyzed by using Transmission electron microscopy (TEM), high-resolution TEM, scanning transmission electron microscopy (STEM) and related EDX mapping, field-emission measurements and cathodoluminescent spectroscopy. Potential applications of these novel nanostructures, using as nanothermometers, nanoscale optoelectronics, electronics, were also demonstrated.

1. INTRODUCTION

One-dimensional (1D) nanostructures play an important role in the nanotechnology due to their utilization in building blocks of nanoscale circuits, optoelectronic, electrochemical and electromechanical devices [1]. It has been reported that semiconducting nanowires can often exhibit novel functional electronic and optical properties owing to their unusual structural one-dimensionality and related quantum-mechanical confinement effects. They are expected to demonstrated novel properties such as nanowire lasing, phonon and electron transport, chemical sensing, ultimate field emission (flat panel displays) and gate-driven conductivity (field-effect transistors). Some pioneering works on the fabrication of semiconducting nanowires and evaluation of their useful functions in photonics and electronics are promoting the successful fabrication of various ultimate devices, such as logic gates, single-electron transistors, superconductors, light-emitting diodes and lasers [2-7].

1D nanostructures such as rods, wires, belts and tubes can be fabricated through a number of advanced nanolithographic techniques, thermal evaporation and solution-based methods [8-15]. Carbon nanotube can also serve as nanoscale containers to host foreign elements, compounds and/or substances. 1D nanomagnets, nanocables and other functional nanomaterials may be synthesized through taking advantages of tiny nanotube channel dimensions and related confinement effects, and functional properties of guest species encapsulated within, e. g. ferromagnetism, piezoelectricity, electro- and thermoconductivity etc [16]. Focusing on new properties and a wide selection of nano-building blocks for the growing demands of nanoscale devices on the nanosized electrical and optical lines, it has attracted much current attention to assemble known functional materials in either radial or axial directions within a single nanoscale building block to form heterostructures, namely coaxial nanocables. It is important to obtain a potential barrier between adjacent constituting components in a nanoscale device showing the rectifying current-voltage characteristics. This can be realized through the creation of various heterostructures including *p-n* junctions, metal-oxide-semiconductor junctions or metal-semiconductor contacts, which allow signal processing to be reliably performed [17]. Diverse heterostructures assembled in either radial or axial directions within a single nanometer-scale building block have been fabricated [18-27]. These have included the axial block-by-block group III-V (GaAs/GaP, InAs/InP) and group IV (Si/Ge) semiconductor nanowires, epitaxial Si/Ge core-shell nanowire heterostructures, hetereojunctions of carbon nanotubes and silicon nanowires, and NiSi/Si nanowire heterostructures. Semiconductor-core nanoscale heterostructures including SiC-C, Si/SiO_2-C and SiC/SiO_2-BN-C nanocables have also been synthesized via a reactive laser ablation or coating silicon carbide nanowires with carbon.

Silicon has long been considered to be the prime material for the needs of information technology electronics [28]. Recently, it has also been found valuable for photonics [29, 30]. The Si nanowires are particularly important for the miniaturization of the Si integrated circuits (IC). Various approaches, including vapor-liquid-solid (VLS) growth, solution-liquid-solid (SLS) growth and oxide-assisted (OA) growth, have been utilized for the synthesis of Si nanowires [31-34]. Metallic gallium melts at nearly room temperature and has one of the widest liquid ranges among all metals. Gallium possesses unusual crystallization behaviour [35]. Similarly to water, metallic Ga expands as it freezes. Ga displays rather complex phase diagram exhibiting many stable and metastable phases. At ambient pressure, α–Ga(I) phase is orthorhombic with eight atoms in the unit cell of a *Cmca* space group lattice. In this structure a Ga atom is bonded to the nearest neighbouring Ga atom in a manner of a diatomic molecule. With respect to the second closest neighbors located at the distances 10-13% longer than the first coordination sphere, the atom retains metallic properties. It is suggested that covalent and metallic bonds coexist in the Ga metal. In addition, a number of metastable phases including β, γ, δ have been identified. At low temperature Ga metal becomes a strong-coupling superconductor with a well-defined superconducting transition temperature, as confirmed by tunneling experiments [36]. Due to low melting point (29.8℃) and high reactivity, the preparation of isolated metallic Ga nanomaterials is a difficult task. Metallic Indium has similar properties as Ga. Its melting point is around 156 ℃. Gallium oxide (β-Ga_2O_3) is a wide-band-gap compound with a band gap of approximately 4.9 eV at room temperature, promising to act as transparent conductor for the coming generation of optoelectronic devices in the deep ultraviolet wavelength region [37]. The magnetism due to conduction electron spins in this material can exhibit an original memory effect within a temperature range of 4-293 K. Its n-type conduction is thought to be controlled by oxygen vacancies which act as shallow donors with binding energies around 30 meV [38]. One-dimensional crystalline β-Ga_2O_3 nanostructures, including nanowires, nanorods, nanobelts and nanoribbons, have been produced by thermal evaporation [39-42], laser ablation [43], arc-discharge [44, 45], microwave plasma [46, 47], and metal-catalyzed growth [48, 49]. A Vapor-liquid-solid (VLS) or vapor-solid (VS) growth mechanism have been generally attributed to their growth.

Our group has been devoted to the fabrication of nanowire heterostructures [50-61]. The present contribution is focused on the recent progress of our group with respect to the synthesis, detailed structural and crystallographic analysis and optical property measurements of inorganic nanowire heterostructures. These nanostructures including carbon nanotube or silicon-based heterostructures, namely, gallium-filled carbon nanotubes, Ga_2O_3-C core-shell 1D nanostructures, Si-ZnS side-by-side composite nanowires, Si-Cd_4SiS_6 composite nanowires and Si-In end-to-end nanowire heterostructures. These materials are thoroughly analyzed by using Transmission electron microscopy (TEM), high-resolution TEM, scanning transmission electron microscopy (STEM) and related EDX mapping, field-emission measurements and cathodoluminescent spectroscopy.

2. EXPERIMENTAL

2.1. Synthesis

The nanowire heterstructures were produced within our Laboratory by using general thermal evaporation processes in a vertical high-frequency (21 MHz) induction furnace. In brief, the furnace consists of a fused-quartz tube and an induction-heating cylinder made of high-purity graphite coated with a graphite fiber thermo-insulating layer. The length of an induction-heating cylinder can change with different experiments. The system had inlet and outlet graphite pipes for the introduction of the reactive and carrier gases. Figure 1 is a photograph of the working induction furnace used for the synthesis of gallium-filled carbon nanotubes (nanothermometers). Figure 2 depicts a schematic diagram of the vertical induction furnace used for the synthesis.

Figure 1. photograph of the working induction furnace for the synthesis of gallium-filled carbon nanotubes.

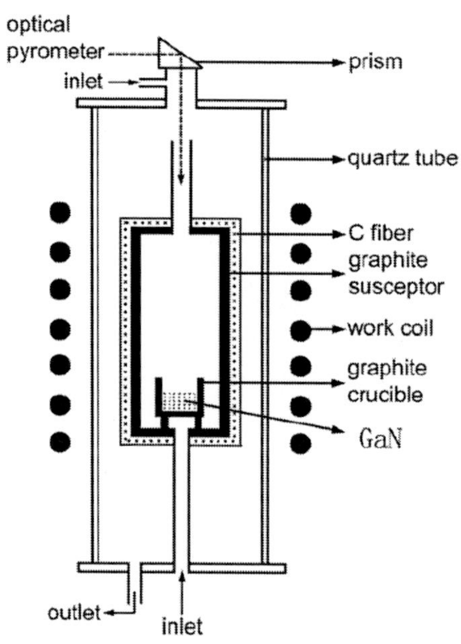

Figure 2. schematic diagram of the vertical induction furnace used for the synthesis.

Mixed precursor powders were put in graphite crucibles placed inside the induction-heated cylinder and heated by the induction currents to 1300-1800 ℃. The precursors sublimated and produced reactive vapors that have been carried by the oncoming carrier gases (N_2/CH_4, N_2 or Ar). Prominent vapor pressure and temperature gradients along and across the hot furnace zone promote one-dimensional nanostructure growth under the continuous condensation of reactive vapors in the low-temperature furnace regions, *e.g.* at outlet pipe and/or crucible or graphite cylinder walls. The fabricated nanomaterials were collected in the powder-like form after furnace cooling, and analyzed.

2.2. Analysis

The product was characterized using X-ray powder diffraction (XRD; RINT 2200) with CuKα radiation. The morphology of the product was observed using a JSM-6700F scanning electron microscope (SEM) operated at 10-20 kV. The product was ultrasonically dispersed in ethanol and transferred to a carbon-coated copper grid for transmission electron microscopy (TEM) observations. A field emission JEM-3000F high-resolution electron microscope operated at 300 kV equipped with an energy-dispersive X-ray spectrometer (EDX) was employed to perform the microanalysis. The detailed study was conducted in a high-resolution field-emission transmission electron microscope JEOL-3100FEF equipped with an energy X-ray (EDX) dispersive system and in-column Omega filter. Spatially-resolved elemental mapping of the nanomaterials was performed using the scanning transmission mode of the instrument (STEM mode, EDX mapping). Field-emission measurements of the nanocables were conducted in a vacuum chamber at a pressure of $\sim 6.0\times 10^{-5}$ Pa. A rod-like aluminum probe with 1 mm^2 in cross section was used as the anode. The as-obtained nanocables were used as the cathode. A dc voltage sweeping from 0 to 1100 V was applied to the sample. Cathodoluminescence (CL) spectra were collected with a high spatial resolution CL system attached to a thermal field-emission scanning electron microscope (TFE-SEM, Hitachi S4200).

3. Results and Discussion

3.1. Gallium-Filled Carbon Nanotubes [50-52]

Metallic gallium is often used in the fabrication of thermometers, for it has a long liquid range from room temperature to 2403℃ and a low vapor pressure even at high temperatures. If the liquid gallium is confined in the cavity of a carbon nanotube, it can be used as a nanotherometer for the application in a wide variety of microenvironments. Through the thermal evaporation of Ga_2O_3/C mixture, carbon nanotubes filled with metallic Ga can be obtained at a low yield. Ga_2O_3 is always the by-product.

In order to obtain high-quality carbon nanotubes filled with Ga and avoid oxide contaminants, an oxygen-free synthetic environment is required. We use the GaN as the gallium precursor and methane as carbon precursor. During the synthetic process, GaN sublimes and decomposes at high temperature (approximately 1300℃) to produce Ga vapor.

This vapor condenses at low temperature (approximately 800℃) to form Ga droplets. When the CH_4 gas flows through the induction furnace at high temperature, it decomposes to produce C clusters. Ga droplets induce the anisotropic growth of C nanotubes, meanwhile the nanotube cavities are filled with liquid Ga. The simultaneous generation of a Ga vapor and carbon clusters promote the uniform filling of C nanotubes. The growth process is schemed in figure 3.

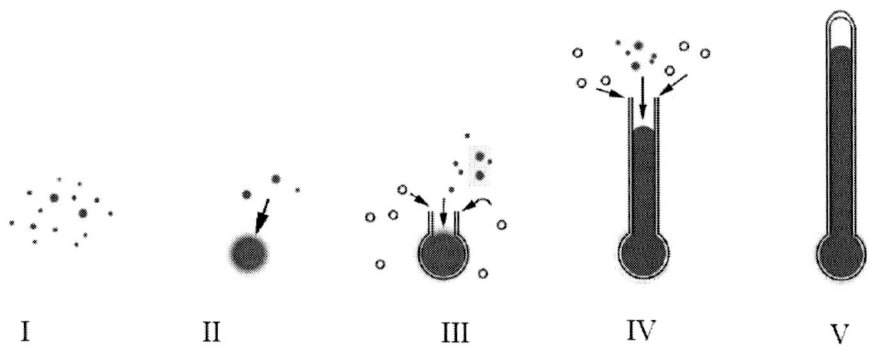

Figure 3. Growth scenario for the formation of liquid gallium columns sheathed with carbon: (I) vaporization of gallium, (II) condensation of gallium vapor to form gallium droplets, (III) nucleation of carbon nanotubes, (IV) growth of carbon nanotubes and in-situ filling with gallium, (V) sealing of carbon nanotubes.

A representative scanning electron microscopy (SEM) image of the product is depicted in figure 4a, which clearly showed that the product is composed of wire-like nanostructures with a length up to several micrometers. High-magnification SEM analysis reveals that the wire-like nanostructures have core-sheath structures. The end of a core-sheath structure is shown in figure 1 (b). Detailed structural and chemical analyses of individual nanostructures were carried out using transmission electron microscopy (TEM), electron diffraction (ED), Energy-dispersive X-ray spectroscopy (EDS). Figure 5a shows a TEM image of a typical as-obtained product, which reveals the uniform core-sheath nature. The diameter of its metallic Ga core is approximately 25nm, whereas its sheath is about 6nm thick. The length of these nanostructures can reach several micrometers. EDS analyses conducted on these nanocables using a nanoprobe reveal that the core is composed of Ga and the sheath is composed of C. Its corresponding EDS spectrum is shown in figure 5b. Selected-area electron diffraction pattern (figure 5c) taken on the Ga-C core-sheath nanostructures only shows one pair of weak (0 0 2) diffraction arcs and additional diffraction rings, which suggests that the carbon sheaths in these Ga-C core-sheath nanostructures are less ordered in contrast with carbon nanotubes made up of cylindrically curved graphene layers. The gallium cores seem in liquid state. Both ends of these Ga-C core-sheath composite nanostructures are enclosed by carbon, as revealed by figure 5 d and e. The cavity on the one end is owing to the voluminal contraction of metallic Ga column confined in the carbon nanotubes when it cools down from a high temperature to room temperature. Besides the straight Ga-C core-sheath composite nanostructures, Y-junction core-sheath nanostructures may also be generated as shown in figure 6. The pyrolysis of organometallic precursor along with thiophene can generate Y-junction carbon nanotubes [62]. At high temperature, Y-junction carbon nanotubes may be generated through pyrolysis of CH_4. Meanwhile, the cavity of carbon nanotubes is filled with gallium to generate Y-junction Ga-C core-shell nanostructures.

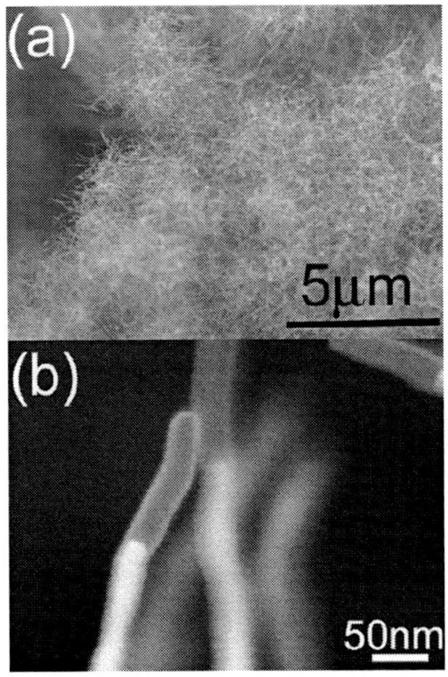

Figure 4. (a) SEM image of the as-obtained liquid gallium columns sheathed with carbon; (b) High-magnification SEM image displaying the tips of the nanostructures. From Ref. [52]

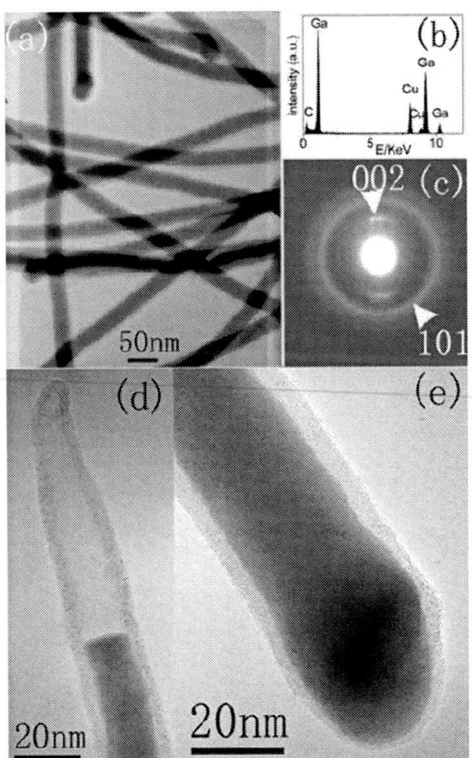

Figure 5. (a) TEM image of Ga nanocolumns sheathed with C; (b) EDS spectrum recorded from a composite nanostructure; (c) Selected-area electron diffraction (SAED) pattern taken from an individual nanostructure; (d) and (f) TEM images displaying the opposite ends of a given nanostructure. From Ref. [52]

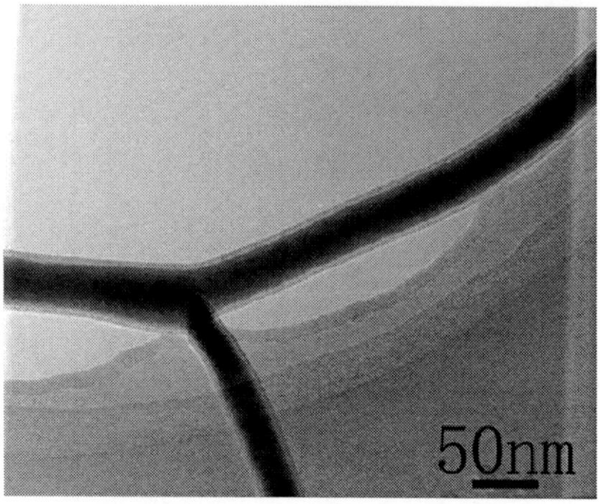

Figure 6. TEM image of a typical Y-junction Ga-C nanostructure. From Ref. [52].

Moreover, modern electron microscopes with field emitters allow the focusing of nanometer-sized electron beam spots on the specimen. Electron-beam irradiation is a powerful tool to manipulate nanostructures locally because atoms can be displaced by collisions with high-energy electrons [63-68]. We demonstrate that metallic Ga nanowires sheathed with carbon can be welded, cut and encapsulated using convergent electron beam. After the reaction was terminated, the furnace cooled down from a temperature as high as 1300℃ to a room temperature in almost half an hour. It is possible that the rapid cooling may cause the discontinuity of gallium nanowires encapsulated in the carbon sheaths. Figure 7a shows a TEM image of a typical separated Ga nanowire displaying a gap of about 70nm. The incident electron energy (E_p) of electron beam can be determined from the equation $E_p = mc^2(1/\sqrt{1-\beta^2}-1)$, where m is the rest mass of the electron, c is the speed of light, v is the velocity of the electron, and $\beta = v/c$. This energy may cause the local temperature rise. When the electron beam focused with a beam spot size about 50nm in diameter moved up and down along the bottom fraction of Ga nanowire shown in figure 7a, it expanded continuously owing to beam heating. Finally, the bottom nanowire was connected with the top one. Figure 7 a-e shows the expanding course of gallium nanowire. The wetting angle of the gallium surface changes a little with the prolonged irradiation. The wetting angle change and expansion of gallium encapsulated in carbon nanotubes may be caused by a thermal effect of the irradiation. Our estimates show that the specimen can typically be heated to almost a hundred centigrade degrees during the irradiation with the present current density, which is far above the melting point of metallic Ga (29.8℃). The top metallic Ga nanowire remains unchanged because it is free from irradiation, which implies that the thermal conductivity of carbon is negligible in contrast to that of metallic Ga. The thermal effect of electron beam only takes place on the irradiated region. On the other hand, the expansion of metallic Ga nanowire is also an indicator for the thermal effect of electron beam.

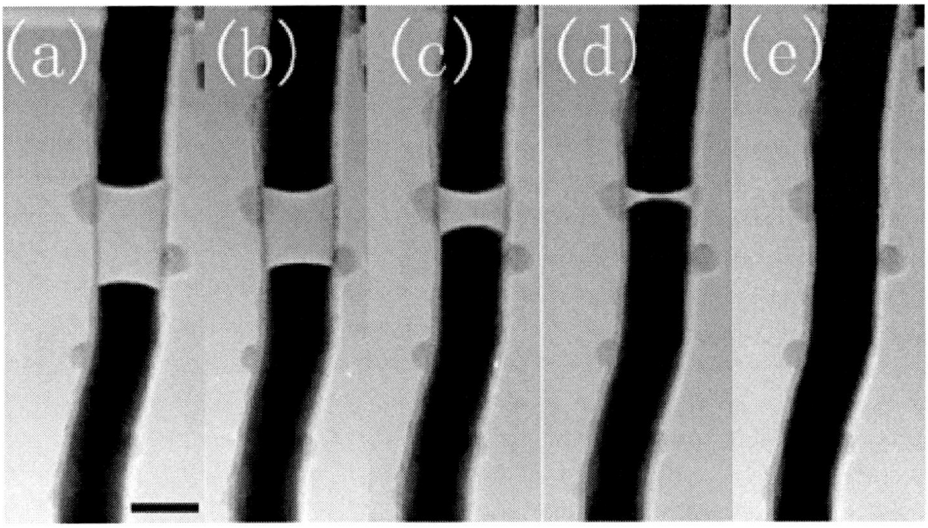

Figure 7. TEM images showing the expanding of metallic Ga nanowire under electron-beam irradiation (a) 0 minute; (b) 2 minutes; (c) 5 minutes; (d) 8 minutes; (e) 10 minutes. From Ref. [52].

When the electron beam strikes a specimen, it is widely known that inelastic and energy-loss electrons are generated with abundant signals including x-rays, inelastic and secondary electrons which are very useful for materials analysis. If the beam intensity is above the threshold, it can cause the damage of a target. The beam damage on carbon and metals is generally dominated by direct displacement of atoms from the specimen to create point defects, namely knock-on displacement. Apart from beam heating, we also demonstrate the ability of high energetic electron beam in cutting Ga nanowires sheathed with carbon. Figure 8 shows an in-situ overall recording that the Ga nanowire sheathed with carbon is cut by a focused electron beam. The beam size is about 5nm in diameter as schemed by the white dot in figure 8a. This convergent electron beam is moved breadthwise back and forth. Figure 8b shows the initial state of the Ga nanowire sheathed with carbon. After an intense beam irradiation for about two minutes (figure 8c), the irradiated region becomes thinner. The image contrast in this irradiation region becomes lighter, suggesting that metallic Ga is partially removed. A further two-minute irradiation can completely remove the metallic Ga from the irradiated region (figure 8d), which implies that cavity of carbon sheath collapses completely. The Ga nanowire sheathed with carbon is finally separated into two ones after prolonged irradiation (figure 8e and f).

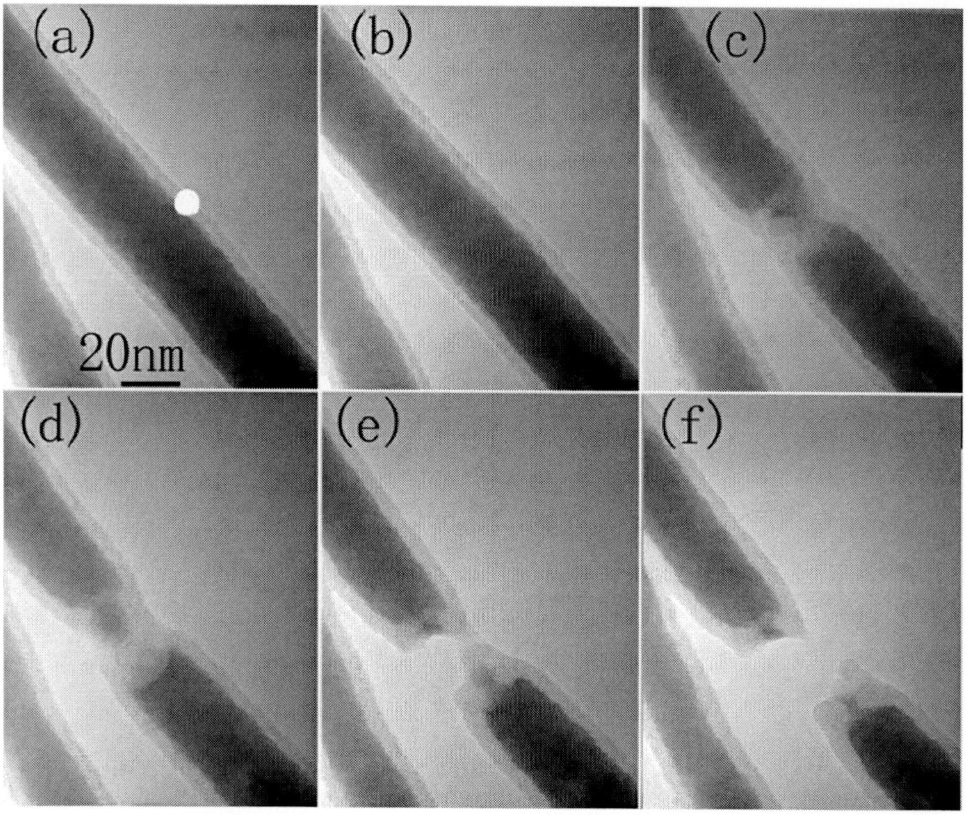

Figure 8. TEM images showing the morphological changes of a Ga-C composite nanowire when it is irradiated with a moved electron beam. (a) the white dot indicating the electron beam; (b) 0 minutes; (c) 2minutes; (d) 4 minutes; (e) 7 minutes; (f) 9 minutes. From Ref. [52].

Detailed insight into structural transitions caused by the electron beam irradiation was carried out using high-resolution electron microscopy (HRTEM). Figure 9a-f show the HRTEM images taken from the irradiation regions under various stages, corresponding to the figure 8b-f respectively. As suggested by the electron diffraction pattern of the Ga nanowire sheathed with carbon, the carbon sheath is close to an amorphous state. HRTEM image of the longitudinal edge displays that the thin carbon sheath has no as well-defined layered structure as carbon nanotubes produced by an arc discharge or chemical vapor deposition with transition metal nanoparticles as catalysts. After two-minute beam irradiation, the crystalline perfection seems a little improved with the shrinkage of the carbon sheath. This result implies that growth defects including undulating, dangling carbon sheets, dislocation edges, and local irregularities may be annealed in situ after a few minutes of electron irradiation in a 300kV electron microscope. In some respects, an intense irradiation resembles a high- temperature effect, allowing interstitial and vacancies to aggregate into clusters to reduce their free energy [69, 70]. Further electron-beam irradiation cause the complete shrinkage of the carbon sheath, meanwhile a nest-like carbon nanostructure form as indicated by the arrow in the figure 9c. An increased irradiation creates a big cleft in this composite nanostructure. Moreover, a semispherical carbon onion of a diameter about 15nm is also generated. Figure 9d shows this semispherical carbon onion is connected the top fraction of composite nanostructure by an 8nm-wide carbon neck. This carbon neck can completely be peeled off by electron-beam irradiation to create two separated composite nanostructures. A spherical carbon onion of diameter about 10nm is left on the top right corner of the bottom composite nanostructure (figure 9e). A similar one is created on the top left corner of the bottom composite nanostructure. The top composite nanostructure is ended with a sharp tip, as shown by figure 9f. Figure 9a-f clearly demonstrates that the encapsulated metallic Ga nanowire can not hinder the reconstruction of the carbon sheath under intense electron-beam irradiation. Once the carbon sheath shrinks, the liquid Ga nanowire is ready to recede and separate into two owing to its liquid state. The electron-beam irradiation can anneal the carbon sheath and cause the knock-on damage. Moreover, self-organization phenomena also take place to form spherical carbon onions on the tip of a separated Ga-C composite nanostructure during the irradiation because irradiation can create a situation where the solid is far from thermal equilibrium and allow a degree of structural fluidity. The spherical carbon onion seems more stable than the semispherical one under irradiation.

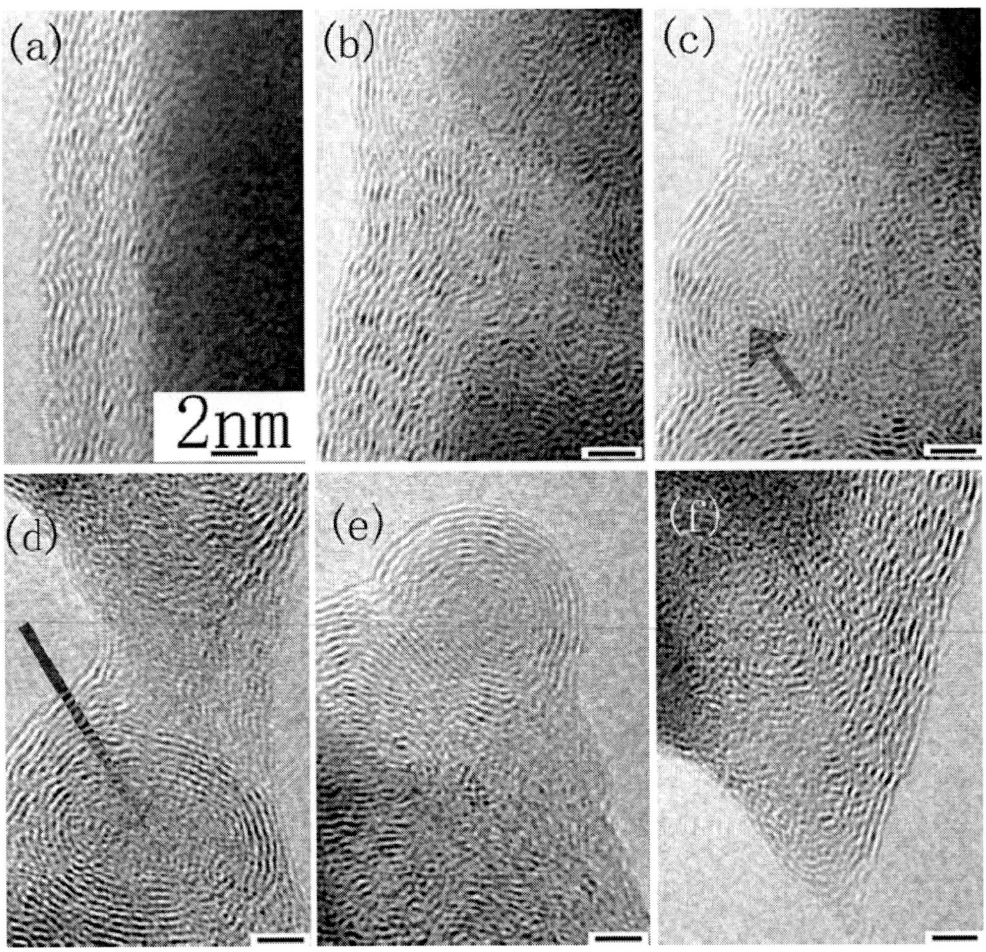

Figure 9. HRTEM images showing the structural changes of the Ga-C composite nanowire under electron-beam irradiation, corresponding to figure 4b, c, d, e and f, respectively. (a) 0 minutes; (b) 2minutes; (c) 4 minutes; (d) 7 minutes; (e) 9 minutes, showing the end of the lower Ga-C composite nanowire and (f) 9 minutes, showing the end of the upper Ga-C composite nanowire. From Ref. [52].

Generally a cavity is left in one end of the 1D Ga-C composite nanostructure at room temperature (10a). Electron-beam irradiation may eliminate this cavity to make the Ga nanowire completely covered with carbon. Firstly an electron beam with beam size about 5nm irradiates over carbon sheath in front of the Ga nanowire for about 2 minute, which may result in the collapse of carbon sheath of an outer diameter about 22nm to form a short solid carbon rod of a diameter about 18nm (figure 10b). Constant irradiation on one side of the carbon sheath may cause its bending (see supporting information). Knock-on damage make the carbon rod thinner and thinner with the increased electron-beam irradiation. Finally the unwanted carbon sheath can completely be removed, as shown by figure 10 c and d. The HRTEM images may help to understand the structural change of carbon sheath during the irradiation, including the annealing, knock-on damage, separation of unwanted fraction and self-organization into carbon onion-like nanostructures. After the unwanted carbon is separated from the Ga nanowire, a carbon onion head are generated on the tip of 1D Ga-C composite nanostructure. This carbon onion nanostructure may prevent the liquid metallic Ga from spilling.

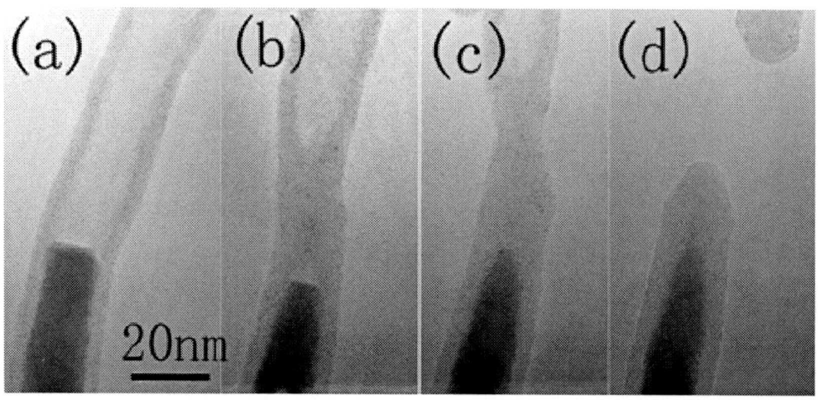

Figure 10. TEM images showing the morphological changes of the carbon sheath in a Ga-C composite nanowire under electron-beam irradiation. (a) 0 minute; (b) 2 minutes; (c) 4 minutes; (d) 8 minutes. From Ref. [52].

3.2. Ga_2O_3-C Core-Shell Nanocables [53]

Although the inner cavity of an open carbon nanotube can be used to accommodate foreign elements or compounds [71-73], it is difficult though to create filling via capillarity over distances more than 100 nm due to the fact that the filling terminates once the inner hydrostatic pressure equilibrates the outer pressure. Herein, we demonstrate the continuous filling of carbon nanotubes with gallium oxide. It may originate from the simultaneous crystallization of gallium oxide and carbon nanotubes in line with the following chemical reactions. These reactions are assumed based on the thermodynamical properties consecutively noted in the parentheses below:

$$Ga_2O_3 + 2C\ (graphite) \rightarrow Ga_2O(g) + 2CO\ (g) \qquad (1)$$

$$(\Delta G^{\circ}_{1600K} = -123.222kJ/mol,\ logK_{1600K} = 4.022)$$

$$Ga_2O(g) + 2CO\ (g) \rightarrow Ga_2O_3 + 2C\ (graphite) \qquad (2)$$

$$(\Delta G^{\circ}_{1000K} = -198.173kJ/mol,\ logK_{1000K} = 10.35)$$

As revealed by the reaction equation (1), the total entropy considerably increases to form gaseous Ga_2O and CO. It is favorable for the reaction (1) to increase its temperature. At a temperature as high as 1600 K, Ga_2O_3 spontaneously reacts with a carbon powder to form gaseous Ga_2O and CO due to negative Gibbs free energy. Being carried by the N_2 gas flow to a lower temperature region, gaseous Ga_2O and CO can recombine to form crystalline gallium oxide and carbon nanotubes in accord with the reaction (2). Ga_2O_3 grows along a direction perpendicular to the $(\bar{2}02)$ plane, meanwhile, carbon is coated on its surface.

The product deposited on a carbon plate placed below the outlet in a vertical induction furnace. Representative SEM image of a sample is shown in Figure 11(a), which indicates that the product is composed of wire-like nanostructures with a length of up to several micrometers. High-magnification SEM image reveals that the wire-like nanostructures have core-shell structures and thus can be defined as nanocables. Figure 11(b) shows the open end of a nanocable, indicated by the arrow. A uniform continuous filling of the nanocable is marked with the "diamond"-like symbols in figure 11(b). The diameter of nanocables is approximately 50 nm.

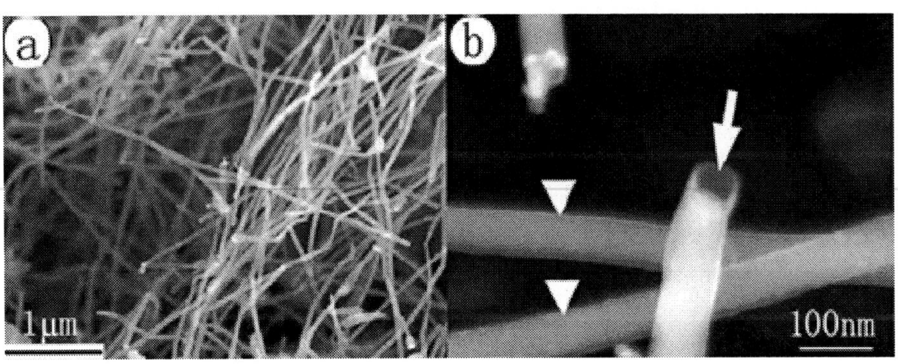

Figure 11. (a) a SEM image of as-obtained Ga_2O_3-C nanocables; (b) a typical SEM image showing the open end of a Ga_2O_3-C nanocable. From Ref. [53].

Detailed structural and chemical analysis of individual composite nanowires was carried out using high-resolution TEM (HRTEM). Figure 12(a), a TEM image of a typical as-obtained Ga_2O_3-C nanocable, displays that its core diameter is approximately 40 nm, whereas its sheath is 6 nm thick. The corresponding selected-area electron diffraction pattern in figure 12(b) suggests that the cable is a single crystal. All the spots on the pattern can be indexed as those natural for the [101] zone axis of β-Ga_2O_3. High resolution TEM images taken from the core and the longitudinal edge of a nanocable are depicted in figures 12(c) and 12(d), respectively. The HRTEM image of the core indicates clearly-resolved interplanar distances of 2.82 Å and 2.68 Å, corresponding to the $(\bar{2}02)$ and $(\bar{1}0\bar{1})$ lattice spacings, respectively. HRTEM analysis also reveals that the crystalline Ga_2O_3 core is free from dislocations or stacking faults. The HRTEM image of the nanocable longitudinal edge displays that the thin wall around the core has a layered structure. The interlayer distance of approximately 3.40 Å coincides with that between adjacent basal planes in graphite crystals. The crystalline Ga_2O_3 core has a sharp interface with the carbon wall. As revealed by the HRTEM observation and SAED pattern, the growth direction of crystalline Ga_2O_3 core is nearly perpendicular to the $(\bar{2}02)$ crystal planes, namely $(\bar{2}02)*$.

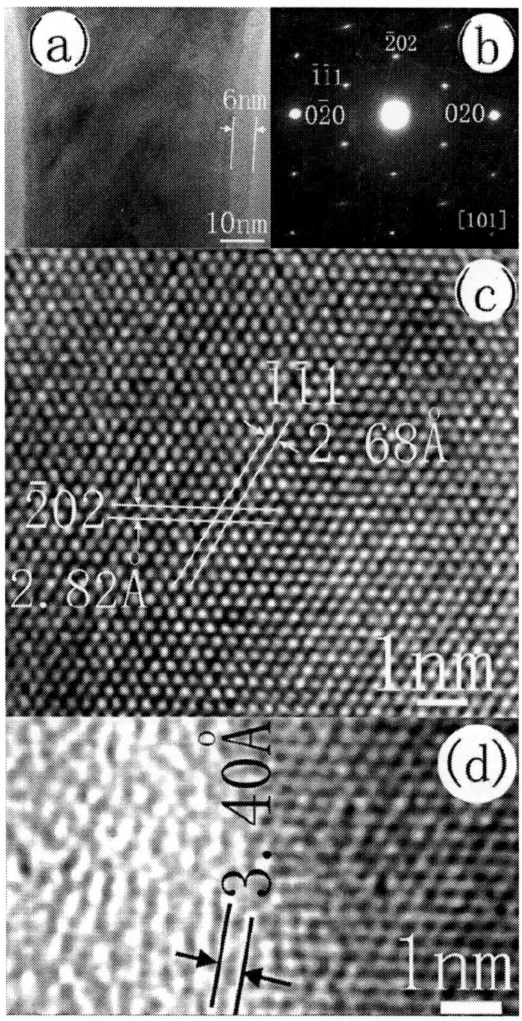

Figure 12. (a) a typical TEM image of an as-obtained Ga_2O_3-C nanocable; (b) Selected-area electron diffraction (SAED) pattern of the nanocable; (c) HRTEM image of the nanocable core; (d) HRTEM image of the nanocable wall. From Ref. [53].

β-Ga_2O_3 is a thermodynamically stable form of gallium oxide in ambient conditions. It has a monoclinic unit cell and belongs to the space group C2/m. The lattice constants of β-Ga_2O_3 are a = 12.227 Å, b = 3.0389 Å, c = 5.8079 Å, β = 103.82°. The O atoms have a distorted cubic close packing in β-Ga_2O_3 crystals, which contain two crystallographically different Ga atoms in the asymmetric unit, one with tetrahedral coordination geometry (Ga1) and the other with octahedral coordination geometry (Ga2). The Octahedrons share edges to form double chains parallel to b-axis, connected by corner-sharing tetrahedrons. Figure 13(a) displays the structural model of β-Ga_2O_3 crystal viewed along the [1 0 1] axis. The $(\bar{2}02)^*$ plane and b-axis are shown in figure 13(a). As documented by the structural model, the double chains of edge-sharing octahedrons and chains of corner-sharing tetrahedrons stack alternately along the $(\bar{2}02)^*$ planes forming corner-sharing connections. The growth direction of crystalline Ga_2O_3 core is dependent on this crystallographic feature. The lattice constant (12.227 Å) of β-Ga_2O_3 is nearly ten times larger than the interplanar distance (1.234 Å) between the (110) planes in graphite. The $(\bar{2}02)$ interplanar distance of β-Ga_2O_3 is nearly two times larger than the sp^2 C-C bond length (1.421 Å) in graphite. Figures 3(b) and 3(c) show the crystallographic relationship between β-Ga_2O_3 and a single graphene sheet.

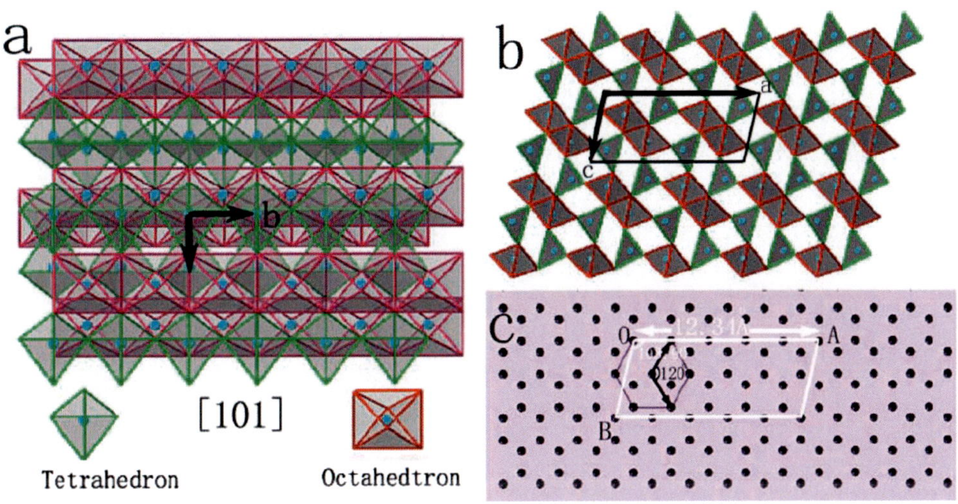

Figure 13. (a) A scheme of β-Ga_2O_3 crystal structure viewed along the [101] direction. (b) Structural model of the (020) crystal plane of β-Ga_2O_3 crystal. (c) Structural model displaying the structural relationship between of β-Ga_2O_3 and a single graphene sheet. The white parallelogram on a graphene sheet corresponds to that in Fig. 13 (b).

Field-emission measurements of the nanocables were conducted in a vacuum chamber at a pressure of ~6.0×10^{-5} Pa. A rod-like aluminum probe with 1 mm^2 in cross section was used as the anode. The as-obtained nanocables were used as the cathode. Figure 14 shows emission current density, J, versus applied field curves at an anode-sample distance of 97 μm. The Fowler-Nordheim (F-N) plot of the data is shown in the inset of figure 14. A turn-on field 7.73 V/μm and a threshold field of 8.45 V/μm were obtained. This value is compatible to that of carbon nanotubes [74-76]. The behavior of the emission current versus the applied voltage was analyzed using the Fowler-Nordheim equation [77],

$\ln\left(J/E^2\right) = -\left(B\phi^{3/2}/\beta\right)E^{-1} + \ln\left(A\alpha\beta^2/\phi\right)$, where J is the current density, E the applied field (V/d), ϕ the work function, B and A are constants, α the effective emission area and β the F-N enhancement factor. The slope of the Fowler-Nordheim (F-N) plot shown in figure 14 inset indicates that the emission is generally consistent with the F-N mechanism by exhibiting linear dependence, but the F-N curve is composed of two discrete straight lines separated at the threshold field. The sharp increase in the emission current at the threshold field may be due to the fact that the interaction between the carbon shells and the gallium oxide cores starts to contribute to the total emission current at a higher field. It has been reported that the carbon nanotube films had an emission current density of 100 mA/cm^2 at an electric field of 10 V/μm [74] and the needle-shaped SiC nanowires had a field emission current density of 30.8mA/cm^2 at an electric field of 9.6 V/μm [78]. The emission current density of the nanocables can be as high as 722.5mA/cm^2 at an electric field of 10 V/μm, indicating that the as-obtained nanocables are a prospective candidate for field-emitters.

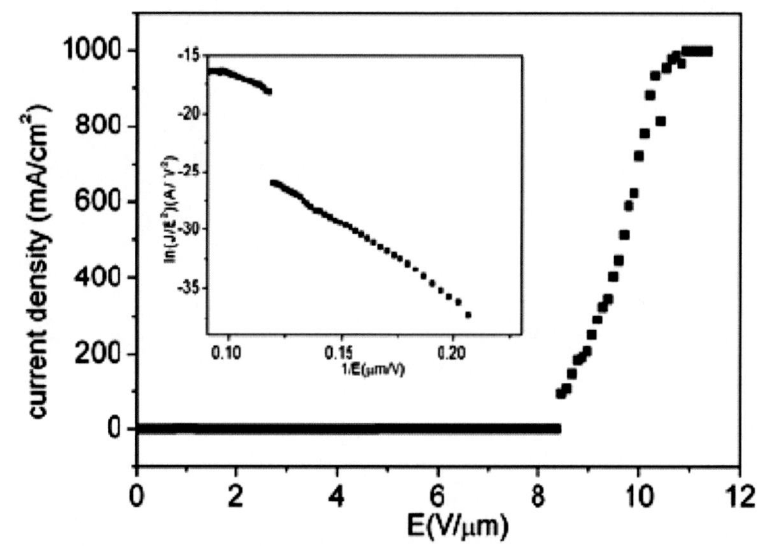

Figure 14. Field-emission I-V curve obtained on the nanocables. The inset depicts the F-N plot, which is composed of two discrete straight lines. From Ref. [53].

3.3. Si/ZnS, Si-Cd_4SiS_6 Composite Nanowires [54, 55]

The vapor-liquid-solid (VLS) growth process has been extensively used for the synthesis of 1D nanostructures. The key feature during VLS growth is the existence of an alloy droplet capping one end of a one-dimensional nanostructure. This promotes the anisotropic crystal growth [79, 80]. In a VLS growth process, the metal catalysts are rationally selected according to a given phase diagram. A catalyst may act as a preferential site for the absorption of vapors to form a liquid alloy. Once an absorbed reactant is supersaturated within a liquid alloy at a given temperature, it nucleates from a supersaturated liquid droplet. Further condensation of the reactant vapors results in the axial growth of a 1D structure.

Generally, a diameter of the nanowires grown via a VLS process is confined with a diameter of the liquid alloy droplets. A VLS process has widely been used to produce 1D nanostructures from various inorganic solids including elemental semiconductors (Si, Ge), Ⅲ-Ⅴ semiconductors and Ⅱ-Ⅵ semiconductors using Au or Fe nanoparticles as catalysts [81,82]. The melting point of elemental tin is approximately 232℃. Tin can form a liquid alloy with silicon and zinc sulfide at a high temperature, but, by contrast, its solubility in solid-state silicon and zinc sulfide is negligible at room temperature. Therefore it is reasonable to assume that an elemental tin can be used to simultaneously catalyze anisotropic growth of zinc sulfide and silicon nanowires. If two substances are evaporated simultaneously, tin nanoparticles may catalyze growth of composite nanowires. Herein, we demonstrate a typical example of bimorph composite nanowires via a one-catalyst-confined growth. ZnS/Si and Si/Cd_4SiS_6 nanowire heterostructures were generated through the novel tin-catalyzed growth.

Side-by-side ZnS/Si composite nanowires have been obtained through co-thermal evaporation of ZnS and Si powders with a small amount of tin sulfide as an additive. During the synthetic process, a mixture of zinc sulfide and silicon powders was thermally co-evaporated at a high temperature (approximately 1100℃) with a small amount of tin sulfide as an additive. Then silicon and zinc sulfide vapor condense on metallic tin droplets at a low temperature (approximately 600℃), which results in the anisotropic growth of silicon and zinc sulfide simultaneously. A powder XRD pattern of a product, shown in figure 15a, suggests the coexistence of silicon and zinc sulfide with zinc-blende (ZB) structures. The XRD diffraction peaks labeled with a "diamond" mark in figure 15a could be assigned to ZnS of the wurtzite (W) structure. SEM image, shown in figure 15b, indicates that the product is composed of one-dimensional nanostructures of high aspect ratio and each nanowire is capped with a spherical catalytic particle.

Detailed structural and chemical analysis of individual composite nanowires was carried out using high-resolution TEM (HRTEM). Figure 16a, a TEM bright-field image of a typical nanowire, clearly shows that the wire is composed of two distinct fragments and capped with a spherical tip. The diameter of the composite nanowire is approximately 90nm. Energy-dispersive X-ray spectroscopic (EDS) analyses using a nanoprobe were conducted on the tip and the two distinct wire fragments. The corresponding EDS spectra are shown in figure 16 b, c and d, respectively. It is evidenced that the spherical cap is made of metallic tin. The part exhibiting a brighter contrast is composed of silicon, whereas that of a darker contrast is composed of zinc sulfide with a composition of Zn (49.4%) and S (50.6%). High-resolution TEM images of a ZnS domain, a Si domain, and the interface between them are displayed in figure 17a, b and c, respectively. As seen in these images, both fragments are structurally uniform and have no dislocations or other defects. The clearly resolved fringes with a 3.10Å separation are peculiar to the {111} d~spacing of a diamond-like Si and ZB ZnS. The interface between ZnS and Si fragments is homogeneous and atomically sharp. It has been reported that ZnS thin films can epitaxially grow on a silicon substrate [83]. In fact, the similarity between ZB ZnS and diamond-like Si lattice constants results in only marginal lattice mismatch of ~0.5% (ZB ZnS: a = 5.4060Å, JCPDS card 05-0566; diamond-like Si: a = 5.43088, JCPDS card 27-1402). It is worth noting that both fragments in the ZnS-Si composite nanowires have a twin-like crystallographic relationship, as displayed on the

HRTEM images: $(1\bar{1}1)_{ZnS}$ is parallel to $(1\bar{1}1)_{Si}$, while $(\bar{1}11)_{ZnS}$ and $(\bar{1}11)_{Si}$ are mirror planes on $(\bar{1}11)_{Si}$. The growth direction of composite nanowires is close to $[\bar{1}12]_{Si}$. Figure 17d is the [110] projection of ZB ZnS and diamond-like Si crystal lattices, which illustrates the structural relationship between the two fragments in HRTEM images.

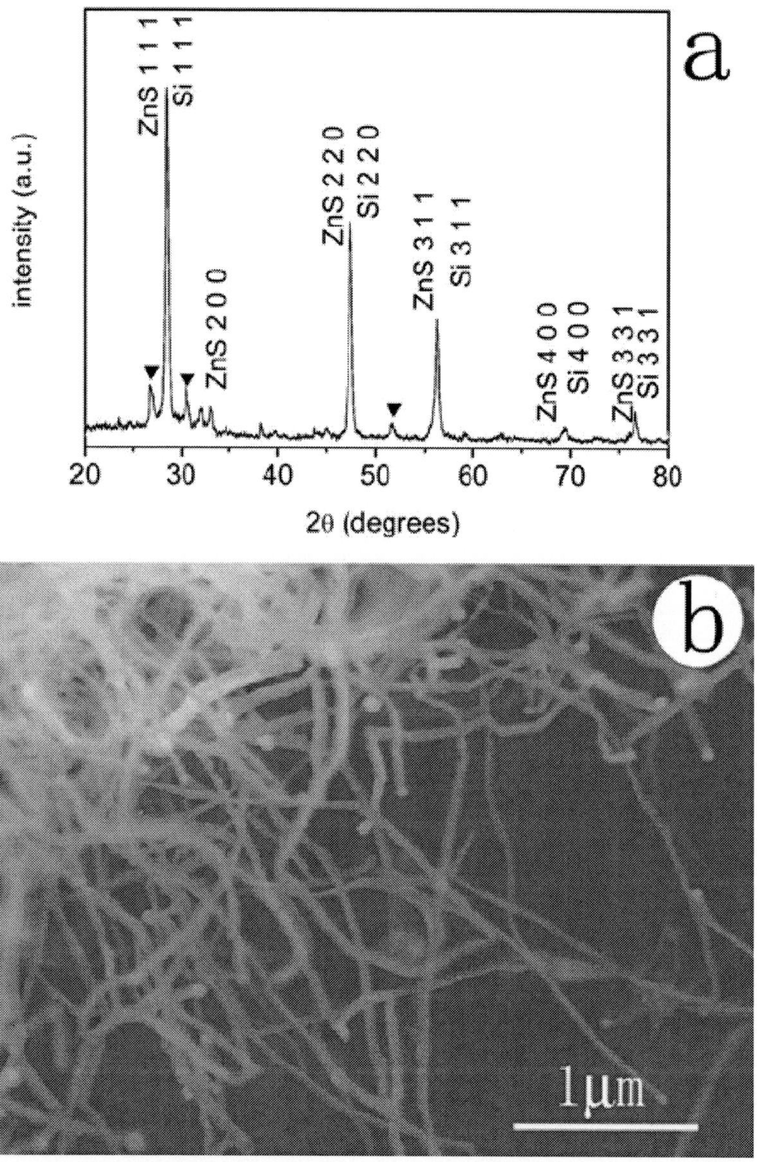

Figure 15. a) XRD pattern of an as-obtained product. b) SEM image of the composite nanowires. From Ref. [54].

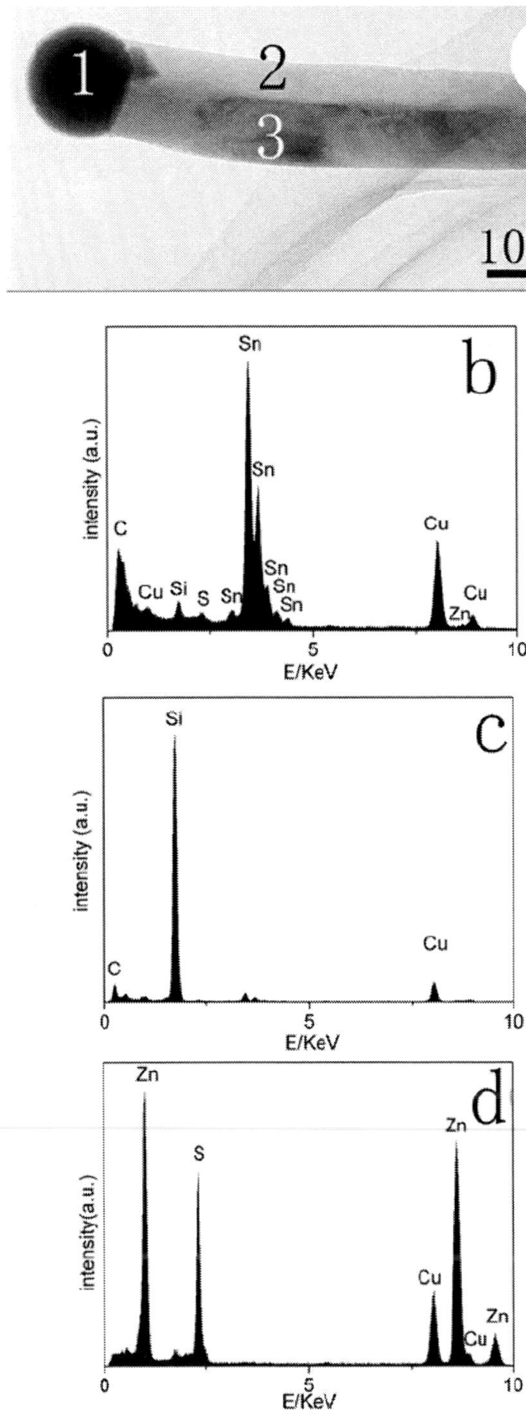

Figure 16. a) TEM image of a typical composite nanowire. b), c) and d) nano-beam EDS spectra from the regions labeled with 1, 2, and 3 on the nanowire, respectively. From Ref. [54].

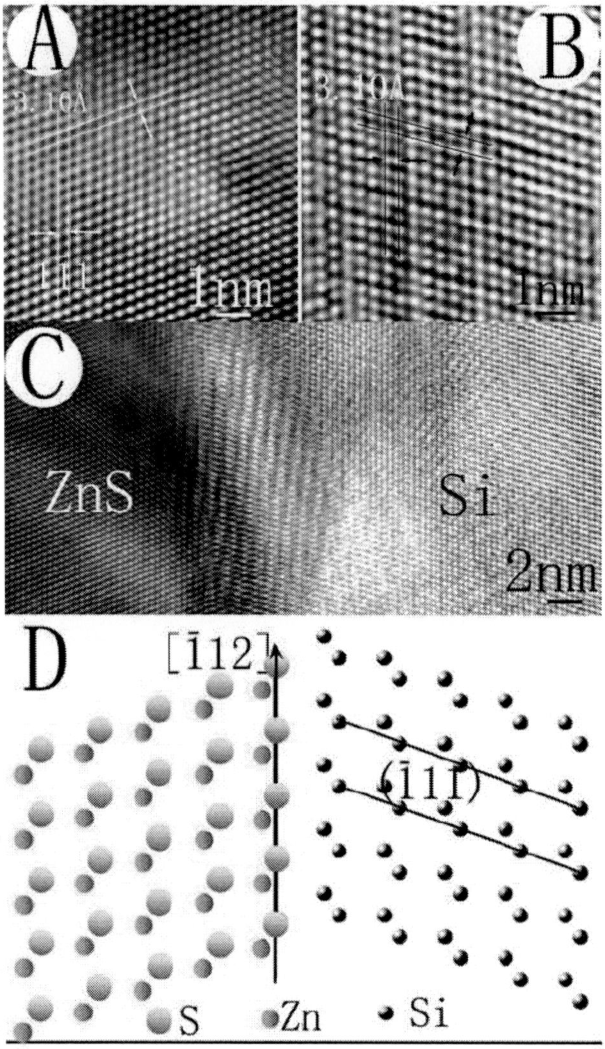

Figure 17. High-resolution TEM images of a) a ZnS wire fragment b) a Si wire fragment c) an interface between two fractions. d) structural model showing the crystallographic relationship between the ZnS and Si fragments. From Ref. [54].

70% of the nanowires displayed diameters of approximately 90 nm and 30% of them have diameter approximately 15 nm. The composite nanowires exhibiting nearly similar diameter frequently aggregate together, as shown in figure 18. The TEM image displays that these nanowires are also capped with spherical particles. Figure 19a, TEM image of a typical composite nanowire of ~18 nm in diameter, also reveals two distinct fragments. The nano-beam EDS analyses on thin composite nanowires indicate the ZnS/Si side-by-side compositional characteristics and metallic tin spherical tips, similarly to the thick wires discussed above. Further HRTEM studies provide detailed structural information on ZnS and Si wire fragments. Figure 19b, HRTEM image of the interface between ZnS and Si fractions, verifies that ZB ZnS epitaxially grows on a diamond-like Si, in contrast to the observed twin-like crystallographic relationship in the thicker composite nanowires. The epitaxial growth is schemed in figure 19c, for the [110] projection of diamond-like Si and ZB ZnS. The preferred growth orientation of a thin composite nanowire is close to $[1\bar{1}0]_{Si}$. The Si fragment is structurally homogeneous, whereas the bent contrast within the ZnS domain (figure 19a) suggests its structural inhomogeneity. The HRTEM studies reveal the coexistence of ZB ZnS and wurtzite ZnS (W-ZnS). Figure 20 is a HRTEM image recorded from a ZnS wire fragment exhibiting the bent contrast. A region between the two dark lines in figure 20 consists of a hexagonal phase oriented along the [100] direction, as illustrated in the figure 20 inset. The existence of W-ZnS phase in thin composite nanowires may account for the presence of low-intensity diffraction peaks of W-ZnS marked with "diamond" symbols in the XRD pattern (figure 1a). In the thin composite nanowires, $(001)_{W-ZnS}$ is parallel to $(1\bar{1}1)_{ZB-ZnS}$ and $(1\bar{1}1)_{Si}$. Both ZB-ZnS and W-ZnS phases can be considered as a stacking of ZnS_4 tetrahedrons. The ABCABC stacking forms the ZB-ZnS, whereas the ABAB stacking forms the W-ZnS. The two structures can transform by changing stacking sequences along the $[111]_{ZB-ZnS}$ or $[001]_{W-ZnS}$ directions.

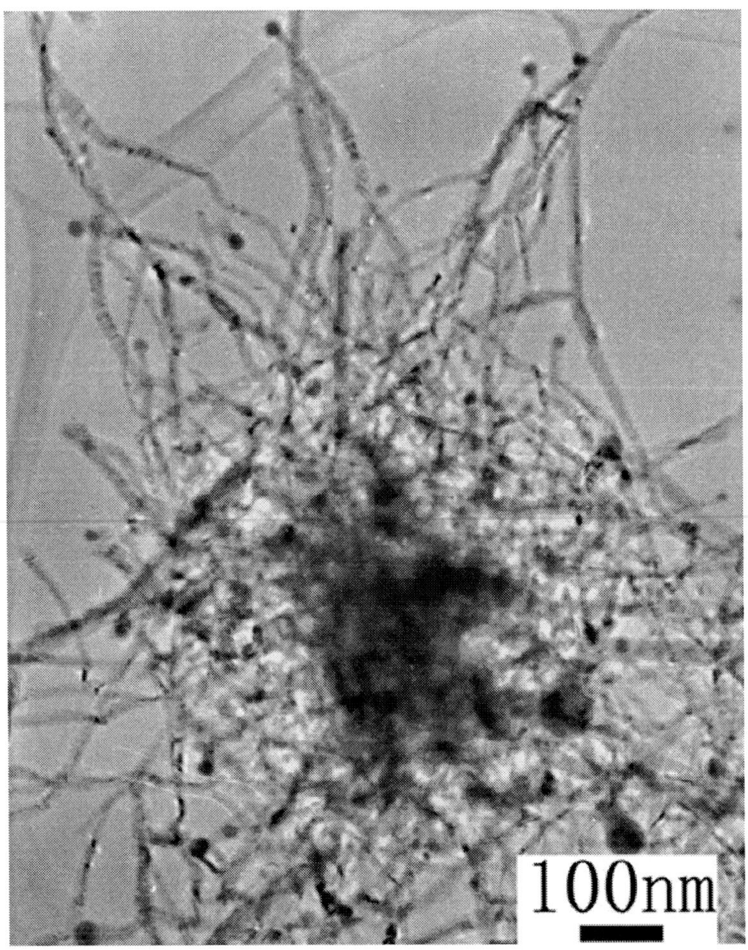

Figure 18. a) TEM image of thin composite nanowires.

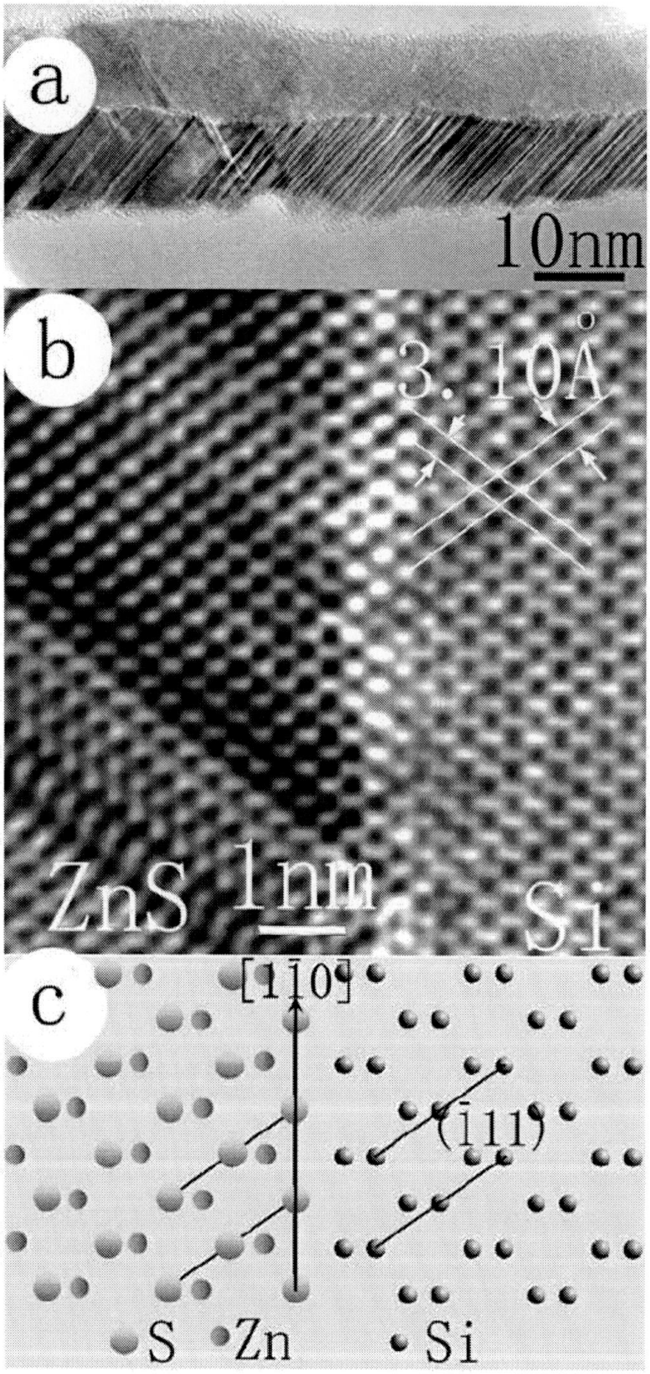

Figure 19. a) TEM image of a single thin composite nanowire b) High-resolution TEM image displaying the interfacial structure of the composite nanowire. c) structural model showing the crystallographic relationship between the ZnS and Si wire fragments in the nanowire. From Ref. [54].

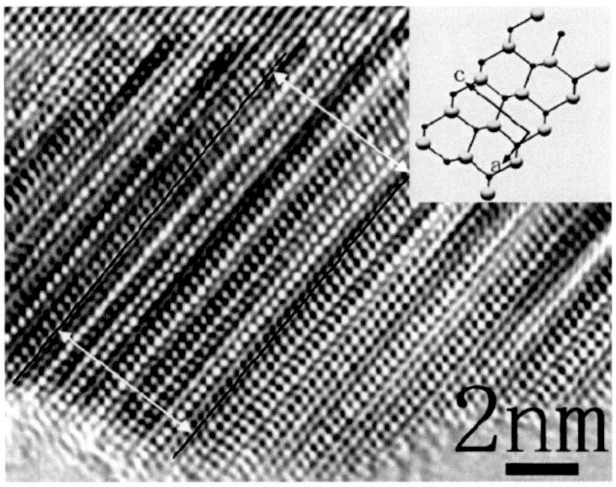

Figure 20. a) High-resolution TEM image confirming the existence of wurtzite structure in a thin composite nanowire. The inset depicts the [100] projection of a wurtzite structure.

The high-spatial-resolution CL spectra (figure 21) collected from either thick or thin composite nanowires always exhibit two peaks. The emission peak at approximately 685 nm (1.81eV) may be attributed to the Si fragments within the composite nanowires, which is consistent with the literature data [84, 85]. The peaks at 465 nm (2.67eV) in figure 21a and at 452nm (2.7eV) in figure 21b may be attributed to the ZnS fragments within the nanowires. These peaks originate from the native defects in ZnS including vacancies and dislocations. The thinner composite nanowires contain much higher defect densities than the thicker ones, therefore they naturally display much more intense blue emission.

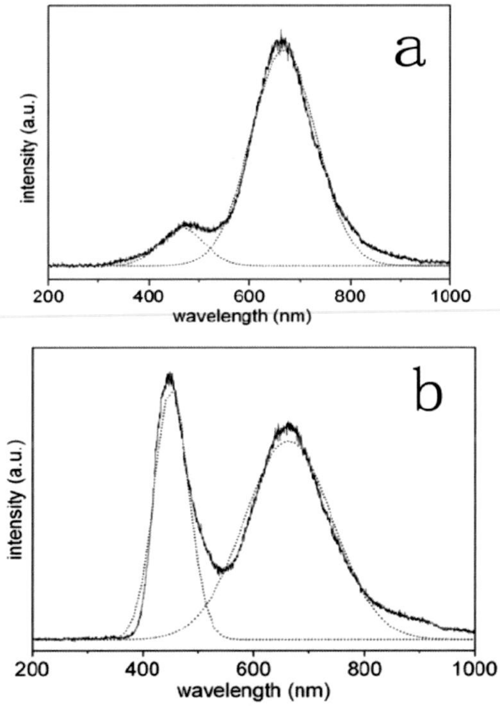

Figure 21. CL spectra recorded at 17K from composite nanowires with an approximate diameter of a) 90nm; b) 15nm. From Ref. [54].

During the synthetic process, tin sulfide plays an important role in the thermal evaporation of silicon and zinc sulfide to finally generate composite nanowires. The as-obtained composite nanowires are capped with metallic tin droplets, which is the key feature of the VLS growth mechanism. The growth process is schemed in figure 22. Most importantly, in the absence of tin sulfide, the nanowires can not be generated. Based on thermodynamics [86], the involved chemical reactions may be proposed as follows:

$$Si(s) \rightarrow Si(g) \quad (1)$$
$$(\Delta G^{\circ}1400K = 243.306 kJ/mol, \log K1400K = -9.078)$$
$$ZnS(s) \rightarrow Zn(g) + 1/2S2(g) \quad (2)$$
$$(\Delta G^{\circ}1400K = 109.016 kJ/mol, \log K1400K = -4.067)$$
$$Si + ZnS(s) \rightarrow Zn(g) + SiS(g) \quad (3)$$
$$(\Delta G^{\circ}1400K = 29.774 kJ/mol, \log K1400K = -1.110)$$
$$Zn(g) + SiS(g) \rightarrow Si(s) + ZnS(s) \quad (4)$$
$$(\Delta G^{\circ}900K = -138.783 kJ/mol, \log K_{900K} = 8.055)$$
$$SiS(g) \rightarrow Si(s) + 1/2SiS_2(s) \quad (5)$$
$$(\Delta G^{\circ}900K = -62.3205 kJ/mol, \log K_{900K} = 3.617)$$

Figure 22. Growth scenario for the formation of 1D composite nanostructures. (I) formation of liquid Sn droplets. (II) condensation of ZnS and Si vapors on Sn droplets. (III) Simultaneous crystal nucleation of ZnS and Si. (IV) anisotropic crystal growth of ZnS and Si.

From the thermodynamic point of view, an equilibrium vapor pressure of $P_{Zn(g)}$ and $P_{SiS(g)}$ created in accord with the reaction (3) at 1400K is as high as 211.7 mm Hg. On the other hand, the self-evaporation of ZnS in line with the reaction (2) is generally preferred, owing to the fact that the reaction between Si and ZnS must overcome a diffusion barrier. The self-evaporation of Si is negligible, for the equilibrium vapor pressure $P_{Si(g)}$ created in accord with the reaction (1) at 1400 K is as low as 6.35×10^{-7} mm Hg. SnS melts at 1155K. Liquid SnS can considerably reduce the diffusion energy for the reaction between solid-state Si and ZnS; thus the reaction (3) becomes preferential. From the kinetic viewpoint, SnS can accelerate the reaction (3). At a lower temperature, Zn and SiS vapors are inclined to recombine with the production of Si and ZnS, in line with the reaction (4). Although SiS (g) can possibly decompose to produce Si and SiS_2 in line with the reaction (5), the reaction (4) is considerably preferred in the presence of Zn (g). Moreover the decomposition of SnS results in the formation of a tin vapor. The tin vapor condenses at a lower temperature to produce liquid tin droplets which then absorb ZnS and Si vapors, and catalyze the anisotropic growth of ZnS/Si composite nanowires. Through a similar co-thermal evaporation process, Cd_4SiS_6/Si composite nanowires were obtained from a mixture CdS and Si powders with a small amount of tin sulfide as an additive. Figure 23 a and d depict side-by-side and core-shel Cd_4SiS_6/Si composite nanowires, respectively.

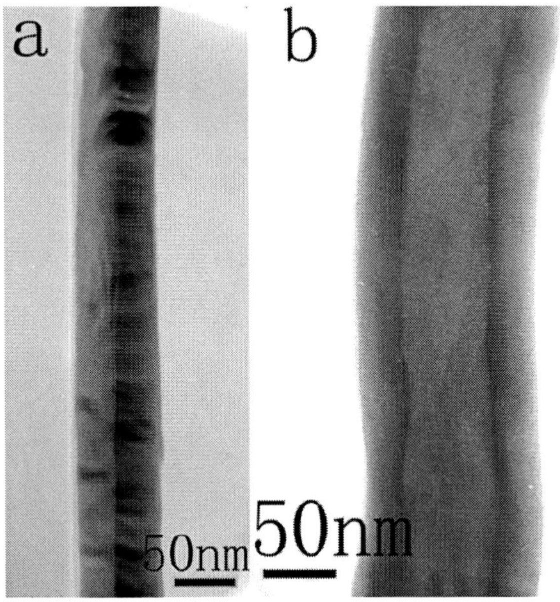

Figure 23. TEM image of Cd_4SiS_6/Si composite nanowires: (a) side-by-side; (b) core-shell.

3.4. Si/In End-to-End Nanowire Heterostructures [56]

If a metallic nanowire is connected to a semiconducting Si nanowire, a unique metal-semiconductor nanoscale contact can be generated. Then it may become a basic component of a novel miniaturized semiconductor device. We demonstrate the fabrication of In-Si end-to-end nanowire contacts and thoroughly analyze their morphological and structural characteristics. The metal-semiconductor nanowire junctions are sheathed with insulating silicon oxide.

Thermal evaporation of a In and SiO powder mixture at around 1400℃ in an induction furnace results in the generation of a gray-color product in the outlet of a graphite induction-heated cylinder. A powder XRD pattern of the product initially suggests the coexistence of diamond-cubic Silicon (JCPDS file 27-1402) and tetragonal Indium (JCPDS file 05-0642). Detailed structural and chemical analyses of the products are further carried out using TEM, HRTEM, electron-diffraction (ED), and Energy-dispersive X-ray spectroscopy (EDS). Figure 24a displays TEM micrograph of a typical nanowire junction. The nanostructure is composed of two distinct end-to-end segments. The length of the bottom gray-contrast part is of several micrometers. The length of the upper black-contrast part may reach dozens of micrometers.

The diameter of the 1D nanostructure is ~200 nm. Energy-dispersive X-ray spectra (EDS) using an electron nanoprobe were collected from the three spots located within a tube-like segment, a black-contrast segment and a gray-contrast segment, as indicated by the corresponding white dots in figure 24a. The results are shown in figures 24b, 24c and 24d, respectively. It is documented that the tube is composed of silicon oxide with a composition of O (68.6 at.%) and Si (31.4 at.%). The dark-contrast branch is composed of Indium, whereas that of a gray contrast is made of silicon. The O peaks in figures 24c and 24d may originate from a Si oxide sheath over the In and Si nanowires, respectively, whereas the C and Cu signals come from a C-coated Cu TEM grid. EDS elemental mapping sheds an additional light on the nanowire junction chemistry. Figure 25a is a scanning transmission electron microcopy (STEM) image of an In-Si end-to-end nanowire heterojunction. This reveals a clear interface between the two domains. The In, Si and O elemental maps are shown in figures 25 b, 25c and 25d, respectively. They demonstrate well-defined composition variations and the sharp interface between the In and Si segments. The uniform Oxygen scattering originates from the amorphous silicon oxide sheathing the entire nanowire.

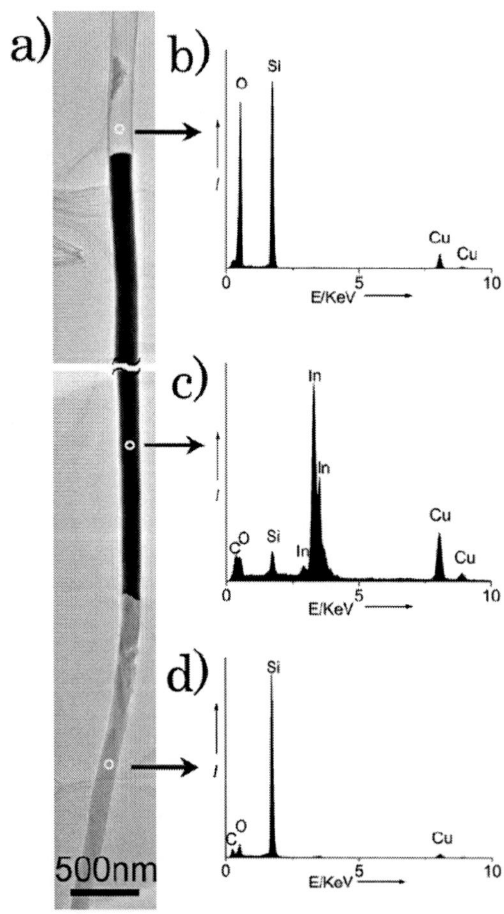

Figure 24. a) TEM image of a typical In-Si nanowire heterojunction. b), c) and d) EDS spectra taken from the selected regions marked with the white circles on the nanowire image, respectively. From Ref. [56].

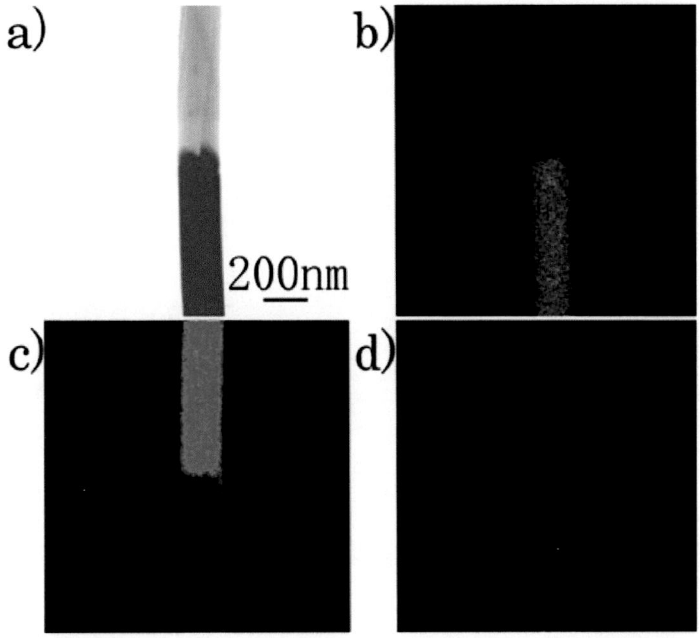

Figure 25. a) STEM image depicting a Si (light) and In (dark) domains within a In-Si nanowire junction. b) In; c) Si; and d) O elemental maps demonstrating the elemental spatial distribution. From Ref. [56].

Detailed TEM investigations allow us to gain further insight into the heterojunction microstructure. Figure 26a depicts a representative In-Si nanowire heterojunction. Figure 26b presents its magnified view. It is visible that the whole nanowire junction is uniformly coated with a 8 nm thick silicon oxide sheath. Selected-area electron diffractions (SAED) were taken from a Si nanowire segment (light contrast), a In nanowire segment (dark contrast) and the boundary between the two. The SAED patterns are shown in figures 26c, 26d and 26e, respectively. All the diffraction spots in figures 26c and 26d can be indexed to those of the [110] zone axis of diamond-like Si and the [110] zone axis of tetragonal In, respectively. Streaking of the reflections in the underfocused electron diffraction patterns, shown in the right-hand-side parts of figures 26c and 26d, implies that the preferred growth directions of the Si and In nanowires are close to the $[\bar{1}13]$ and $[1\bar{1}4]$ orientations, respectively. High-resolution TEM image taken from the Si domain is depicted in figure 27a. This indicates the clearly-resolved 3.10 Å interplanar fringes corresponding to the {111} lattice distances. The inset in figure 27a is the [110] projection of a diamond-like Si crystal lattice. In addition, HRTEM analysis reveals that the crystalline Si nanowire is free from planar defects and coated with an amorphous sheath. The pattern taken from the boundary region between the Si and In segments, figure 26e, implies that the (002) plane of Si (marked with a circle) is nearly parallel to the $(1\bar{1}2)$ plane of In (marked with a square). Figure 27b shows the structural model of diamond-like Si on the (002) plane and that of tetragonal In on the $(1\bar{1}2)$ plane. The square dimensions on the Si (002) plane are close to those of the rectangle on the In $(1\bar{1}2)$ plane. Thus it is reasonable to suggest that the Si and In nanowire growth is crystallographically related.

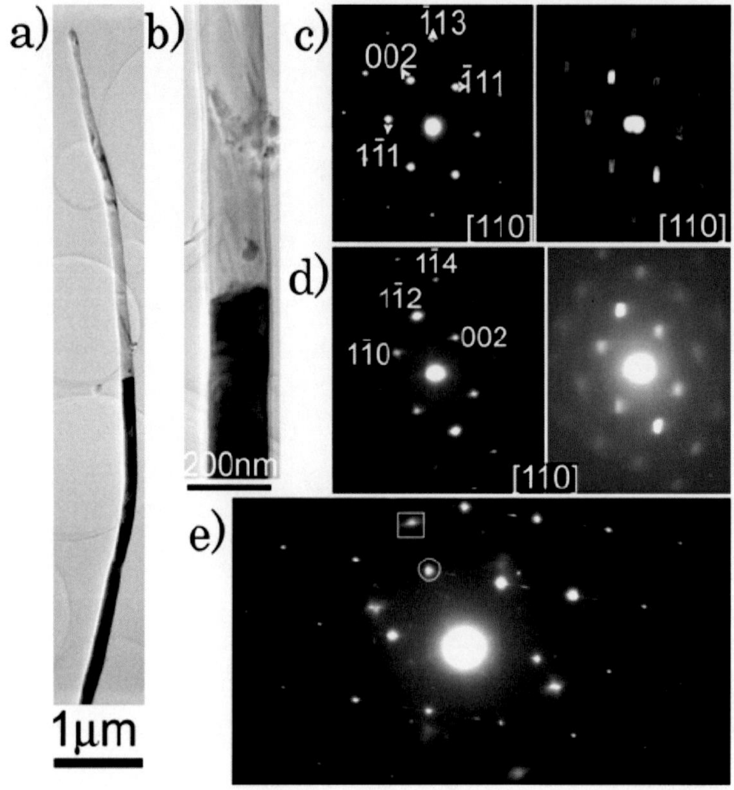

Figure 26. a) TEM image of a single In-Si nanowire junction. b) Magnified TEM image displaying that the In-Si nanowire junction is uniformly sheathed with silica. c), d) and e) SAED patterns taken from the selected Si, In and In-Si boundary regions respectively. From Ref. [56].

Figure 28a, HRTEM image of a Si branch, clearly shows that a Si crystalline core is sheathed by a ~8 nm-thick amorphous shell. The spots on the corresponding SAED pattern (figure 28a, inset) can be indexed to the [112] zone axis of a diamond-like Si crystal. The clearly resolved interplanar fringes of 3.10 Å and 1.90 Å are seen in HRTEM image (figure 28b). These respectively coincide with the (111) and (220) lattice spacings of a diamond-like Si crystal. The HRTEM image is consistent with the projection of the diamond-like Si crystal along the [112] orientation, as shown in the inset of figure 28b. The $[2\bar{2}0]$ preferential growth orientation is also documented by the SAED pattern and the HRTEM image. Some other preferential growth orientations such as $[1\bar{1}1]$ and $[3\bar{3}1]$ are additionally demonstrated in supplemental materials.

Vapor-liquid-solid (VLS) or oxide-assisted (OA) mechanisms are generally attributed to the anisotropic growth of Si nanowires. During a VLS growth process, a metal nanoparticle acts as a preferential site for the vapor absorption to form a liquid alloy. Once an absorbed reactant is supersaturated within a liquid alloy at a given temperature, it nucleates from the supersaturated liquid droplet. Further condensation of the reactant vapors on the alloy droplets may result in the continuous axial growth of Si nanowires. It is crucial for the VLS growth that one end of a Si nanowire is capped with an alloy droplet. The Si nanowires generated *via* a VLS process are generally oriented along the [111] axis. In contrast, the Si atom supersaturation during the OA growth is relatively negligible or moderate.

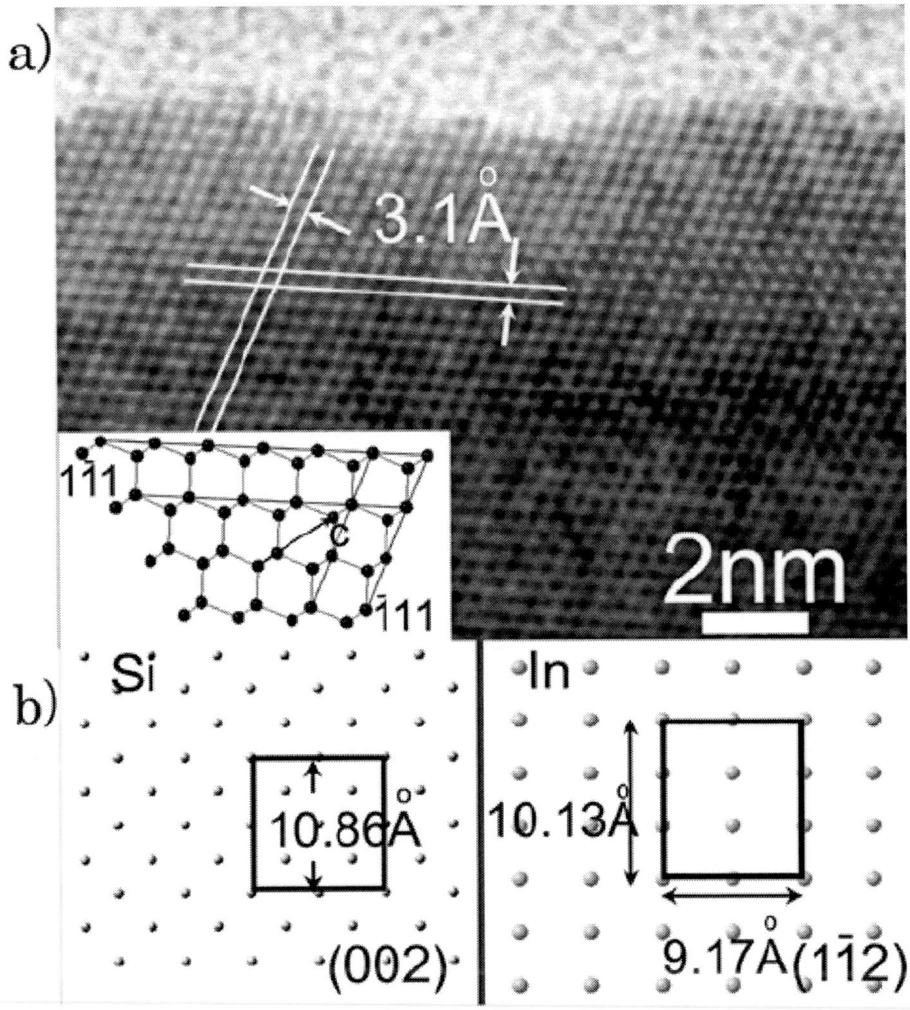

Figure 27. a) High-resolution TEM image of a Si subnanowire. Two sets of {111} (d_{111} = 3.135Å) planes are marked with the double lines. The inset is the scheme of a diamond-like Si crystal structure viewed along the [110] direction. b) Structural model showing the (002) plane of diamond-like Si and the (1$\bar{1}$2) plane of tetragonal In. From Ref. [56].

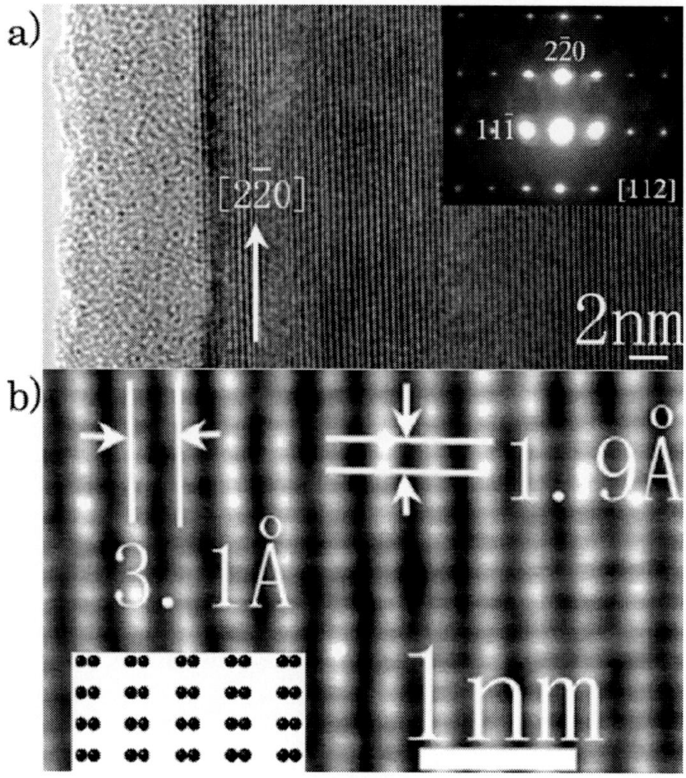

Figure 28. a) TEM image of a Si subnanowire, the inset is corresponding SAED pattern. b) High-resolution TEM image of the Si subnanowire growing along the $[1\bar{1}0]$ direction ($d_{11\bar{1}}$ = 3.13Å, $d_{2\bar{2}0}$ = 1.92Å, as indicated by double lines), the inset is the scheme of diamond-like Si crystal structure viewed along the [112] direction. From Ref. [56].

The present Si nanowires are sheathed with an amorphous oxide over their entire bodies. The presence of oxygen causing the amorphous silicon oxide formation has a considerable effect on the surface stability of crystalline Si. This may promote the anisotropic growth of Si nanowires. Various growth directions have been documented for the OA Si nanowires. Considering the fact that various orientations are possible for the anisotropic growth of Si nanowires sheathed with amorphous silica, a OA growth process is assumed to be involved in their growth. Silicon suboxide clusters are stable as a gas phase at high temperatures [87]. They facilitate the nucleation and growth of Si nanostructures. Disproportionation of SiO to Si and SiO_2 can result in the growth of Si nanowires. During the synthetic process, a SiO powder may sublimate to generate a SiO vapor. At 1400℃, the equilibrium pressure of metallic In is as high as ~1000 Pa. Being carried by a Ar gas flow to a lower temperature region (~900℃), a In vapor condenses to produce liquid In droplets. Gaseous SiO deposits over the In droplets and subsequently disproportionates to generate Si nanocrystals and an amorphous silicon oxide sheath. Anisotropic growth of Si nanocrystals *via* OA process results in the formation of Si nanowires sheathed with an amorphous silicon oxide. At ~ 900℃, the mobility of atoms is high enough to let newly arriving atoms rapidly assemble themselves within the growing edges and promote the fast growth of Si oxide nanotubes over In droplets.[24] During the growth of Si nanowires, Si oxide nanotubes may grow in the opposite direction. Meanwhile, a condensed In vapor is sucked into the nanometer-sized cavity of Si nanotubes due to the capillary force.[25] No In oxide is observed since the bonding energy of Si-O (798.8 kJ/mol) is much higher that of In-O (360.21 kJ/mol).[26] Consequently, a uniform SiO_2 sheath over the entire In-Si contact is formed. When the furnace temperature becomes below the melting point of metallic In (156.6℃), the liquid In columns confined with Si oxide nanotubes transform to solid In nanowires. The crystalline Si nanowires may have a certain impact on the crystallization of In nanowires, since both materials have closely-related crystallographic features, as demonstrated in figure 26. Owing to a volume decrease of an In column upon crystallization, an empty amorphous silica nanotube may be left above a In-Si heterojunction, as shown in figure 24.

Finally, the thermal expansion behavior of a In column confined with a silica nanotube and adjacent to a Si segment was investigated by means of a Gatan heating holder attached to TEM. Figure 29 shows consecutive TEM images of a In column during heating. Firstly, the In column abruptly jumps under its melting (m.p. 156.6°C) due to significant density difference between the solid and liquid phases. Secondly, the In column expands linearly upon heating in accord with the thermal expansion behavior of a bulk metallic In. The expansion of the liquid In column, being a part of a In-Si 1D heterojunction sheathed with silica, is envisaged to make possible the smart design of a temperature-driven switch and/or sensor within an electronic device.

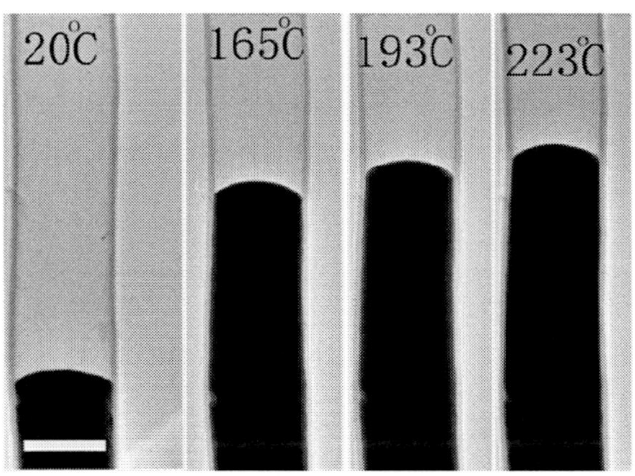

Figure 29. Consecutive TEM images showing melting and thermal expansion of an In column (the part of a In-Si heterojunction) confined with a silica tube during *in-situ* TEM heating. From Ref. [56].

4. Summary and Prospects

One-dimensional nanowire heterostructures made of different materials were introduced in this chapter. These nanostructures were generally obtained through one-step thermal evaporation. The mixed precursors were evaporated at a higher temperature. At a lower temperature, phase separation generated the modulated nanostructures along either the radial direction or axial directions with a distinction from other multelemental nanowires such as compound nanowires. Phase distribution in core-shell one dimensional nanostructures, such as Ga-filled carbon nanotubes and Ga_2O_3-C core-shell nanowires, is generally homogeneous along the axial directions. When two solids have a specific crystallographic relationship, unconventional side-by-side one dimensional nanostructures are obtained. A typical example is demonstrated on ZnS/Si nanowire heterostructures. The formation process of In-Si end-to-end nanowire contact is a little complex. The capillarity effect of silica nanotube sheathed on the silicon nanowires seems to play an important role on the condensation of In vapor. The growth of silicon nanowires and condensation of In vapors in silicon nanotubes took place simultaneously, which may be attributed to formation of unique In-Si end-to-end nanowire heterostructures.

In contrast with the compound nanowires, nanowire heterostructures provide a large possibility to take advantage of distinct functions and properties of different materials within an individual nanosized component. The gallium column encapsulated in carbon nanotubes can expand upon heating or shrink upon cooling, which give a possibility to record the temperate changes in a variety of microenvironments. Moreover, the liquid-gallium-filled carbon nanotubes posses a unique combination of two types of conductivity: low-resistantive Ga-filled nanotube segments and relatively high resistive empty nanotube regions [88]. This character allows one to realize an electrically controlled nanoscaled thermal sensor/switch. Zinc sulfide is a direct-band-gap semiconductor (3.68 eV at 300 K), while silicon is an indirect-band gap semiconductor. A single one-dimensional nanostructure composed of ZnS and silicon may stimulate interest in fabrication of novel nanoscaled optoelectronics. Potential barrier may exist in end-to-end metal-semiconductor contact of a Si-In nanowire heterstructure, which allows reliable signal processing. Linear thermal expansion of an In branch (after its melting under heating) makes possible the design of a temperature-driven switch and/or sensor within an electronic device. A thin SiO_2 outer layer will not only provide electrical insulation but also to protect the functional nanowire from potential chemical damage.

REFERENCES

[1] Xia, Y.; Yang, P.; Sun, Y.; Wu, Y.; Mayers, B.; Gates, B.; Yin, Y.; Kim, F.; Yan, Y.; *Adv. Mater.* 2003, *15*, 353-389.

[2] Duan, X. F.; Huang, Y.; Agarwal, R.; Lieber, C. M. *Nature* 2003, *421*, 241-245.

[3] Cui, Y.; Lieber, C. M. *Science* 2001, *291*, 851-853.

[4] Dai, H. *Acc. Chem. Res.*2002, *35*, 1035-1044.

[5] Han, W. Q.; Fan. S. S., Li, Q. Q.; Hu, Y. D. *Science* 1997, *277*, 1287-1289.

[6] Huang, H. M.; Mao, S.; Feick. H.; Yan, H. Q.; Wu, Y.Y.; Kind, H.; Weber, E.; Russo, R.; Yang, P.D. *Science* 2001, *292*, 1897-1899.

[7] Law, M.; Sirbuly, D.; Johnson, J.; Goldberger, J.; Saykally, R.; Yang, P. *Science* 2004, *305*, 1269-1273.

[8] Dai, H. J.; Wong, E. W.; Lu, Y. Z.; Fan, S. S., Lieber, C. M. *Nature* 1995, *375*, 769-772.

[9] Hu, J.; Odom T. W.; Lieber, C. M. *Acc. Chem. Res.* 1999, *32*, 435-445.

[10] Rao, C. N. R.; Deepak, F. L.; Gundiah, G.; Govindaraj, A. *Progr. in Solid State Chem.* 2003, *31*, 5-147.

[11] Zhan, J.; Yang, X.; Wang, D.; Li, S.; Xia, Y.; Qian, Y. *Adv. Mater.* 2000, *12,* 1348-1341

[12] Patzke, G. R.; Krumeich, F.; Nesper, R. *Angew. Chem. Int. Ed.* 2002, *41*, 2446-2461.

[13] Pan, Z. W.; Dai Z. R.; Wang, Z. L. *Science* 2001, *291*, 1947-1949.

[14] Dai, Z. R.; Pan, Z. W.; Wang, Z. L., Adv. Funct. Mater. 2003, *13*, 9-24.

[15] Zhang, R. Q.; Lifshitz, Y.; Lee, S. T. *Adv. Mater.* 2003, *15,* 635-640.

[16] Sloan, J.; Kirkland, A. I.; Hutchison, J. L.; Green, M. L. H. *Chem. Comm.* 2002, 1319-1332.

[17] Sze, S. M. *Physics of Semiconductor Devices*, John Wiley and Sons, New York, 1981.

[18] Wu, Y.; Fan, R.; Yang, P. *Nano Lett.* 2002, *2*, 83-86.

[19] Bjork, M. T.; Ohlsson, B. J.; Sass, T.; Persson, A. I.; Thelander, C.; Magnusson, M. H.; Deppert, K. Wallenberg, L. R.; Samuelson, L. *Nano Lett.* 2002, *2*, 87-89.

[20] Gudiksen, M. S.; Lauhon, L. J.; Wang, J.; Smith, D. C.; Lieber, C. M. *Nature* 2002, *415*, 617-620.

[21] Lauhon, L. J.; Gudiksen, M. S.; Wang, D.; Lieber, C. M. *Nature* 2002, *420*, 57-61.

[22] Hu, J.; Ouyang, M.; Yang, P.; Lieber, C. M. *Nature* 1999, *399*, 48-51.

[23] Wu, Y.; Xiang, J.; Yang, C.; Lu, W.; Lieber, C. M. *Nature* 2004, *430*, 61-65.

[24] Shi, W. S.; Peng, H. Y.; Xu, L.; Wang, N.; Tang, Y. H. H.; Lee, S. T. *Adv. Mater.* 2000, *12,* 1927-1930.

[25] Li, Y.; Bando, Y.; Golberg, D. *Adv. Mater.* 2004, *16,* 93-96.

[26] Kim, H. Y.; Park, J.; Yang, Y. *Chem. Comm.* 2003, 256-257.

[27] Zhang, Y.; Suenaga, K.; Colliex, C.; Iijima, S. *Science* 1998, *281*, 973-975.

[28] Luryi, S.; Xu, J. M.; Zaslavsky, A. *Future Trends in Microelectronics: The Road Ahead*, John Wiley and Sons, New York, 1999.

[29] Liu, A.; Jones, R.; Liao, L.; Samara-Rubio, D.; Rubin, D.; Cohen, O.; Nicolaescu, R.; Paniccia, M. *Nature* 2004, *427*, 615-618;

[30] Mathur, N. *Nature* 2002, *419*, 573-575.

[31] Wu, Y.; Cui, Y.; Huynh, L.; Barrelet, C. J.; Bell, D. C.; Lieber, C. M. *Nano Lett.* 2004, *4*, 433-436.

[32] Holmes, J. D.; Johnston, K. P.; Doty, R. C.; Korgel, B. *Science* 2000, *287*, 1471-1473.

[33] Lee, S. T.; Zhang, Y. F.; Wang, N.; Tang, Y. H.; Bello, I.; Lee, C. S. *J. Mater. Res.* 1999, *14*, 4503-4507;

[34] Gundiah, G.; Deepak, F. L.; Govindaraj, A.; Rao, C. N. R. *Chem. Phys. Lett.* 2003, *381*, 579-583.

[35] Kenichi, T.; Kazuaki, K.; Masao, A. *Phys. Rev. B* 1998, *58,* 2482-2486.

[36] Wühl, H.; Jackson, J. E.; Briscoe, C. V. *Phys. Rev. Lett.* 1968, *20,* 1496-1499.

[37] Binet, L.; Gourier, D. *J. Phys. Chem. Solids* 1998, *59*, 1241-1249.

[38] Lorenz, M. R.; Woods, J. F.; Gambino, R. J. *J. Phys. Chem. Solids* 1967, *28*, 403-408.

[39] Dai, Z. R.; Pan, Z. W.; Wang, Z. L. *J. Phys. Chem. B* 2002, *106*, 902-904.

[40] Chun. H. J.; Choi, Y. S.; Bae, S. Y.; Seo, H. W.; Hong, S. J.; Park, J.; Yang, H. *J. Phys. Chem. B* 2003, *107*, 9042.

[41] Gao, Y. H.; Bando, Y.; Sato, T.; Zhang, Y. F. *Appl. Phys. Lett.* 2002, *81*, 2267-2269.

[42] Yu, D. P.; Bubendorff, J. L.; Zhou, J. F.; Wang , Y. L.; Troyon, M. *Solid State Comm.* 2002, 124, 417-421.

[43] Hu, J. Q.; Li, Q.; Meng, X. M.; Lee, C. S.; Lee, S. T. *J. Phys. Chem. B* 2002, *106,* 9536-9539.

[44] Han, W. Q.; Kohler-Redlich, P.; Ernst, F.; Rühle, M. *Solid State Comm.* 2000, *115*, 527-529.
[45] Park, G. S.; Choi, W. B.; Kim, J. M.; Choi, Y. C.; Lee, Y. H.; Lim, C. B. *J. Crystal Growth* 2000, *220*, 494-500.
[46] Graham, U. M.; Sharma, S.; Sunkara, M. K.; Davis, B. H. *Adv. Funct. Mater.* 2003, *13*, 576-581.
[47] Sharma, S.; Sunkara, M. K. *J. Am. Chem. Soc.* 2002, *124*, 12288-12293.
[48] Chang, K. W.; Wu, J. J. *Adv. Mater.* 2004, *16*, 545-549.
[49] Liang, C. H.; Meng, G. W.; Wang, G. Z.; Wang, Y. W.; Zhang, L. D.; Zhang S. Y. *Appl. Phys. Lett.* 2001, *78*, 3202-3204.
[50] Gao, Y. H.; Bando, Y. *Nature* 2002, *415*, 599-599.
[51] Liu, Z.W; Bando, Y.; Mitome, M.; Zhan, J.H. *Phys. Rev. Lett.* 2004, *93*, 095504
[52] Zhan, J.H.; Bando, Y.; Hu, J.Q.; Sekiguchi, T.; Golberg, D. Nakanishi, H. *J. Phys. Chem. B.* 2005, *109*, 11580-11584.
[53] Zhan, J.H.; Bando, Y.; Hu, J.Q.; Li, Y. B.; Golberg, D. *Chem. Mater.* 2004, *16*, 5158-5161
[54] Zhan, J.H.; Bando, Y.; Hu, J.Q.; Sekiguchi, T.; Golberg, D. Adv. Mater. 2005, *17*, 225 – 230.
[55] Zhan, J.H.; Bando, Y.; Hu, J.Q.; Golberg, D. *J. Electron Microscopy* 2005, *54*, 485-491.
[56] Zhan, J.H.; Bando, Y.; Hu, J.Q.; Liu, Z.W.; Yin, L.W.; Golberg, D. *Angew. Chem. Int. Ed.* 2005, *44*, 2140-2144.
[57] Hu, J.Q.; Bando, Y.; Liu, Z. W.; Sekiguchi, T.; Liu, Q.L.; Golberg, D.; Zhan, J. H. *J. Am. Chem. Soc.* 2003, *125*, 11306- 11313
[58] Hu, J.Q.; Bando, Y.; Zhan, J. H.; Golberg, D Adv. Mater. 2005, *17*, 1964-1967.
[59] Hu, J.Q.; Bando, Y.; Zhan, J. H.; Golberg, D Appl. Phys. Lett. 2004, *85*, 2932-2934.
[60] Hu, J.Q.; Bando, Y.; Zhan, J. H.; Zhi, C. Y.; Golberg, D. *Nano Lett.* 2006, *6*, 1136-1140
[61] Zhan, J. H.; Bando, Y.; Hu, J. Q.; Golberg, D.; Kurashima, K. Small 2006, *2*, 62-65.
[62] Satishkumar, B. C.;Thomas, P. J.; Govindaraj, A.; Rao, C. N. R. *Appl. Phys. Lett.* 2000, 77, 2530-2532.
[63] Banhart, F.; Ajayan, P.M. *Nature* 1996, 382, 433-435.
[64] Banhart, F. *Rep. Prog. Phys.* 1999, 62, 1181-1221.
[65] Li, J.; Banhart, F. *Nano. Lett.* 2004, 4, 1143-1146.
[66] Terrones, M.; Terrones, H.; Banhart, F.; Charlier, J. C.; Ajayan, P. M. *Science* 2000, 288, 1226-1229.
[67] Chopra, N.; Ross, F. M.; Zettl, A. *Chem. Phys. Lett.* 1996, 256, 241-245.
[68] Crespi, V. H.; Chopra, N. G.; Cohen, M. L.; Zettl, A. Radmilovic V. *Appl. Phys. Lett.* 1998, *73*, 2435-2437.
[69] Goldberg, D.; Bando, Y.; Eremets, M.; Takemura, K.; Kurashima, K.; Tamiya, K.; Yusa, H. *Chem. Phys. Lett.* 1997, *279*, 191-196.
[70] Ugarte, D. *Nature* 1992, *359*, 707-709.
[71] Ajayan, P. M.; Ebbesen, T. W.; Ichihashi, T.; Iijima, S.; Tanigaki, K.; Hiura, H. *Nature* 1993, *362*, 522-525.
[72] Tsang, S. C.; Chen, K.; Harris, P. J. F.; Green, M. L. H. *Nature* 1994, *372*, 159-162.
[73] Pham-Huu, C.; Keller, N.; Estournès, C.; Ehret, G.; Ledoux, M. J. *Chem Comm.* 2002, 1882-1883.

[74] Xu, X. P.; Brandes, G. R. *Appl. Phys. Lett.* 1999, *74*, 2549-2551.
[75] Fan, S. S.; Chapline, M. G.; Franklin, N. R.; Tombler, T. W.; Cassell, A. M.; Dai, H. J. *Science* 1999, *283*, 512-514.
[76] Collins, P. G.; Zettl, A. *Phys. Rev. B* 1997, *55*, 9391-9399.
[77] Marcus, R. B.; Ravi, T. S.; Gmitter, T.; Busta, H. H.; Niccum, J. T.; Chin, K. K.; Liu, D. *IEEE Trans. Electron Devices* 1991, *38*, 2289-2293.
[78] Wu, Z. S.; Deng, S. Z.; Xu, N. S.; Chen, J.; Zhou, J.; Chen, J. *Appl. Phys. Lett.* 2002, *80*, 3829-3831.
[79] Wu, Y.; Yang, P. *J. Am. Chem. Soc.* 2001, *123*, 3165-3166.
[80] Levitt, A. P. *Whisker Technology*, Wiley-Interscience, New York, 1970.
[81] Duan, X. F.; Lieber, C. M. *Adv. Mater*. 2000, *12*, 298-302.
[82] Wang, D.; Dai, H. *Angew. Chem. Int. Ed.* 2002, *41*, 4783-4786.
[83] Tran, N. H.; Hartmann, A. J.; Lamb, R. *J. Phys. Chem. B* 2000, *104*, 1150-1152.
[84] Lee, S. T.; Zhang, Y. F.; Wang, N.; Tang, Y. H.; Bello, I.; Lee, C. S. *J. Mater. Res.* 1999, *14*, 4503-4507.
[85] Gundiah, G.; Deepak, F. L.; Govindaraj, A.; Rao, C. N. R. *Chem. Phys. Lett.* 2003, *381*, 579-583.
[86] Barin, I. *Thermochemical data of pure substances*, VCH Publishers, New York, 1989.
[87] Brewer, L. *Chem. Rev.* 1953, *52*, 1-75.
[88] Dorozhkin, P. S.; Tovstonog, S. V. ; Golberg, D.; Zhan, J.; Ishikawa, Y.; Shiozawa, M.; Nakanishi, H.; Nakata, K.; Bando, Y. *Small* 2005, *1*, 1008-1093

In: Progress in Advanced Materials Research
Editor: Nicolas H. Voler, pp. 221-236
ISBN: 978-1-60021-893-4

Chapter 8

PROGRESS IN SPM-BASED ULTRAHIGH DENSITY INFORMATION STORAGE

Yongqiang Wen and Yanlin Song*
Organic Solid Key Laboratory, Center for Molecular Science,
Institute of Chemistry, Chinese Academy of Sciences, Beijing 100080; China

ABSTRACT

With the explosive growth in data volume in this digital age, higher and higher requirements on information storage techniques have been put forward. In recent years, novel storage techniques have welled up continuously. Among them, SPM-based information storage techniques, including scanning tunneling microscopy (STM), atomic force microscopy (AFM), scanning near-field optical microscopy (SNOM), etc., have become one of the most active research fields. The development of data storage technique depends to a large degree on the development of storage medium. Storage medium with high storage density, fast response speed and long lifetime is the critical factors for ultrahigh-density data storage. This review gives an overview of SPM-based nanoscale data storage on different medium, including inorganic materials, polymers, and organic small molecules films. Particular emphasis is placed on the novel organic materials. The related storage mechanism and principals are introduced and the development foreground is predicted.

I. INTRODUCTION

The 21st century is an era of high-tech with economy informationization and information digitalization. In this digital age, information technology infiltrates various aspects of economy, politics, military and society, and influences greatly all trades and professions, supports industrial development and contributes to lifestyles, economies and society. With the explosive increasing in data volume, higher and higher requirements on information storage

* Email: ylsong@iccas.ac.cn (Prof. Y. L. Song)

techniques have been put forward, and people need to learn how to save, manage, and use more and more data availably. Currently, microelectronic-based magnetic and optical data storage, from the point of storage mechanism, have basically come to or approach to the superparamagnetic effect and optical diffraction limit [1, 2] and are difficult to satisfy the increasing demand of information storage. In order to overcome these limits, novel storage techniques have been developed continually to increase the data storage density in recent years. Among them, SPM-based information storage techniques, including scanning tunneling microscopy (STM), [3] atomic force microscopy (AFM), [4] scanning near-field optical microscopy (SNOM), [5] etc., have become one of the most attractive research fields.

The development of data storage technique depends to a large degree on the progress of novel storage medium. In addition to high storage density, fast response speed and long lifetime are also the pursuing goals for development of information storage technology. To meet these demands, developing novel high-powered storage medium and high-quality storage thin film has also become a principal challenge for ultrahigh-density data storage device. In this article, we give an overview of SPM-based nanoscale data storage based on different storage mechanism on various media, including inorganic materials, polymers, and organic small molecule films. The advantages and disadvantages are discussed and the development foreground is predicted.

II. Probe Storage Devices and Advantages

Since STM was invented by Binnig et al. at IBM research center in 1982, [6] it has been proved to be one of the most powerful tools for investigating local electronic surface structure and has also led to a number of related scanning tunneling microscopy technologies, such as AFM, SNOM, MFM, friction force microscopy (FFM), kelvin force microscope (KFM), etc. A SPM, which can be used not only to observe the surface morphology of materials but also to locally modify surfaces on nanometer scale, has already displayed huge potential in ultrahigh-density data storage. The progress in SPM technologies has made it possible to manipulate a single molecule or even an atom.

Figure 1 shows a comparison of the storage areal density and data rate for various tip-based storage modes and conventional data storage devices.[7] From this comparison, it can be clearly seen that the attained information density of SPM-based storage technology is 3-6 orders of magnitude higher than microelectronic technique-based magnetic and optical data storages, although limitations in storage rate exist. This apparent predominance of SPM-based storage mode in data density has made it a powerful tool in future information storage development.

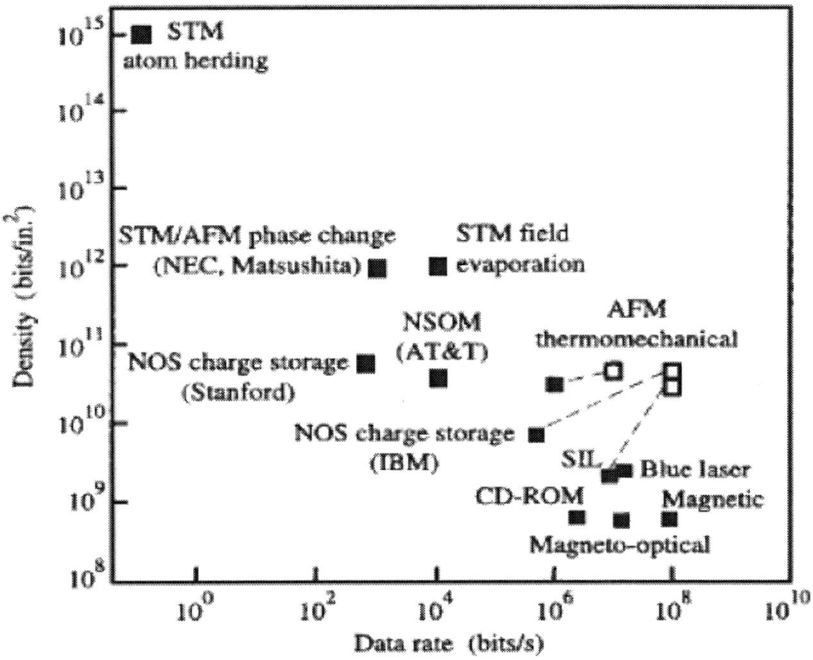

Figure 1. Comparison of storage capabilities and rates between SPM-based demonstrations and conventional data storage devices. NOS: nitride-oxide-semicon doctor; SIL: solid immersion lens. NSOM: near-field scanning optical microscopy. Copyright 1995 by IBM Corporation, republished with permission from the IBM Journal of Research and Development [7].

III. Storage Medium and Principles of Probe Storage

The development of storage medium is also one of the most important factors for the development of data storage technique. When the size of information dots arrives at nanometer scale, the traditional materials and techniques would face an enormous challenge. Thus, studies of novel SPM-based materials and devices, as one of the central issues of nanoelectronics, have become an important topic and would provide a solid theoretical and experimental foundation for the future development of information technology. At the earlier stage, the representative media for data storage using SPM are inorganic materials.[3, 4] Compared with inorganic materials, organic materials including small organic molecules, polymers, organic complexes (including organic/inorganic or organic/metal complexes), have been paid much attention in various fields [8-11] in recent years due to their unique properties, such as easily processing, lower price, special responsive characters to light, electric, and/or magnetic field, controllable molecular structure and capability, etc.

The key to realize data storage is that the storage media should have two distinguishable stable states (i.e. states of “0” and “1”) and can transform from one state to another under certain conditions. Combining the traditional storage principle with SPM technology, SPM can be used to modify materials’ surface and realize the data “write” on the surface by applying local electrical, optical, magnetic, or thermal actions between the SPM tip and the substrate to bring the property transformation, such as phase change, charge transfer, chemical reaction, etc.; and physical or chemical changes in nanometer scale such as electrical, optical, magnetic properties and morphology can be observed by virtue of their high resolving ability, i.e. "readout" the corresponding information. The tools of SPM-based storage mainly include: STM, AFM, SNOM, MFM, scanning capacitance microscopy (SCM), electrostatic force microscope (EFM), etc. Accordingly, the storage mechanism of SPM-based data storage can be generally divided into two kinds. (1) "write" and "read" information are based on the change of surface morphology. (2) "write" and "read" information are based on the change of materials property (electrical, optical or magnetic properties, etc.) while the surface morphology remains unchanged. Figure 2 presents the schematic diagrams of the principles of several SPM-based information storage.

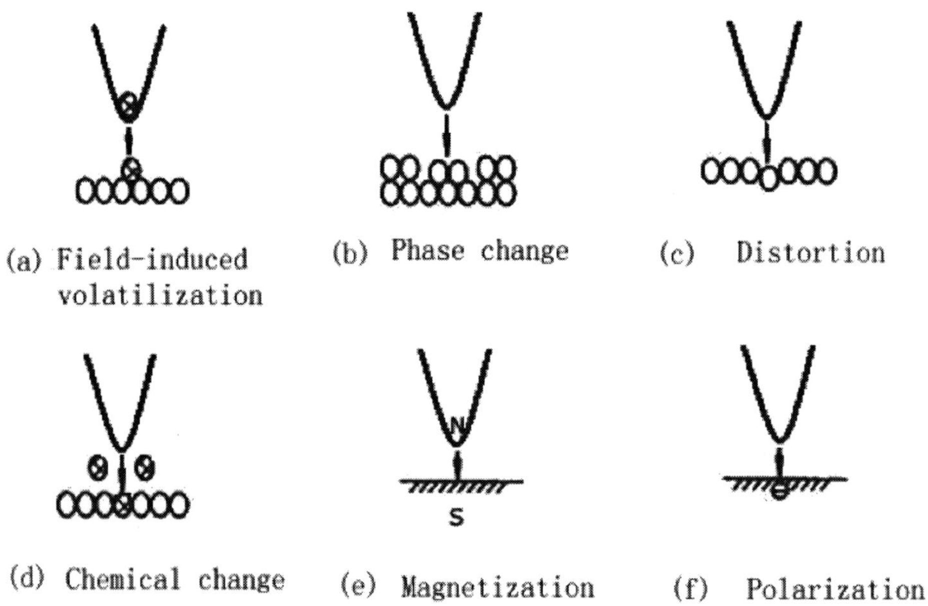

Figure 2. Schematic diagrams of the principles of SPM-based information storage.

1. Mechanical Storage

Here, mechanical storage refer to the recording who causes morphology change in the material surface due to thermomechanical tip impression or thermal-induced volatilization, phase change, etc., where tiny pits or holes on the material surface represent the digital data.

In the early 1990s, Mamin and Rugar [12] at the IBM Almaden Research Center pioneered the use of an AFM tip for writing and reading of topographical features on a rotating polycarbonate substrate for data storage. Data was formed by a thermomechanical process of applying local heating and force to the polymer through the tip. In this way, density of up to 30 Gb/in^2 was achieved, representing a significant advance compared to the densities of that day. Afterwards, Binnig et al [13] found that the size of the recording marks was influenced by the thermal conductivity of the substrate. Then they modified their technique to increase data rate and storage density by using a line polymer layer on a silicon substrate as the recording medium. Because Si is a much better heat conductor than the polymer, the achieved bit sizes decreased to the range of 10 to 50 nm and the storage density was of the order of 500 Gbit/in^2. In 1999, Weidenhof et al [14] studied rewritable phase transitions induced by local laser heating in an inorganic material. Reversible switching between the amorphous and crystalline states of $Ge_2Sb_2Te_5$ thin films, which is accompanied by a considerable change in optical properties, was exploited to store bits of information. Morphology changes were detected with AFM, which allows a precise characterization of the bit size and bit depth.

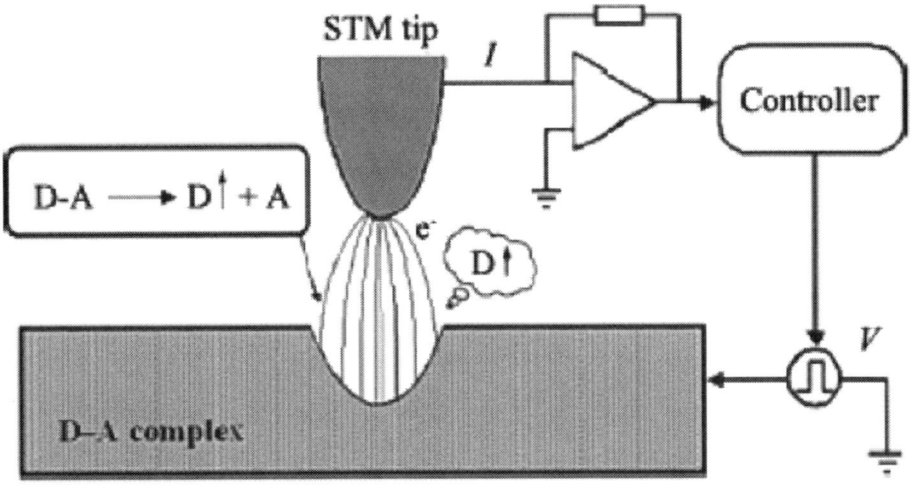

Figure 3. Diagram of the STM-based THB process. D: low boiling-point component, A: high boiling-point component.

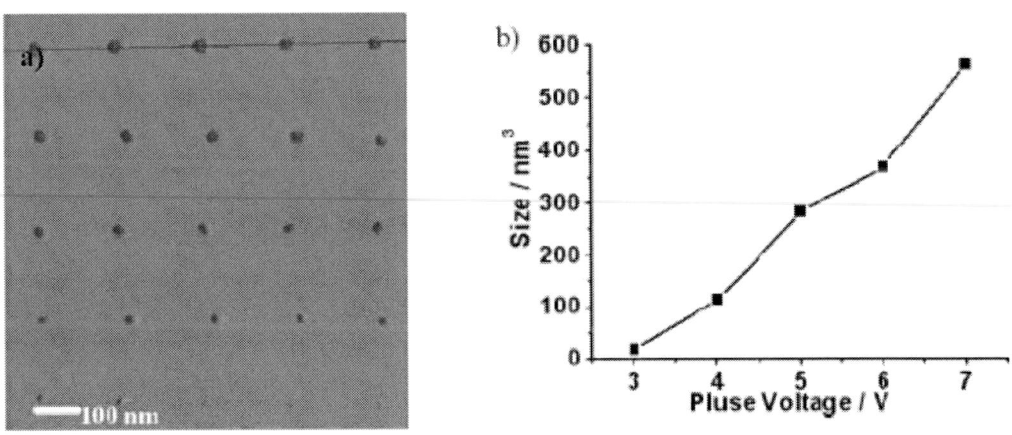

Figure 4. Dependence of the hole size on the pulse voltage and polarity of the imaging bias. a) A 5 × 5 array formed by voltage of 3, 4, 5, 6 and 7 V at a fixed duration of 300 µm, corresponding to columns from bottom to top, respectively. b) The size of the hole as a function of the pulse voltage. Figure 3 and figure 4 were reprinted from Advanced Materials, Vol. 17, Peng, H. L.; Ran, C. B.; Yu, X. C.; Zhang, R.; Liu, Z. F. "Scanning-tunneling-microscopy based thermochemical hole burning on a new charge-transfer complex and potential for data storage" Pages 459-464. Copyright (2005), with permission from Wiley-VCH.

In recent years, Liu et al have presented the concept of ultrahigh density thermochemical hole burning (THB) memory (figure 3), in which information bit was recorded as a hole. A STM tip was used to induce a localized THB effect on a series of charge transfer (CT) complexes crystals, [15-17] which is attributed to the localized thermochemical decomposition of the complex induced by the current heating effect. The decomposition reaction evolves the low boiling point decomposition components of the charge transfer complex, leaving a nanometer-sized hole on the crystal surface. Figure 4 shows a STM image of recording holes array on the surface of a CT crystalline complex, dibutylammonium bis-7,7,8,8-etracyanoquinodimethane [DBA $(TCNQ)_2$]. With the increasing of pulse voltage, the hole size was found to increase.[15]

2. Electrical Storage

Electrical storage generally involves changes in electrical property of storage media due to phase change, charge transfer, chemical reaction, etc., which usually is caused by applying pulse voltage between the SPM tip and the substrate. In 1993, Sato et al [3] demonstrated the possibility of STM being used for reversible recording and erasing of marks about 10 nm in size at ambient temperature and pressure on the surface of a composite medium, β-$Na_xV_2O_5$. Instead of using changes in surface morphology for recording as previously reported, changes in tunneling current were produced due to locally induced phase transitions between crystalline and glassy forms.

Since the middle of last decade, a research group from Chinese Academy of Science and Peking University has made great progress in STM-based electrical information storage with organic materials. [18-21] Several organic materials have been designed from the point of improving the film quality and the properties for STM-based information storage. By investigating the recording properties of the organic thin films and their relationship with molecular structures, they further studied the corresponding storage mechanisms.

In 1996, they designed and synthesized an organic complex made of m-nitrobenzole malonitrile (m-NBMN) and diamine benzene (DAB), and succeeded in writing marks with a diameter of 1.3 nm. [18] In 2000, they further demonstrated that conductance transition in m-NBMN/pDA (see figure 5) crystalline organic complex induced by local electric field could be reversed by the application of reverse pulses.(see figure 6) [19] Phase changes between amorphous and crystalline state were proposed as the reason for such reversible transition. Considering the difficulty for organic complex to prepare mono-dispersed thin film, an organic molecule combining two functional groups (electron donor and electron acceptor), N-(3-nitrobenzyllidene)-p- phenylene-diamine (NBPDA), was further designed and used for preparing thin film by vacuum evaporation method.[20] Ultrahigh density data storage (marks with 0.7 nm in diameter) was achieved and electrical bistability induced by charge transfer was measured on crystalline NBPDA films. However, because of the absent of strong interactions among the molecules, it was still difficult to obtain a large and smooth surface.

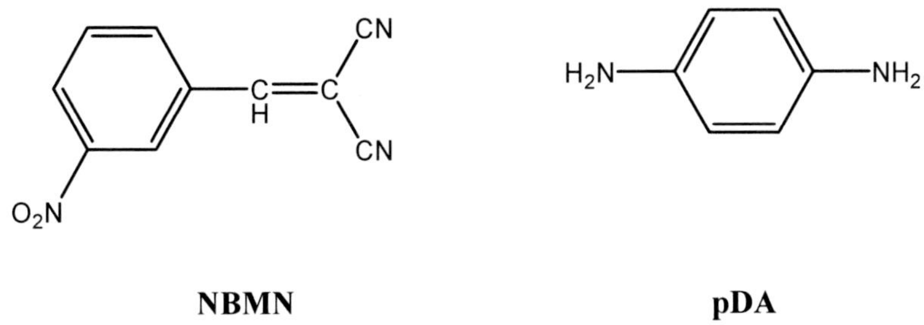

Figure 5. The molecular structure of NBMN and pDA.

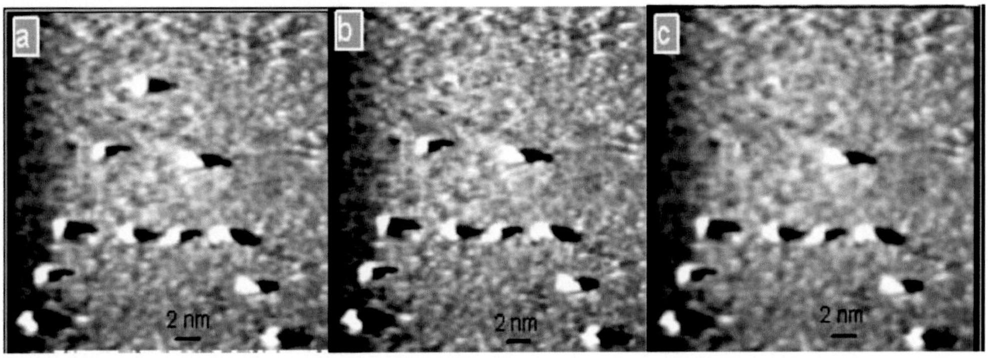

Figure 6. STM images of the NBMN-pDA film in the information recording experiments. (A) a "A" recorded pattern (B), (C) after erasing one mark each time. Reprinted with permission from Gao, H. J.; Sohlberg, K.; Xue, Z. Q.; Chen, H. Y.; Hou, S. M.; Ma L. P.; Fang, X. W.; Pang, S. J.; Pennycook S. J. Physical Review Letters Vol. 84, 1780-1783. Copyright (2000) by the American Physical Society.

In contrast to small organic molecules, supramolecular materials were often easier to be synthesized and processed due to their excellent self-assembling property originated from the strong intermolecular noncovalent interactions. In 2005, stable, reversible conductance transitions and memory effects were observed in solid-state supramolecule rotaxane-based Langmuir-Blodgett (LB) thin films by Feng et al.[21] Taking advantage of the switching property, they have achieved reproducible nanometer-scale recording dots on rotaxane LB films via voltage application using STM. The stable and easily controlled conductance switching made the rotaxanes a hopeful candidate for nanorecording.

Song and Diederich et al provided a new rout to improve the performance of organic storage materials by molecular self-assembly during vacuum deposition. They designed a stable conjugated organic molecule with strong push-pull electron groups, 1,1,2-tricyano-2-[(4-dimethylaminophenyl)ethynyl] ethane (TDMEE) (see figure 7a).[22] The high-quality crystalline thin films were prepared based on the self-assembly of TDMEE molecules by intermolecular π-π and donor-acceptor interactions using the vacuum deposition method. The molecules undergo π-stacking in the crystalline thin films, and the packing adopts regular alternate donor/acceptor stacks. Experimental and theoretical studies indicated that the recording was based on an electric-field-induced intermolecular charge transfer, which was strongly influenced by the ordered, antiparallel packing mode of the dipolar donor-acceptor molecules in the film.

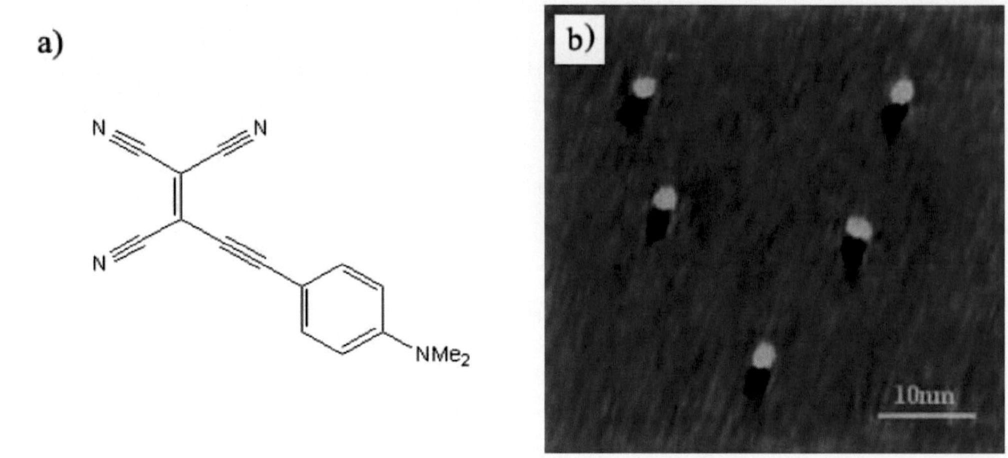

Figure 7. a) The molecular structure of TDMEE and b) The STM image of a recorded pattern. The average size of marks is about 2.1 nm in diameter. Reprinted from Advanced Materials, Vol. 17, Jiang, G. Y.; Michinobu, T.; Yuan, W. F.; Feng, M.; Wen, Y. Q.; Du, S. X.; Gao, H. J. Jiang, L.; Song, Y. L.; Diederich, F.; Zhu, D. B. "Crystalline Organic Thin Film for Nanoscale Data Recording Through Intermolecular Charge Transfer interactions" Pages 2170-2173. Copyright (2005), with permission from Wiley-VCH.

Besides STM, conductive AFM can also be used for nanometer-scale recording. Kado [23] have demonstrated this technique with an amorphous $GeSb_2Te_4$ film as a phase change medium. Data are recorded by changing the local electric properties of the thin film using a conductive AFM probe. The conductance of the film can be increased by more than one hundred times by applying a pulse voltage between the probe and the film. The recorded data can be read by detecting the change in the conductance using the probe. The smallest possible recording region was 10 nm in diameter, which corresponds to a data storage density of 1 Tbit/cm^2. The data can be erased by applying a negative voltage to the probe. The mechanism of the reversible conductance change in the film was attributed to an electric-field-induced Schottky barrier tunneling.

Conductive polymer is also a kind of important materials for ultrahigh density data storage. Yoshida et al [24] have studied the reversible electrical modification of a conductive polymer based on an AFM. Using a surface-initiated polymerization method, a thin polyaniline self-assembled monolayer (SAM) was formed on Au surface. With the transformation of the polarity of the applied voltage, the conductivity of the SAM was reversibly changed. The reversible electrical modification was considered to originate in the oxidation-reduction reaction or protonation- deprotonation reaction of PANI from x-ray photoelectron spectroscopy measurement. In 2005, the possibility of electrochemical nanowriting on a electroactive precursor polymer thin film of poly(vinylcarbazole) (PVK) was demonstrated by Advincula et al.,[25] in which nanopatterns were created by the application of a voltage-biased AFM tip through electropolymerization and crosslinking between the carbazole units of a PVK precursor that is spin coated onto a conducting substrate.

3. Near-Field Optical Storage

Classical optics describes the laws of diffraction in the so-called far field. The maximum resolving power of a far-field optical microscope is given by the Rayleigh criterion, which is also referred as the diffraction limit:

$$d_{min} = 0.61\ \lambda/n\sin\theta$$

where λ is the wavelength of light and $n\sin\theta$ is the numerical aperture. SNOM allows the investigation of optical properties beyond diffraction limit. The spatial resolution depends on the aperture diameter and the distance of tip to sample surface. The illuminated subwavelength aperture acts as a nanometer size light source and is scanned over the sample surface within the optical near-field. [5, 26] During the past few years, more and more attention has been aroused to this technique that shows enormous potential for imaging, sensing and modification at near-molecular resolution, and the potential storage density is far beyond currently available digital storage. [27, 28]

SNOM recording can be realized by virtue of magneto-optical change, phase change or chemical change of the recording film for writing, and polarized or reflected light detection for reading. One major advantage of SNOM, besides high density and mass storage capacity, is that in principle many existing storage techniques, for example, flying head technology in hard disk drivers and magneto-optical material and phase change material storage techniques, could be modified for SNOM data storage, which would greatly reduce the time and cost required for research and development. [27] So far there are three methods to realize the near-field recording: probe scanning microscopy (PSM), super resolution near-field structure (Super-RENS) and solid immersion lens (SIL). [27, 29]

In 1992, Betzig group [5] firstly performed near-field recording on magneto-optical medium using SNOM with a fiber tip about 20 nm in diameter as recording probe. The SNOM can control the spacing between the probe and the sample, maintaining constant by using a shear force for non-contact AFM operation. In their experiment, recording marks of 60 nm and storage density of 45 Gb/in^2 were obtained. However, low efficiency of throughput of the probe limited the speed of recoding and read-out.

In 1990, a method of SIL was proposed by Kino and Mansfield of Stanford University to improve the resolution of optical microscope,[30] and used for optical data storage later. [31] With this method, the size of recording marks can be decreased by a factor of the refractive index “n” of the SIL. By combining the SIL with optical data storage technology, it may dramatically increase the areal density while retaining erase ability and high data rate. In 1994, Terris et al used a SIL (made from n = 1.83 glass) near-field optical recording system and got a recording mark of 350 nm in magneto-optic material.[32] Later they further developed a SIL flying optical head technique. Writing and reading were achieved at a density of 3.8×10^8 bits/cm^2 with a data rate of 3.3×10^6 bits/s. [33]

Recently, a new technique called Super-RENS to generate and control optical near-field effects without a probe like SNOM was developed for optical near-field recording. Super-RENS is usually composed of a SiN/Sb/SiN/$Ge_2Sb_2Te_5$/SiN multilayered structure on a substrate disk, where $Ge_2Sb_2Te_5$ is used for a recording layer and Sb for an aperture layer. The characteristic of this technique is that a near-field aperture is fabricated in the aperture layer in the disk and a distance between the aperture and the recording film is always fixed at a few tens of nanometers with a SiN dielectric thin film. Tominaga et al successfully developed this technique to achieve nanoscale optical data storage. Optical resolution on optical marks with a size about 81 nm has been achieved on a phase change medium. The mark size was far beyond the diffraction limit of an optical pickup with a wavelength of 635 nm and a numerical aperture of 0.6.[34]

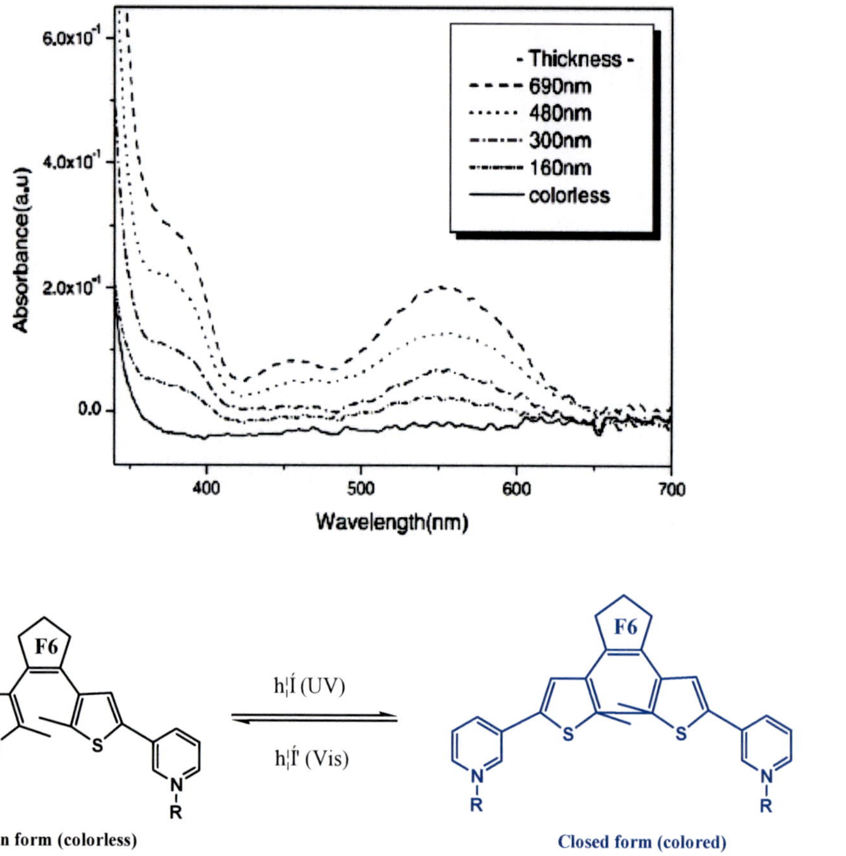

Figure 8. The changes in the absorption spectra of the diarylethene compound before and after UV irradiation with different thickness. The solid line shows the spectrum of a fresh diarylethene film, the upper spectra show the absorption spectra after UV irradiation with different thickness. In the lower part of the figure, the photochromic reaction of the diarylethene is shown.

Reversible data storage is an important goal for future information storage, which depends primarily on the development of rewritable recording media. Photochromic materials, which can reversibly isomerize between two discrete structures upon absorption of different wavelength light, without doubt, are a series of promising materials. Using them, near-field nanowriting and 3D-rewritable media have been developed. Up to now, a serial of photochromic materials, such as azobenzenes, fulgides, 9-chloroanrhracene, diarylethene, have been used for SNOM-based data storage. [35-38] In 2003, with SNOM, sub-micron scale optical read, write, erase on azo-polymethacrylate thin films have been realized by Labardi et al.[35] In the experiment by Lee et al,[39] near-field writing and reading were realized using the 514 nm line of an argon ion laser on deposited films of diarylethene. For writing, a laser power of 5 mW was coupled into the SNOM probe. A normal film of diarylethene is transparent. Upon UV-irradiation, the absorption spectrum changed (see figure 8). An absorption peak appeared at around 550 nm, and the film became red. Upon irradiation with a wavelength in this region of the absorption spectrum, bleaching will occur, thus leaving transparent marks in the otherwise red film. Using deposited diarylethene films, memory bits with 1.4 μm diameter size was clearly observed (see figure 9).

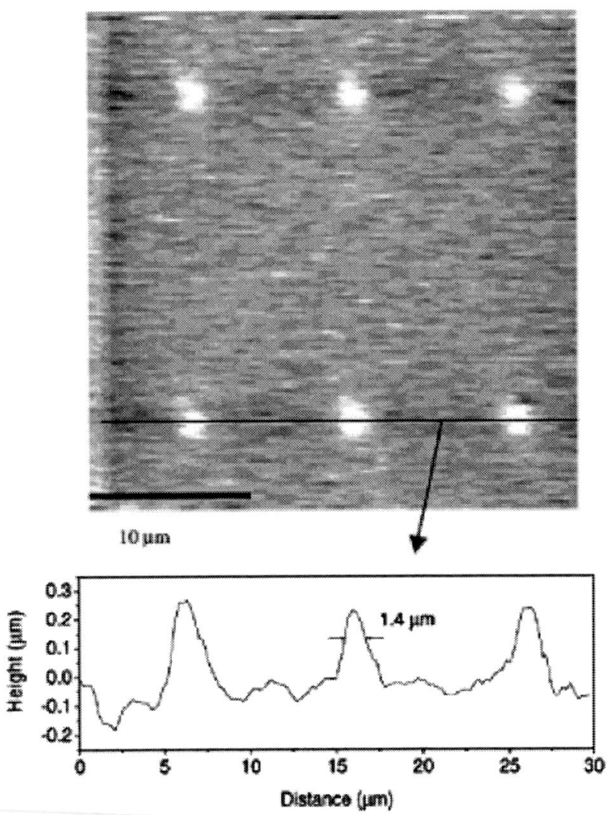

Figure 9. Bits formation on a diarylethene photochromic thin film as read with the same setup which was used for writing the bits. The cross section analysis (lower figure) shows that the size of the bits is 1.4 μm in diameter. Figure 8 and figure 9 were reprinted from Optical Materials, Vol. 21, Lee, H. W.; Kim, Y. M.; Jeon, D. J.; Kim, E.; Kim, J.; Park, K. "Rewritable organic films for near-field recording" Pages 289-293. Copyright (2002), with permission from Elsevier.

4. Magnetic Storage

Since magnetic force microscopy (MFM) was developed in 1987, [40] it has emerged as a powerful tool for studying a wide variety of local magnetic phenomena. By measuring changes of the interaction force between a magnetized probe and the local stray magnetic field from the sample, MFM allows direct visualization of magnetic domains for reading and provides the experimental basis for theoretical modeling. The probe is typically a cantilever made of silicon or silicon nitride, with a ferromagnetic tip on the free end. MFM is capable of detecting details with diameters down to 30 nm,[41, 42] and the writing process can be realized by applying additional external fields [43] or locally heating the sample [44] when the tip is nearly in contact with the sample.

In recent years, great progress has been made in magnetic recording with MFM. Much work has been done in preparation of advanced MFM probes, micromagnetic simulations and system design to achieve information storage with high density and low SNR.[45-49] In 1997 Kong et al. of the University of Minnesota described data storage with MFM, and error-free results were obtained up to 7.5 Gbits/in^2. In their experiment, a complicated two tips architecture was used in the writing and reading process, one for erasing and writing, and another for reading; both were magnetically "hard".[50] Carley et al of Carnegie Mellon University suggested an alternative magnetic storage system that relies on magneto-optic storage materials with a "soft" magnetic tip coupled to a write coil and giant magnetoresistive read structures. Comb-drive micro electro mechanical system (MEMS) motors were used in this magnetic memory that can store over 2 GB of data in 2 cm^2 of die area.[51] In 2005, they investigated the feasibility of using a magnetic force microscopy for ultrahigh-density information storage, and an areal density of up to 0.3 Tb/in^2 was achieved with thermally stable magnetic marks and a SNR of 20-25 dB in the existence of additional electronic noise.[46]

Figure 10 shows an MFM image of MFM-based storage, which displays a switched magnetic domains of a Pt/Co multilayered film. The dimensions of the information dots were 60 nm. In the experiment, the current induced heating was used for writing and the MFM observation for reading. The MFM probe consisted of SiO_2 base, and a PtCoCr-alloyed film was used for a magnetic probe and a carbon film to make the probe conductive.[44, 52]

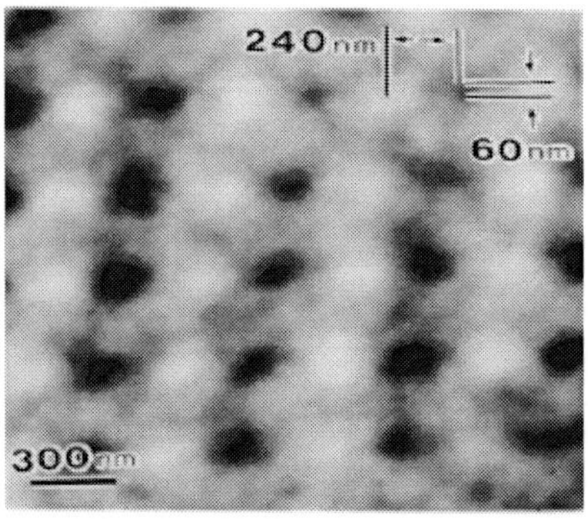

Figure 10. MFM image of written domains induced by local heating (writing). Reused with permission from S. Hosaka, H. Koyanagi, A. Kikukawa, M. Miyamoto, Ryo Imura, and J. Ushiyama, Journal of Vacuum Science & Technology B, 13, 1307-1311. Copyright (1995), with permission from AVS The Science & Technology Society.

IV. Future Perspectives of SPM-Based Storage

Although great progress in data density have been made, major limitations and challenges still remain on SPM-based data storage. For practical application, a high readout speed is necessary. However, the readout speeds proposed for SPM-based storage were by far several orders of magnitude slower than practical applications (see figure 1). For example, currently a single AFM operates at best on microsecond time scale, while conventional magnetic storage operates at best on nanosecond time scale. Besides the readout rate, some other issues must be addressed for a practicable SPM-based storage, such as a systematic engineering, a system reliability, etc.

Hosaka [52, 53] summarized the readout speed of various SPM-based storage and predicted that AFM and SNOM based storages have greater possibility to be the candidates for future storage, which have the potential to reach a high-speed readout of over 1 MHz with ultrahigh recording densities of 1 Tb/in^2 and 170 Gb/in^2, respectively.

In recent years, an attractive technology with high density and high data transfer rate has been proposed to overcome the limitation of the storage rate, in which an array of nano-scaled tips are used for writing and reading.[54-57] Figure 11 shows the schematic diagram of a 2D cantilever array data storage system proposed by researchers of IBM, called "Millipede".[57] It is based on a mechanical parallel x/y scanning of either the cantilever array chip or the storage medium for writing/reading through feedback-controlling the tip-medium contact. It is worth noting that the Millipede approach is not based on individual z-feedback for each cantilever, but uses a feedback control for entire chip. In principle, using many cantilevers working in parallel, one can overcome low data transfer rates according to the result of IBM researchers. At the first step a 5 × 5 and then a 32 × 32 array chip were designed and fabricated to test the basic Millipede concept. With a 32 × 32 AFM cantilever arrays, data rates of several hundreds Mbit/s can be realized.[55] Figure 12 shows the SEM image of the first parallel writing/reading result. The dots are about 50 nm in diameter, which results in an areal densities of 100-200 Gb/in^2 depending on the ability to separate the bit indentations by a distinguishable amount.[54]

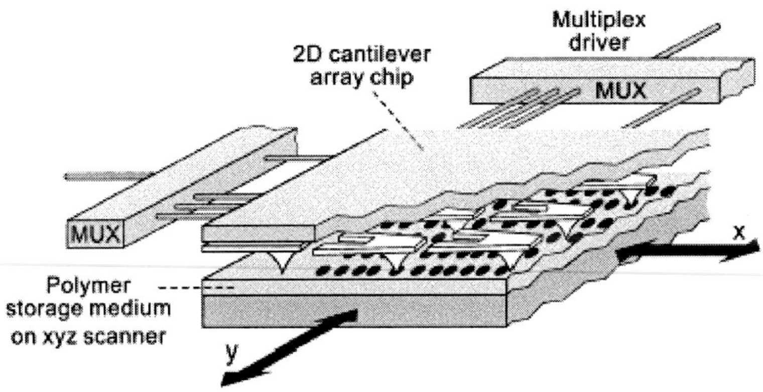

Figure 11. Illustration of the Millipede concept. Reprinted from IEEE Transactions on Magnetics Vol. 39, Eleftheriou, E.; Antonakopoulos, T.; Binnig, G. K.; Cherubini, G.; Despont, M.; Dholakia, A.; Dürig, U.; Lantz, M. A.; Pozidis, H.; Rothuizen, H. E.; Vettiger, P. "Millipede-A MEMS-Based Scanning-Probe Data-Storage System" Pages 938-945. Copyright (2003), with permission from IEEE.

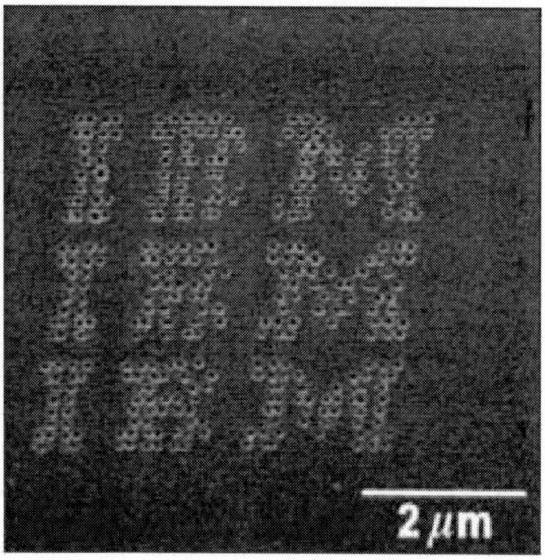

Figure 12. SEM image of x/y/z scanner Millipede writing result (50-nm-thick PMMA medium). Copyright 2000 by IBM Corporation, republished with permission from the IBM Journal of Research and Development [54].

V. Conclusion

SPM has been proved to be a powerful tool for ultrahigh density data storage. It would increase several orders of the data density compared with microelectronic technique-based magnetic and optical data storages. In the past few years, SPM-based information storage has become an important research field in information science, and great progress has been made in relative techniques. Meanwhile, many novel materials have been developed, and organic materials has especially been paid much attention due to their special properties and advantages. Although some critical challenges still exist, in principle, the main limitation of readout rate would be solvable in a massively parallel system when an array of recording/reading tips is used. In this way, the concept of high-density and high rate storage operation with 2D AFM array chip has already been demonstrated. With the further development of millipede array chip fabrication technique, and with the development of novel materials with excellent performance, the commercial application of SPM-based storage with high density, high rate, and long time can be expected in the future.

References

[1] Thompson, D. A.; Best, J. S. *IBM J. Res. Develop,* 2000, 44, 311-322.

[2] Betzig, E.; Trautman, J. K. *Science.* 1992, 257, 189-195.

[3] Sato, A.; Tsukamoto, Y. *Nature.* 1993, 363, 431-432.

[4] Kado, H.; Tohda, T. M. *Appl. Phys. Lett.* 1995, 66, 2961-2963.

[5] Betizg, E.; Trautman, J. K.; Wolfe, R.; Gyorgy, E. M.; Finn P. L.; Kryder, M. H.; Chang, C. H. *Appl. Phys. Lett.* 1992, 61, 142-144.

[6] Binnig, G.; Rohrer, H. *Helv. Phys. Acta.* 1982, 55, 726-735.

[7] Mamin, H. J.; Terris, B. D.; Fan, L. S.; Hoen, S.; Barrett, R. C.; Rugar, D. *IBM J. Res. Develop.* 1995, 39, 681-700.

[8] Joachim, C.; Gimzewski, J. K.; Aviram, A. *Nature.* 2000, 408, 541-548.

[9] Ma, L. P.; Liu, J.; Yang, Y. *Appl. Phys. Lett.* 2002, 80, 2997-2799.

[10] Carroll, R. L.; Gorman, C. B. Angew. *Chem. Int. Ed.* 2002, 41, 4378-4400.

[11] Liu, Z. M.; Yasseri, A. A.; Lindsey, J. S.; Bocian, D. F. *Science.* 2003, 302, 1543-1545.

[12] Mamin, H. J.; Rugar, D. *Appl. Phys. Lett.* 1992, 61, 1003-1005.

[13] Binnig, G..; Despont, M.; Drechsler, U.; Haberle, W.; Lutwyche, M.; Vettiger, P.; Mamin H. J.; Chui, B. W.; Kenny, T. W. *Appl. Phys. Lett.* 1999, 74,1329-1331.

[14] Weidenhof, V.; Friedrich, I.; Ziegler, S.; Wuttig, M. *J. Appl. Phys.* 1999, 86, 5879-5887.

[15] Peng, H. L.; Ran, C. B.; Yu, X. C.; Zhang, R.; Liu, Z. F. *Adv. Mater.* 2005, 17, 459-464.

[16] Peng, H. L.; Chen, Z.; Tong, L. M.; Yu, X. C.; Ran, C. B.; Liu, Z. F. J. *Phys. Chem.* B 2005, 109, 3526-3530.

[17] Yu, X. C.; Zhang, R.; Peng, H. L.; Ran, C. B.; Zhang, Y. Y.; Liu, Z. F. *J. Phys. Chem.* B 2004, 108, 14800-14803.

[18] Ma, L. P.; Song, Y. L.; Gao, H. J.; Zhao, W. B.; Chen, H. Y.; Xue, Z. Q.; Pang, S. J. *Appl. Phys. Lett.* 1996, 69, 3752-3753.

[19] Gao, H. J.; Sohlberg, K.; Xue, Z. Q.; Chen, H. Y.; Hou, S. M.; Ma L. P.; Fang, X. W.; Pang, S. J.; Pennycook, S. J. *Phys. Rev. Lett.* 2000, 84, 1780-1783.

[20] Ma, L. P.; Yang, W. J.; Xue, Z. Q.; Pang, S. J. *Appl. Phys. Lett.* 1998, 73, 850-852.

[21] Feng, M.; Guo, X. F.; Lin, X.; He, X. B.; Ji, W.; Du, S. X.; Zhang, D. Q.; Zhu, D. B.; Gao, H. J. *J. Am. Chem. Soc.* 2005, 127, 15338-15339.

[22] Jiang, G. Y.; Michinobu, T.; Yuan, W. F.; Feng, M.; Wen, Y. Q.; Du, S. X.; Gao, H. J. Jiang, L.; Song, Y. L.; Diederich, F.; Zhu, D. B. *Adv. Mater. 2005*, 17, 2170-2173.

[23] Kado, H.; Tohda, T. Jpn. *J. Appl. Phys. Part 1* 1997, 36, 523-525.

[24] Yoshida, S.; Ono, T.; Oi, S.; Esashi, *M. Nanotechnol.* 2005, 16, 2516-2520.

[25] Jegadesan, S.; Advincula, R. C.; Valiyaveettil, S. *Adv. Mater.* 2005, 17, 1282-1285.

[26] Betzig, E.; Trautman, J. K.; Harris, T. D.; Weiner, J. S.; Kostelak, R. L. *Science.* 1991, 251, 1468-1470.

[27] Heinzelmann, H.; Pohl, D.W. *Appl. Phys. A: Mater. Sci. Process.* 1994, 59, 89-101.

[28] Dunn, R. C. *Chem. Rev.* 1999, 99, 2891-2928.

[29] Hong, T.; Wang, J.; Li, D. C. *Opt. Tech.* 2001, 27, 255-259.

[30] Mansfield, S. M.; Kino, G. S. *Appl. Phys. Lett.* 1990, 57, 2615-2616.

[31] Mansfield, S. M.; Studenmund, W. R.; Kino, G. S.; Osato, K. *Opt. Lett.* 1993, 18, 305-307.

[32] Terris, B. D.; Mamin, H. J.; Rugar, D.; Studenmund, W. R.; Kino, G. S. *Appl. Phys. Lett.* 1994, 65, 388-390.

[33] Terris, B. D.; Mamin, H. J.; Rugar, D. *Appl. Phys. Lett.,* 1996, 68, 141-143.

[34] Nakano, T.; Sato, A.; Fuji, H.; Tominaga. J.; Atoda, N. *Appl. Phys. Lett.* 1999, 75, 151-153.

[35] Labardi, M.; Coppede, N.; Pardi, L.; Allegrini, M.; Giordano, M.; Patane, S.; Arena, A.; Cefali, E. *Mol. Cryst. Liq. Cryst.* 2003, 398, 33-43.

[36] Rath, S.; Mager, O.; Heilig, M.; Strauss, M.; Mack, O.; Port, H. *J. Lumin.* 2001, 94, 157-161.

[37] Kaupp, G.; Herrmann, A.; *Ultramicroscopy.* 1998, 71, 383-388.
[38] Kim, J.; Song, K.B.; Park, K.H.; Lee, H.W.; Kim, E. *Jpn. J. Appl. Phys.,* Part 1 2002, 41, 5222-5225.
[39] Lee, H. W.; Kim, Y. M.; Jeon, D. J.; Kim, E.; Kim, J.; Park, K. *Opt. Mater.* 2002, 21, 289-293.
[40] Martin, Y.; Wickramasinghe, H. K. *Appl. Phys. Lett.* 1987, 50, 1455-1457.
[41] Wadas, A.; Dreyer, M.; Lohndorf, M.; Wiesendanger, R. *IEEE Trans. Magn.* 1997, 33, 4050-4052.
[42] El-Sayed R. T.; Carley, L. R. *IEEE Trans. Magn.* 2002, 38, 1892-1894.
[43] Ohkubo. T.; Kishigami, J.; Yanagisawa, K.; Kaneko, R. *IEEE Trans. Magn.*1991, 27, 5286-5288.
[44] Hosaka, S.; Koyanagi, H.; Kikukawa, A.; Miyamoto, M.; Imura, R.; Ushiyama, J. *J. Vac. Sci. Technol.,* B 1995, 13, 1307-1311.
[45] Koblischka, M. R.; Hewener, B.; Hartmann, U.; Wienss, A.; Christoffer, B.; Persch-schuy, G.. Appl. Phys. A: *Mater. Sci. Process.* 2003, 76, 879-884.
[46] El-Sayed, R. T.; Carley, L. R. *IEEE Trans. Magn.* 2005, 41, 1209-1217.
[47] El-Sayed, R. T.; Carley, L. R. *IEEE Trans. Magn.* 2004, 40, 2326-2328.
[48] Wood, R. *IEEE Trans. Magn.* 2000, 36, 36-42.
[49] Carley, L. R.; Ganger, G.; Guillou, D. F., Nagle, D. *IEEE Trans. Magn.* 2001, 37, 657-662.
[50] Kong, L. S.; Zhuang, L.; Chou, S. Y. *IEEE Trans. Magn.* 1997, 33, 3019-3021.
[51] Carley, L. R.; Bain, J. A.; Fedder, G. K.; Greve, D. W.; Guillou, D. F.; Lu, M. S. C. Mukherjee, T.; Santhanam, S.; Abelmann, L.; Min, S. *J. Appl. Phys.* 2000, 87, 6680-6685.
[52] Hosaka, S. *IEEE Trans. Magn.* 1996, 32, 1873-1877.
[53] Hosaka, S.; Kikukawa, A.; Koyanagi, H.; Shintani, T.; Miyamoto, M.; Nakamura, K.; Etoh, K. *Nanotechnol. 1997*, 8, A58-A62.
[54] Vettiger, P.; Despont, M.; Drechsler, U.; Dürig, U.; Häberle, W.; Lutwyche, M.; Rothuizen, H.; Stutz, R.; Widmer, R.; Binnig, G. *IBM J. Res. Develop.* 2000, 44, 323-340.
[55] Vettiger, P.; Cross, G.; Despont, M.; Drechsler, U.; Dürig, U.; Gotsmann, B.; Häberle, W.; Lantz, M. A.; Rothuizen, H. E.; Stutz, R.; Binnig, G. K. *IEEE Trans. Nanotech.* 2002, 1, 39-55.
[56] Vettiger, P.; Brugger, J.; Despont, M.; Drechsler, U.; Dürig, U.; Haberle, W.; Lutwyche, M.; Rothuizen, H.; Stutz, R.; Widmer, R.; Binnig, G. *Microelectron. Eng.* 1999, 46, 11-17.
[57] Eleftheriou, E.; Antonakopoulos, T.; Binnig, G. K.; Cherubini, G.; Despont, M.; Dholakia, A.; Dürig, U.; Lantz, M. A.; Pozidis, H.; Rothuizen, H. E.; Vettiger, P. *IEEE Trans. Magn.* 2003, 39, 938-945.

INDEX

C

F

G

H

I

N

O

P

Q

R

S

T

U

V

W

X

Y

Z